# Student Study G

*for use with*

# Inquiry into Life

## Tenth Edition

## Sylvia S. Mader

Boston   Burr Ridge, IL   Dubuque, IA   Madison, WI   New York   San Francisco   St. Louis
Bangkok   Bogotá   Caracas   Lisbon   London   Madrid
Mexico City   Milan   New Delhi   Seoul   Singapore   Sydney   Taipei   Toronto

# McGraw-Hill Higher Education

*A Division of The McGraw-Hill Companies*

Student Study Guide for use with
INQUIRY INTO LIFE, TENTH EDITION
SYLVIA S. MADER

Published by McGraw-Hill Higher Education, an imprint of The McGraw-Hill Companies, Inc.,
1221 Avenue of the Americas, New York, NY 10020. Copyright © The McGraw-Hill Companies,
Inc., 2003, 2000, 1997. All rights reserved.

This book is printed on recycled, acid-free paper containng 10% postconsumer waste.

RECYCLED

1 2 3 4 5 6 7 8 9 0 QPD QPD 0 9 8 7 6 5 4 3 2

ISBN 0-07-239972-4

www.mhhe.com

# CONTENTS

# TO THE STUDENT

The philosophy behind this edition of the *Student Study Guide* for *Inquiry Into Life* is based upon the belief that learning occurs gradually. Therefore students should complete the study questions sequentially according to each portion of the textbook.

The **Chapter Review,** like the summary at the end of each chapter in the textbook, provides a quick overview of the main concepts in the chapter. Key terms, identified in the textbook by bold print, are also found here and used in context.

The **Study Questions** are correlated to each numbered main section of the textbook and to the learning objectives for that section. Page numbers allow you to find the answer quickly. A systematic and thorough study of the textbook makes it easier to study for a test. You can also use the Study Questions section as an excellent review by answering the questions again after having someone else read the statements to you or by covering up the answers that you filled in. Some instructors may have you answer the Study Questions for extra credit. Ideally, you should have completed them before the instructor ever lectures on the topic.

The **Games** are a good way to apply what you have learned in a fun way. If you can successfully do the games you know that you have successfully mastered the material. You may find it especially enjoyable to do the games with a fellow student. Practice in learning and using the terms is also provided in each chapter. The experience is varied; some chapters have a **Wordsearch,** others a **Crossword,** and still others have a **Wordmatch.** You should avail yourself of this opportunity to test your knowledge of the terms.

The **Chapter Test** is designed to allow you to test yourself. The multiple choice and thought questions are similar to those most teachers would ask. They allow you to determine how well you understand the concepts and provide an evaluation of your newly acquired knowledge. The thought questions involve critical thinking, and there may not be a single correct answer.

**Answers** are provided for the Study Questions and the Chapter Test at the end of each chapter. However, be sure to give your own answers first. The questions will be of little help if you simply fill in your *Student Study Guide* with the answers provided.

*Inquiry Into Life,* tenth edition, by Sylvia S. Mader, also has many student aids that you should be sure to use.

**History of Biology End Papers:** The inside front covers list major contributions to the field of biology in a concise, chronological manner. Students may refer to these whenever it is appropriate.

**Text Introduction:** Chapter 1 discusses the characteristics of life and examines the scientific process. This chapter surveys the field of biology as a whole and prepares students for the study of the rest of the text.

**Chapter Concepts:** Each chapter begins with an integrated outline in which the main sections of the chapter are numbered. Instead of chapter concept statements for each section, students are asked questions to direct their study. The questions are page referenced, and the outline enables instructors to assign just portions of the chapter.

**Key Terms:** Key terms are boldfaced in the chapter and are defined in context. The text ends with a glossary, and an even more extensive glossary is located at the *Inquiry* Web site, http://www.mhhe.com/maderinquiry10.

**Internal Summary Statements:** Short internal summary statements appear at the end of major sections and help students focus their study efforts on the basics.

**Readings:** Reading topics correlate to the subject matter of the chapter. Four types of readings are included in the text:

- *Health Focus* readings review measures to keep healthy, such as the need for exercise ("Exercise, Exercise, Exercise," Chapter 19) and guidelines for preventing cancer ("Prevention of Cancer," Chapter 25).
- *Ecology Focus* readings draw attention to a particular environmental problem, such as acid rain ("The Harm Done by Acid Deposition," Chapter 2).
- *Science Focus* readings, such as "How Memories Are Made" in Chapter 17, are designed to introduce students to the research methods of scientists.
- A *Bioethical Focus* topic is covered in a nonbiased manner at the ends of most chapters. These topics stimulate thought and discussion about today's most relevant bioethical issues.

**Illustrations:** The illustrations in *Inquiry Into Life* are consistent with multicultural educational goals. Integrative illustrations relate micrographs with drawings. This enables students to see an actual structure alongside a diagram drawn similarly to the micrograph. Leaders and labels common to both the micrograph and the drawing allow the student to accurately locate and visualize a particular structure.

The *Visual Focus* illustrations illustrate in depth a main topic, system, or cycle in the chapter. The use of **icons** helps students relate a part to the whole.

**Chapter Summaries:** The summary, called "Summarizing the Concepts," is organized according to the major sections of the chapter, and the content of each section is summarized in a short paragraph or two. Chapter summaries offer a concise review of the material in each chapter. Students may read them before beginning the chapter to preview the topics of importance, and they may also use them to refresh their memories after they have a firm grasp of the concepts presented in each chapter.

**Testing Yourself** consists of 25–30 multiple choice objective questions. Answers are supplied in Appendix A.

**Appendices and Glossary:** The appendices contain optional information. Appendix A is the answer key to the questions found at the end of each chapter. Appendix B is an expanded table for the classification of organisms. Appendix C is the metric system. Appendix D is an expanded table of chemical elements.

The glossary defines the boldfaced terms in the text. These terms are the ones most necessary for the successful study of biology.

**Index:** The text also includes an index in the back of the book. By consulting the index, it is possible to determine on what page or pages various topics are discussed.

**e-Learning Connection:** New to this edition, each chapter ends with a page that organizes the relevant technological material by major sections, helping to create a stronger association between available study activities at the *Inquiry Into Life* Web site. This feature is expanded on the Online Learning Center, where students can easily find the appropriate learning experience. Go to http://www.mhhe.com/maderinquiry10.

# Helpful Study Hints

The following list tells you how to study more productively.

**Have a positive attitude.**

Feel that you can and will be successful and this will lead you to taking steps, such as those listed here, that lead to success.

**Be motivated.**

The desire to do well has to come from within yourself. To be committed means that you will study even when you do not feel like studying, and even when you are having personal problems.

**Attend class.**

Make every possible effort to attend lecture. Try to read the chapter in your textbook before your instructor starts to lecture on that chapter. Even if you don't understand the material yet, just having read the chapter will help you understand the lecture. Be attentive during class and ask questions.

**Study regularly.**

Set aside several hours at a certain time on particular days to go over old and new material. Study in a quiet, well lit room away from distractions. Complete all parts of the *Study Guide* for each chapter.

**Be interactive.**

Use your *Study Guide* to ask yourself the right questions in order to master the material. Always try to answer the questions first before looking up the answer in the back of the guide. After answering the questions for a section, summarize to yourself what you have learned. Make up ways for you to remember factual material.

**Get help.**

If particular concepts are difficult for you to understand, get help from your instructor. Ask another person, perhaps a family member, to ask you questions from the *Study Guide* when the answers are already filled in, or cover up the answers and try them again by yourself.

**Be a good test taker.**

If you have done all the above, it will be possible for you to get a good night's sleep before a test day. When taking the test, go slowly and read the questions carefully before giving your answer. Know that your *Study Guide* has prepared you to choose the correct answer and don't choose any answer that is unfamiliar to you.

# 1

# THE STUDY OF LIFE

Although living things are diverse, they share certain characteristics. The organization of living things is exhibited by the smallest unit of life, the cell. In multicellular organisms, similar cells compose a tissue, tissues form organs, and organs are part of organ systems. Each level of organization has emergent properties that cannot be accounted for by simply adding up the properties of the previous levels.

Living things acquire materials and energy from the environment that are used during **metabolism,** a process that maintains **homeostasis.** Living things respond to the environment. When they **reproduce** and develop, genetic changes are passed on that result in **adaptation** to the environment.

Biologists classify organisms into a particular domain (distantly related), kingdom, phylum, class, order, family, genus, and **species.** Each organism is given a binomial name consisting of the genus and species. Domain Archaea and domain Bacteria contain unicellular prokaryotes, which usually absorb food, and domain Eukarya contains kingdoms Protista (unicellular eukaryotes, various modes of nutrition—e.g., protozoans and algae), Fungi (usually multicellular, absorb food—e.g., mushrooms), Plantae (multicellular **photosynthesizers**), and Animalia (multicellular, ingest food).

All living things belong to a population, and populations interact with each other within a community and with the physical environment, forming an **ecosystem.** Ecosystems are characterized by chemical cycling and energy flow, which begin when a photosynthesizer becomes food (organic nutrients) for an animal. Food chains tell who eats whom in an ecosystem. As one population feeds on another, the energy dissipates but the nutrients do not. Eventually, inorganic nutrient molecules return to photosynthesizers, which use them and solar energy to produce more food.

Human activities have totally altered many ecosystems for their own purposes and are putting stress on most of the others. We now realize that the health of the biosphere is essential to the future continuance of the human species. Therefore, we should do all we can to maintain the health of the biosphere.

When studying the natural world, scientists use the scientific process. Observations, along with previous **data,** are used to formulate a hypothesis. New observations and/or experiments are carried out in order to test the hypothesis. A good experimental design includes a **control** group. The experimental and observational results are analyzed, and the scientist comes to a conclusion as to whether the results support or do not support the hypothesis. Several conclusions in a particular area may allow scientists to arrive at a theory—such as the cell theory, the gene theory, or the theory of evolution. The theory of evolution is a unifying theory of biology.

All persons are responsible for deciding how scientific knowledge should be used for the benefit of humankind.

Study the text section by section. Answer the study questions so that you can fulfill the learning objectives for each section.

## 1.1 THE CHARACTERISTICS OF LIFE (PP. 3–5)

The learning objectives for this section are:
- State the characteristics of life that all organisms share.
- Understand different levels of biological organization.

1. Match these characteristics of life with the situations that follow.
    Living things are organized.
    Living things take materials and energy from the environment.
    Living things respond to stimuli.
    Living things reproduce.
    Living things adapt to the environment.
    Living things are homeostatic.
    Living things grow and develop.

   Frogs have a life cycle that includes an egg, a larva (tadpole) that undergoes the process of metamorphosis, and an adult.  a._____

   Humans immediately remove their hands from a hot object.  b._____

   All living things are composed of cells.  c._____

   A flounder is a flattened fish that lives on the bottom of bodies of water, while a tuna is a streamlined fish that swims in the open sea.  d._____

   Most cells use the sugar glucose as an energy source.  e._____

2. Homeostasis refers to keeping  a._____ relatively stable, such as body
   b._____.

3. List the following levels of biological organization in order, from smallest to largest:

   cell, community, ecosystem, molecule, organ, organism, organ system, population, tissue

   a. _____ (smallest)

   _____

   _____

   _____

   _____

   _____

   _____

   _____

   _____ (largest)

   Different groups of organisms interact with the environment at the level of the  b._____.

4. Label each of the following statements as an example of either the unity of life (U) or the diversity of life (D):
    a. _____ Fungi absorb food; plants carry out photosynthesis.
    b. _____ Homeostasis, metabolism, and evolution are characteristics of living things.
    c. _____ Life began with single cells.
    d. _____ Living things consist of cells.
    e. _____ Maple trees have broad, flat leaves, and pine trees have needlelike leaves.

## 1.2 THE CLASSIFICATION OF LIVING THINGS (PP. 6–7)

The learning objective for this section is:
• Classify living things into categories according to their evolutionary relationships.

5. List the following levels of classification in order, from smallest (most exact) to largest (most general):

class      _____ (smallest)

family      _____

domain      _____

genus      _____

kingdom      _____

order      _____

phylum      _____

species      _____ (largest)

6. Name the kingdoms described in each of the following statements:

a. _____ multicellular; ingest food

b. _____ absorb food; includes molds and mushrooms

c. _____ photosynthesize food; includes ferns

d. _____ includes protozoans and algae

7. Name the domain occupied by each of the following:

a. _____ plants

b. _____ bacteria

c. _____ animals

d. _____ fungi

e. _____ unicellular prokaryotes

## 1.3 THE ORGANIZATION OF THE BIOSPHERE (PP. 8–9)

The learning objectives for this section are:
- State the levels of organization beyond the individual.
- Explain how populations interact in ecosystems.
- Explain how humans impact the biosphere, and why biodiversity is so important.

8. Study the diagram that follows and then answer the questions.

a. Which type of population in a food chain does every other population rely upon?

_____

b. What is the ultimate source of energy for a food chain? _____

c. What type of population in a food chain processes organic remains and returns inorganic nutrients to the ecosystem? _____

d. How do producers interact with the abiotic environment? _____

e. How do producers interact with the biotic environment? _____

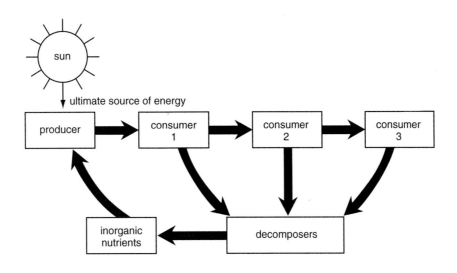

9. Indicate whether these statements are true (T) or false (F).
   a. _____ As more and more ecosystems are converted to towns and cities, fewer of the natural cycles are able to function adequately.
   b. _____ Rain forests play no protective role in the biosphere, and their destruction is of little concern to humans.
   c. _____ The present rate of extinction is normal and about the same as at any other time in the history of the Earth.

## 1.4 THE PROCESS OF SCIENCE (PP. 10–14)

The learning objectives for this section are:
- Understand how the scientific process is used to gather information and to arrive at conclusions.
- List the steps of the scientific method.

10. Place the terms *conclusion, experiment/observations, hypothesis, observation,* and *theory* in the correct boxes of the adjacent diagram:

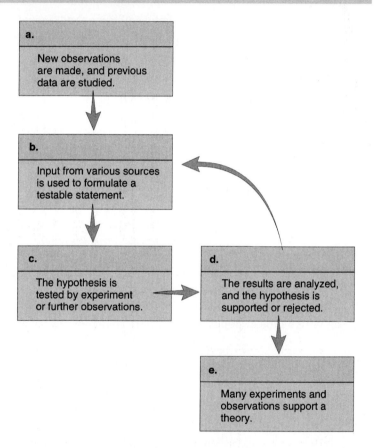

**a.**
New observations are made, and previous data are studied.

**b.**
Input from various sources is used to formulate a testable statement.

**c.**
The hypothesis is tested by experiment or further observations.

**d.**
The results are analyzed, and the hypothesis is supported or rejected.

**e.**
Many experiments and observations support a theory.

11. Each statement demonstrates a step of the scientific method used in a study. Match the steps from question 10 to these examples.

   To spawn, striped bass migrate up a major river. In 1991, a dam was built on this river. Fishers reported catching fewer striped bass in the river in 1993.

   A study of the fish population in the river revealed decreased numbers of striped bass.
   a. _____

   Most likely, the dam is blocking the migration and spawning of striped bass.   b. _____

   A fish ladder is built along the dam. Scientists observe that the ladder helps striped bass clear the dam and spawn upriver.   c. _____

   Survival of striped bass requires free migration of fish upriver to spawn.   d. _____

   Construction of the fish ladder improves fishing. More fish are caught the following year, and studies indicate that the number of striped bass is increasing.   e. _____

12. Biologists are doing a study on athletes' intake of steroids. Which of these would be the experimental variable (E), and which would be the dependent variable (D)?
    a. _____ average gain in body weight of a group of athletes
    b. _____ intake of anabolic steroids by an athlete

13. In the study in question 12, how would a control group of athletes differ from the experimental group? _____
    _____

14. Over a six-month period, the weight gain of the athletes in the experimental group in question 12 averaged 11%. Members of the control group experienced a gain of 4%.

    What is the benefit of mathematical data like this? a. _____
    _____

    State the conclusion of this experiment b. _____
    _____

15. Place the appropriate letter(s) next to each statement. (Some are used more than once.)
    O—observational research only
    E—experimental research only
    E, O—both experimental and observational research

    a. _____ Steps of the scientific method are used.
    b. _____ A control group is employed.
    c. _____ Much of the data required are purely observational.
    d. _____ A hypothesis is disproven or supported.
    e. _____ Observations are made.

16. Match the descriptions to these theories:
    cell theory      theory of biogenesis      theory of evolution      gene theory

    a. _____ Common descent with modification of form.

    b. _____ All organisms are composed of cells.

    c. _____ Organisms inherit coded information.

    d. _____ Life comes only from life.

## 1.5 SCIENCE AND SOCIAL RESPONSIBILITY (P. 14)

The learning objective for this section is:
• Realize that all persons have the responsibility to decide how scientific information can best be used to make ethical or moral decisions.

In questions 17–18, choose the best answer.

_____ 17. Who is responsible for deciding how we use the fruits of science?
    a. scientists
    b. lay people
    c. ministers
    d. All of the above are correct.
    e. Only *b* and *c* are correct.

_____ 18. Which of these is a value judgment?
    a. If the biodiversity of plants declines, we may miss the opportunity to find cures for various illnesses.
    b. It's wrong to kill plants because they are defenseless against humans.
    c. Due to the use of cars for transportation, air pollution has increased.
    d. People shouldn't use their cars because it increases air pollution.
    e. Only *b* and *d* are value judgments.

Review key terms by completing this crossword puzzle, using the following alphabetized list of terms:

*biodiversity*
*biosphere*
*cell*
*control*
*data*
*ecosystem*
*evolution*
*homeostasis*
*hypothesis*
*kingdom*
*reproduce*
*theory*
*vertebrates*

*Across*
1  The total number of species within the biosphere.
3  A major principle in biology, after much experimentation.
7  Community plus its nonliving habitat.
8  Species _____ to make offspring.
10  A theory in biology that suggests that all organisms arose from earlier forms.
11  What a scientist tests in an experiment.

*Down*
1  The layer of life surrounding this planet.
2  What the scientist gathers when making observations.
4  The _____ group receives no experimental treatment in an experiment.
5  Organ systems help the body maintain _____, or steady state.
6  Humans and other organisms with a nerve cord protected by bone are _____.
9  The classification group below domain but above phylum.
12  The basic unit of life of all living things.

# CHAPTER TEST

## OBJECTIVE QUESTIONS

Do not refer to the text when taking this test.

_____ 1. The binomial name *Notorcytes typhlops* refers to taxonomic levels of
   a. class and order.
   b. genus and species.
   c. kingdom and phylum.
   d. order and phylum.

_____ 2. Select the smallest, most exact taxonomic level among the following choices:
   a. class
   b. genus
   c. order
   d. specific epithet

_____ 3. Select the largest, broadest taxonomic level among the following choices:
   a. class
   b. order
   c. domain
   d. phylum

_____ 4. Each is a general characteristic of life EXCEPT
   a. the ability to respond.
   b. growth and development.
   c. organization.
   d. classification.

_____ 5. The lowest level of organization to have the characteristics of life is the _____ level.
 a. atomic
 b. cellular
 c. molecular
 d. population
_____ 6. The term *metabolism* refers best to
 a. chemical and energy transformations.
 b. maintenance of internal conditions.
 c. the ability to respond to stimuli.
 d. the lack of reproduction.
_____ 7. Changes in _____ account for the ability of a
species to evolve.
 a. abiotic factors
 b. ecosystems
 c. genes
 d. sunlight
_____ 8. Through evolution, populations can
 a. adapt to the environment.
 b. change their level of organization.
 c. fail to reproduce.
 d. eliminate cell structure.
_____ 9. Bacteria belong to the domain
 a. Animalia.
 b. Fungi.
 c. Bacteria.
 d. Protista.
_____10. Plants are unique among living things in that
they are
 a. multicellular and absorb food.
 b. unicellular and ingest food.
 c. multicellular and photosynthesize.
 d. All of these are correct.

_____11. Ecosystems
 a. contain only consumers and producers.
 b. need a continual supply of solar energy.
 c. are a part of the biosphere.
 d. prosper if biodiversity is maintained.
 e. All of these are correct.
_____12. Select the incorrect pair.
 a. data—factual information
 b. deductive reasoning—general hypothesis to
 prediction
 c. hypothesis—final conclusion
 d. inductive reasoning—specific data to general
 hypothesis
_____13. Valid scientific results should be repeatable by
other scientific investigators.
 a. true
 b. false
_____14. The statement with the greatest acceptance and
predictive value for scientists is the
 a. hypothesis.
 b. induction.
 c. observation.
 d. scientific theory.

## THOUGHT QUESTIONS

Answer in complete sentences.
 15. Why is any level of organization not a mere sum of its parts?

 16. Testing a hypothesis is the central core of the scientific method. Explain.

**Test Results:** _____ number correct ÷ 16 = _____ × 100 = _____ %

# ANSWER KEY

## STUDY QUESTIONS

**1. a.** grow and develop **b.** respond to stimuli **c.** are organized **d.** adapt to environment **e.** take materials and energy from the environment **2. a.** internal conditions **b.** temperature **3. a.** molecule, cell, tissue, organ, organ system, organism, population, community, ecosystem **b.** emergent properties **c.** cell **d.** ecosystem **4. a.** D **b.** U **c.** U **d.** U **e.** D **5.** species, genus, family, order, class, phylum, kingdom, domain **6. a.** Animalia **b.** Fungi **c.** Plantae **d.** Protista **7. a.** Eukarya **b.** Bacteria **c.** Eukarya **d.** Eukarya **e.** Bacteria and Archaea **8. a.** producer **b.** sun **c.** decomposers **d.** use solar energy and inorganic nutrients **e.** are eaten and supply energy to consumers **9. a.** T **b.** F **c.** F **10. a.** observation **b.** hypothesis **c.** experiment/observations **d.** conclusion **e.** theory **11. a.** observation **b.** hypothesis **c.** experiment/observations **d.** conclusion **e.** observation **12. a.** D **b.** E **13.** Control group would not take steroids. **14. a.** It is objective. **b.** Steroids cause weight gain. **15. a.** E, O **b.** E **c.** O **d.** E, O **e.** E, O **16. a.** theory of evolution **b.** cell theory **c.** gene theory **d.** theory of biogenesis **17.** d **18.** e

## DEFINITIONS CROSSWORD

**Across:**
**1.** biodiversity **3.** theory **7.** ecosystem **8.** reproduce **10.** evolution **11.** hypothesis

**Down:**
**1.** biosphere **2.** data **4.** control **5.** homeostasis **6.** vertebrates **9.** kingdom **12.** cell

## CHAPTER TEST

**1.** b **2.** b **3.** c **4.** d **5.** b **6.** a **7.** c **8.** a **9.** c **10.** c **11.** e **12.** c **13.** a **14.** d **15.** At each level, there are emergent properties not seen at lower levels of organization. For example, the cell, the basic living unit, is more than just a combination of chemicals. **16.** No matter the sequence of steps followed, scientists always test a hypothesis. The testing can take the form of making more observations rather than doing an experiment.

# 2

# THE MOLECULES OF CELLS

All matter is composed of some 92 elements. Each element is made up of just one type of atom. An **atom** has a mass, which is dependent on the number of protons and neutrons in the nucleus, and its chemical properties are dependent on the number of **electrons** in the outer shell.

Atoms react with one another by forming ionic bonds or covalent bonds. **Ionic bonds** are an attraction between charged ions. Atoms share electrons in **covalent bonds,** which can be single, double, or triple bonds.

Oxidation is the loss of electrons (hydrogen atoms), and reduction is the gain of electrons (hydrogen atoms).

Water, acids, and bases are important inorganic molecules. The polarity of water accounts for it being the universal solvent; **hydrogen bonding** accounts for it boiling at 100°C and freezing at 0°C. Because it is slow to heat up and slow to freeze, water is in liquid form at the temperature of living things.

Pure water has a neutral pH; **acids** increase the hydrogen ion concentration [H⁺] but decrease the pH, and **bases** decrease the hydrogen ion concentration but increase the pH of water.

The chemistry of carbon accounts for the chemistry of organic compounds. **Carbohydrates, lipids, proteins,** and **nucleic acids** are macromolecules with specific functions in cells. Macromolecules are polymers that each contain (a) specific monomer(s).

**Glucose** is the six-carbon sugar most utilized by cells for "quick" energy. Like the rest of the macromolecules to be studied, **condensation synthesis** joins two or more sugars, and a **hydrolysis reaction** splits the bond. Plants store glucose as **starch,** and animals store glucose as **glycogen.** Humans cannot digest cellulose, which forms plant cell walls.

Lipids are varied in structure and function. Fats and oils, which function in long-term energy storage, contain glycerol and three fatty acids. Fatty acids can be saturated or unsaturated. The plasma membranes surrounding cells contain phospholipids that have a polarized end. Certain hormones are derived from cholesterol, a complex ring compound.

The primary structure of a polypeptide is its own particular sequence of the possible twenty types of **amino acids.** The secondary structure is often an alpha helix. The tertiary structure occurs when a polypeptide bends and twists into a three-dimensional shape. Proteins can contain several **polypeptides,** and this accounts for a possible quaternary structure.

Nucleic acids (**DNA** and **RNA**) are polymers of **nucleotides,** which have three parts: pentose sugar, nitrogen-containing base, and phosphate. ATP, the high-energy molecule of cells, is a nucleotide with three phosphates.

Study the text section by section. Answer the study questions so that you can fulfill the learning objectives for each section.

## 2.1 BASIC CHEMISTRY (PP. 20–25)

The learning objectives for this section are:
- Name and describe the subatomic particles of an atom, indicating which one accounts for the occurrence of isotopes.
- Draw a simplified atomic structure of any atom with an atomic number less than 20.
- Distinguish between ionic and covalent reactions, and draw representative atomic structures for ionic and covalent molecules.

In questions 1–9, fill in the blanks.

1. List three examples of elements in the human body.

   a. _____  b. _____  c. _____

2. The smallest functional unit of an element is a(n) _____.

3. In the following drawing of an atom, write *P* for protons, *N* for neutrons, and *E* for electrons beside *a* and *b*. Indicate the electrical charge carried by each type of particle.

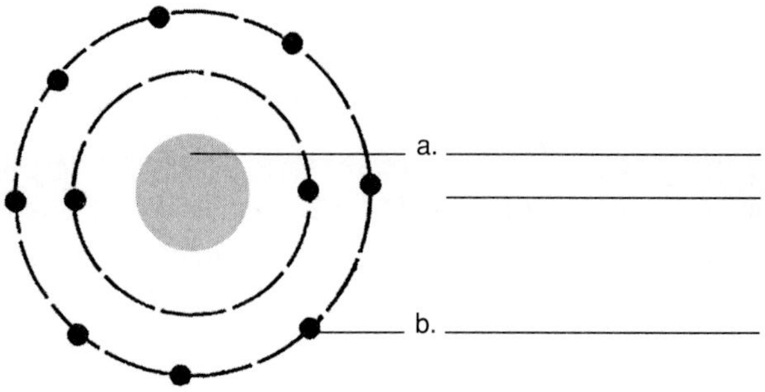

   a. _____
   _____

   b. _____

4. An atom is said to be electrically neutral when it has equal numbers of ᵃ·_____ and ᵇ·_____.

5. Stable, unreactive atoms usually have _____ electrons in their outermost shells.

6. The atomic number of an atom is equal to its number of _____.

7. The atomic weight of an atom is equal to its number of ᵃ·_____ and ᵇ·_____.

8. _____ of an element are atoms that differ in their number of neutrons.

9. Atoms of _____ isotopes, such as carbon 14, radiate high-energy particles from the nucleus as they decay to carbon 12.

In questions 10–16, fill in the blanks.

10. A(n) _____ forms when two or more atoms are bonded together.

11. Sodium (Na) has 11 protons. How many electrons will it have when it is electrically neutral? ᵃ·_____ How many electrons will be in its outermost shell? ᵇ·_____

12. Chlorine (Cl) has 17 protons. How many electrons will chlorine have when it is electrically neutral? ᵃ·_____ How many electrons are in its outermost shell? ᵇ·_____

13. Based on your answers for questions 11 and 12, what type of bond do you expect will form between sodium and chlorine? _____

14. Why? _____
    _____
    _____
    _____

15. When pairs of electrons are shared between atoms, _____ bonds result.

16. In a triple bond, _____ pairs of electrons are shared.

## 2.2 WATER AND LIVING THINGS (PP. 26–30)

The learning objectives for this section are:
- Describe the chemical properties of water, and explain their importance for living things.
- Define an acid and base; describe the pH scale, and state the significance of buffers.

In questions 17–23, fill in the blanks.

17. One type of inorganic molecule, _____, makes up 60–70% of all living things.

18. What type of bond occurs *within* water molecules? _____

19. What type of bond occurs *between* water molecules? _____

20. The following diagram illustrates the dissociation of water molecules.

$$H - O - H \rightleftarrows H^+ + OH^-$$

water        hydrogen   hydroxide
ion      ion

The pH scale is based on the relative abundances of hydrogen ions and hydroxide ions.

When there is more $H^+$ and less $OH^-$, is the pH acidic or basic? a._____

When there is less $H^+$ and more $OH^-$, is the pH acidic or basic? b._____

21. Technically speaking, a(n) a._____ gives off hydrogen ions in a solution, while a(n) b._____ takes up hydrogen ions or gives off hydroxide ions.

22. Because of hydrogen bonding, water sticks to itself and is called a._____. Is this property due to hydrogen bonding or to polarity of the water molecule? b._____

23. A(n) _____ prevents rapid changes in pH by taking up either hydrogen or hydroxide ions.

## 2.3 ORGANIC MOLECULES (P. 31)

The learning objective for this section is:
- Explain and demonstrate the bonding patterns of carbon and organic molecules.

In questions 24–29, fill in the blanks.

24. Carbon shares its electrons with as many as _____ different atoms at once.

25. Organic molecules, like the one illustrated here, usually have a "backbone" made out of the element _____.

26. Organic molecules also contain _____.

27. Are organic molecules characteristic of living or nonliving matter? _____

28. Organic molecules can often attain large sizes and are thus called a._____ or b._____.

29. Polymers are built of smaller units, called _____.

## 2.4 CARBOHYDRATES (PP. 32–33)

The learning objectives for this section are:
- Distinguish between the formation of polymers by condensation synthesis and their breakdown by hydrolysis.
- Give examples of monosaccharides, disaccharides, and polysaccharides, and state their functions.

In questions 30–36, fill in the blanks.

30. The monomer (building block) of carbohydrates is a simple sugar, known technically as a a._____. The simple sugar b._____ is one example.

31. Carbohydrates are the primary source of _____ in the diet.

32. Table sugar, sucrose, is an example of a(n) _____, being built from two sugar molecules.

33. Large organic molecules are built through a reaction called a._____ _____ that removes molecules of b._____.

34. During digestion, organic molecules are broken down into component parts by adding a._____ molecules through a reaction referred to as b._____.

35. When many glucose molecules are joined, the macromolecule (polymer) _____ results.

36. Humans store glucose in the form of a polymer known as a._____ in their livers. Another polysaccharide, b._____, lends structural support to plant cell walls and is not digestible by humans.

## 2.5 LIPIDS (PP. 34–35)

The learning objectives for this section are:
- Give examples of various lipids, and state their functions.
- Recognize the structural formula for a saturated and an unsaturated fatty acid.

In questions 37–43, fill in the blanks.

37. Lipids function as a._____ and keep us warm; they provide long-term storage of b._____; and they form a protective c._____ around our internal organs.

38. Dietary fats are most commonly in the form of a._____, which are formed from one b._____ molecule and three c._____ _____ molecules. Since they repel water and are nonpolar, they are called d._____ fats.

39. Write the word *saturated* or *unsaturated* on the line below the appropriate structure.

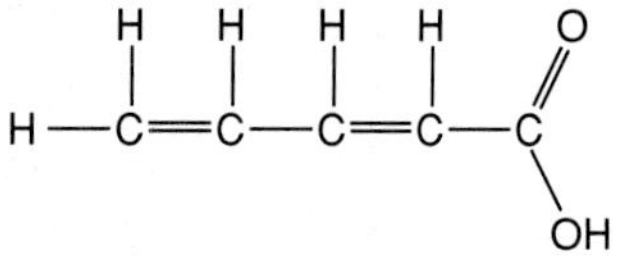

a. 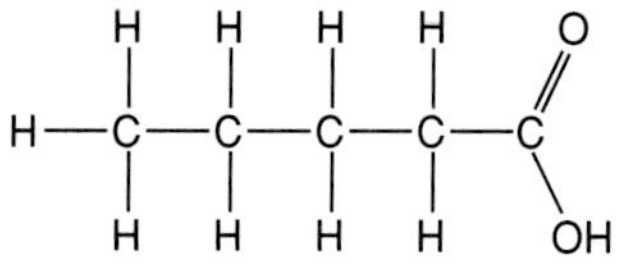 _____        b. _____

40. What does saturated mean? _____

_____

41. Emulsifiers break up fats through a process called a._____ in which the emulsifier surrounds individual fat molecules. b._____ from the gallbladder functions in the same way in the digestive system.

42. A type of lipid with a polar phosphate group, called a(n) a._____, is a major component of plasma membranes.

43. Cholesterol is the precursor of several lipids, known as a._____, that function as hormones in the body.

## 2.6 PROTEINS (PP. 37–38)

The learning objectives for this section are:
- Give examples of proteins, and state their functions.
- Recognize an amino acid and demonstrate how a peptide bond is formed.

In questions 44–48, fill in the blanks.

44. Certain proteins can function as a._____ to catalyze chemical reactions in the cell. Fingernails contain another important protein, called b._____. Another protein, called c._____, is an important structural component of connective tissue.

45. The monomer (building block) of proteins is the _____.

46. In the following diagram of an amino acid, —$NH_2$ is the a._____ group.

—COOH is the b._____ group.

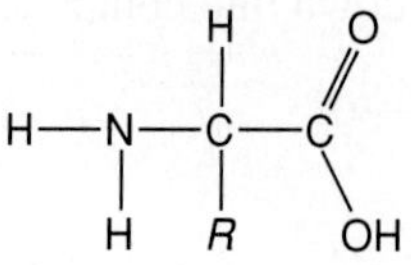

c. How does a peptide bond form between two adjacent amino acids?

_____

_____

_____

47. a._____ structure in proteins refers to the sequence of amino acids. b._____ structure is any twisting of the chain of amino acids. Folding up of the chain is considered to be c._____ structure. d._____ structure exists when two or more chains are folded together, as in the hemoglobin molecule.

48. When proteins are exposed to heat, chemicals, or extremes in pH, they undergo a change in shape. They no longer function as usual, so we say they have been _____.

## 2.7 NUCLEIC ACIDS (PP. 40–41)

The learning objectives for this section are:
- Give examples of nucleic acids, and state their functions.
- State the components of a nucleotide, and tell how these monomers are joined to form a nucleic acid.
- Compare the structures of DNA and RNA.
- Describe the structure and function of ATP.

In questions 49–52, fill in the blanks.

49. The monomers (building blocks) of nucleic acids are _____.

50. The nucleic acid a._____ stores the genetic information for the cell. Another nucleic acid, b._____, works with DNA to specify the order of amino acids during the synthesis of proteins.

51. Refer to the following diagram of a strand of nucleotides to answer questions a–d.
   a. What molecule is represented by S? _____
   b. What molecule is represented by B? _____
   c. How many different types of B are in DNA? _____
   d. What type of bond is represented by the lines? _____

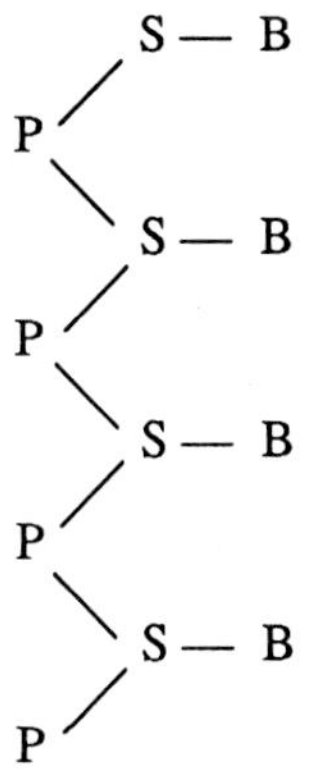

```
           S — B
         P/
          \
           S — B
         P/
          \
           S — B
         P/
          \
           S — B
         P/
```

52. One nucleotide, a._____, functions as an energy carrier inside cells. Energy is released from this molecule when a(n) b._____ group is removed.

53. Complete this summary table.

| | Monomers | Function |
|---|---|---|
| Polysaccharides | | |
| Proteins | | |
| Lipids | | |
| Nucleic acids | | |

In questions 54–55, fill in the blanks.

54. Associate the molecules in the table in question 53 with this diagram.

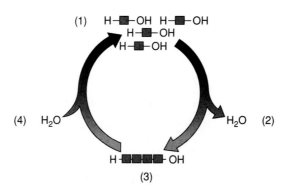

a. What monomers should be associated with (1) in the diagram?

_____

b. What polymers should be associated with (3)? _____

_____

c. Which number in the diagram stands for hydrolysis? _____

d. Which number in the diagram stands for condensation synthesis? _____

55. Of the molecules listed in the first column of the table in question 53,

a. Which are *most* concerned with energy? _____

b. Which one forms enzymes? _____

c. Which one makes up genes? _____

## DEFINITIONS CROSSWORD

Review key terms by completing this crossword puzzle, using the following alphabetized list of terms:

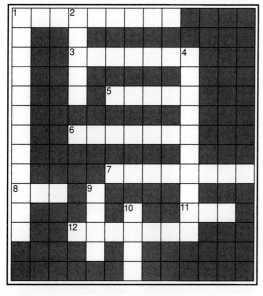

*atom*
*ATP*
*base*
*carbohydrate*
*cellulose*
*DNA*
*electron*
*glucose*
*ionic*
*isotope*
*lipid*
*nucleotide*
*protein*

7   A negatively charged subatomic particle.
8   Nucleotide that serves as the energy carrier for the cell.
11  A nucleic acid that serves as the cell's genetic material.
12  An atom that differs in its number of neutrons.

*Down*
1   Organic compound type that includes mono- and polysaccharides.
2   Insoluble organic compound, such as fats.
4   Building block of nucleic acid.
9   Molecule that releases OH⁻ in solution.
10  Functional unit of an element.

*Across*
1   Complex polysaccharide found in plant tissue that we cannot digest.
3   Organic compound made up of long chains of amino acids.
5   Type of bond formed between oppositely charged ions.
6   Monosaccharide that serves as a building block for starch.

Do not refer to the text when taking this test.

_____ 1. The atomic weight of an atom is determined by the number of
   a. protons.
   b. neutrons.
   c. electrons.
   d. protons and neutrons.

_____ 2. Isotopes differ due to the number of
   a. protons.
   b. neutrons.
   c. electrons.
   d. protons and neutrons.

_____ 3. When an atom either gains or loses an electron, it becomes
   a. electrically neutral.
   b. an ion.
   c. stable.
   d. unreactive.

_____ 4. Chlorine has 17 protons. When chlorine becomes the chloride ion ($Cl^-$), it has
   a. gained an electron.
   b. lost an electron.
   c. gained a proton.
   d. lost a neutron.

_____ 5. When sodium interacts with chlorine, sodium loses an electron while chlorine gains one. This interaction forms
   a. an ionic bond.
   b. a condensation synthesis.
   c. a condensation.
   d. a covalent bond.

_____ 6. Bonds between carbon and hydrogen or oxygen and hydrogen are generally
   a. hydrogen bonds.
   b. ionic bonds.
   c. covalent bonds.
   d. weak and highly transient.

_____ 7. When a strong acid, such as hydrochloric acid, is added to water,
   a. hydrogen ions are taken up.
   b. hydroxide ions are released.
   c. hydrogen ions are released.
   d. the pH stays the same.

_____ 8. A pH of 11.5 is considered to be
   a. slightly acidic.
   b. strongly acidic.
   c. strongly basic.
   d. about neutral.

_____ 9. Which of these is true of organic molecules?
   a. usually ionic bonding
   b. always contain carbon and hydrogen
   c. mainly associated with nonliving matter
   d. are found only in organisms with organ systems

_____ 10. Polymers are
   a. chains of building-block molecules.
   b. formed by dehydration.
   c. broken down by hydrolysis.
   d. All of these are correct.

_____ 11. Which pair of molecules is mismatched?
   a. amino acid—protein
   b. fatty acid—lipid
   c. glucose—starch
   d. glycerol—nucleic acid

For questions 12–16, match the statements to these terms.
   a. triglyceride   b. unsaturated fatty acid
   c. saturated fatty acid   d. phospholipid

_____ 12. Fat or oil.

_____ 13. Has one or more double bonds along the fatty acid chain.

_____ 14. Made up of a glycerol and three fatty acid molecules.

_____ 15. Has hydrogen at every position along the fatty acid chain.

_____ 16. A major component of plasma membranes.

_____ 17. Of the following structures, which molecule is unsaturated?

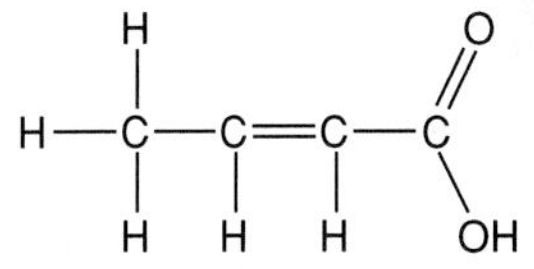

_____ 18. The primary structure of a protein refers to its
   a. three-dimensional shape.
   b. order of amino acids.
   c. order of nucleic acids.
   d. orientation of the amino acids.

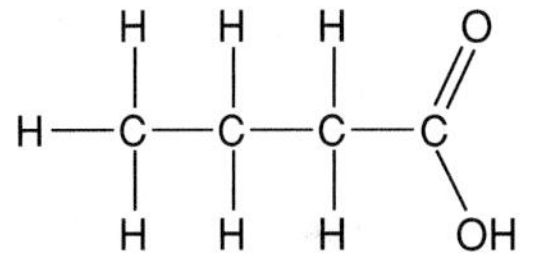

_____ 19. Proteins are polymers of _____, which sometimes function to _____.
   a. amino acids, catalyze chemical reactions
   b. glucose, build muscle strength
   c. nucleotides, synthesize proteins
   d. ribosomes, produce quick energy

_____20. In any amino acid, —NH$_2$ is the _____ group, and —COOH is the _____ group.
    a. carboxyl; amino
    b. amino; carboxyl
    c. peptide; fatty acid
    d. hydroxyl; carboxyl

_____21. A bond forming between the —NH$_2$ group of one amino acid and the —COOH group of another amino acid is called a
    a. double bond.
    b. ionic bond.
    c. hydrogen bond.
    d. peptide bond.

_____22. Twisting of a chain of amino acids into an α-helix is termed the _____ structure of a protein.
    a. primary
    b. secondary
    c. tertiary
    d. quaternary

In questions 23–26, match the statements to these terms.
    a. RNA  b. DNA  c. nucleotide  d. ATP

_____23. Building block of the nucleic acids.
_____24. Functions with DNA to specify the order of amino acids in a protein.
_____25. The genetic material of the cell.
_____26. The energy currency of the cell.

## THOUGHT QUESTIONS

Answer in complete sentences.

27. Give one example of how radioactive isotopes are used in research or medicine.

28. Describe two ways in which the properties of water allow the existence of life.

**Test Results:** _____ number correct ÷ 28 = _____ × 100 = _____%

## ANSWER KEY

### STUDY QUESTIONS

**1. a.** carbon **b.** hydrogen **c.** oxygen **2.** atom **3. a.** protons (positive charge), neutrons (no charge) **b.** electrons (negative charge) **4. a.** protons **b.** electrons **5.** eight **6.** protons **7.** protons, neutrons **8.** Isotopes **9.** radioactive **10.** molecule **11. a.** eleven **b.** one **12. a.** seventeen **b.** seven **13.** ionic **14.** Once sodium gives up the electron in its outermost shell, it is a stable atom. Chlorine also becomes stable by accepting the electron from sodium, now with eight electrons in its outermost shell. Since the chlorine atom now bears a net negative charge, and sodium a net positive charge, the two atoms (now ions) are attracted to each other and held together by an ionic bond. **15.** covalent **16.** three **17.** water **18.** polar covalent bond **19.** hydrogen bond **20. a.** acidic **b.** basic **21. a.** acid **b.** base **22. a.** cohesive **b.** hydrogen bonding **23.** buffer **24.** four **25.** carbon **26.** Hydrogen **27.** living **28. a.** macromolecules **b.** polymers **29.** monomers **30. a.** monosaccharide **b.** glucose **31.** energy **32.** disaccharide **33. a.** condensation synthesis **b.** water **34. a.** water **b.** hydrolysis **35.** polysaccharide **36. a.** glycogen **b.** cellulose **37. a.** insulation **b.** energy **c.** padding **38. a.** triglycerides **b.** glycerol **c.** fatty acid **d.** neutral **39. a.** unsaturated **b.** saturated **40.** The carbon chain has all the hydrogens it can hold; there are no double bonds between carbon atoms. **41. a.** emulsification **b.** Bile **42.** phospholipid **43.** steroids **44. a.** enzymes **b.** keratin **c.** collagen **45.** amino acids **46. a.** amino **b.** carboxyl **c.** When the carboxyl (acidic) group of one amino acid reacts with the amino group of another amino acid, a molecule of water is removed, and the peptide bond forms. **47. a.** Primary **b.** secondary **c.** tertiary **d.** Quaternary **48.** denatured **49.** nucleotides **50. a.** DNA **b.** RNA **51. a.** sugar **b.** base **c.** four **d.** hydrogen **52. a.** ATP **b.** phosphate

**53.**

|  | Monomers | Function |
|---|---|---|
| Polysaccharides | Glucose | Energy storage |
| Proteins | Amino acids | Enzymes speed chemical reactions; structural components (e.g., muscle proteins) |
| Lipids | Glycerol, 3 fatty acids, phosphate group | Long-term energy storage, plasma membrane structure |
| Nucleic acids | Nucleotide with deoxyribose sugar Nucleotide with ribose sugar | Genetic material Protein synthesis |

**54. a.** glucose, amino acids, glycerol and fatty acids, nucleotides **b.** fat, polysaccharide, polypeptide, nucleic acid **c.** 4 **d.** 2 **55. a.** carbohydrates and lipids **b.** protein **c.** nucleic acids

## DEFINITIONS CROSSWORD

**Across:**
1. cellulose   3. protein   5. ionic   6. glucose   7. electron   8. ATP   11. DNA   12. isotope

**Down:**
1. carbohydrate   2. lipid   4. nucleotide   9. base   10. atom

## CHAPTER TEST

1. d   2. b   3. b   4. a   5. a   6. c   7. c   8. c   9. b
10. d   11. d   12. a   13. b   14. a   15. c   16. d   17. b
18. b   19. a   20. b   21. d   22. b   23. c   24. a   25. b
26. d   27. Radioactive isotopes are used to determine the age of fossils. They are also used in many diagnostic medical tests.   28. Because ice is less dense than water, bodies of water freeze from the top down. Aquatic organisms can survive throughout the winter in the cold water beneath the ice. The cohesiveness of water enables water to travel easily through a tube, such as blood, a liquid, does within blood vessels.

# 3

# CELL STRUCTURE AND FUNCTION

Cells are the smallest units displaying the properties of life. Cells normally are measured in micrometers because they are so small. Their small size ensures a suficient amount of plasma membrane to serve the cytoplasm.

All organisms are composed of cells. The two major kinds of cells are prokaryotic and eukaryotic. They differ by the organization of chromosomal DNA and the presence of organelles in the cytoplasm. Prokaryotic cells lack a nucleus and other membranous organelles.

The nucleus of eukaryotic cells (plant and animal) is defined by a nuclear envelope that separates the nucleoplasm from the cytoplasm. The chromosomal material exists as chromatin until the cell divides. The nucleolus in the nucleus contains ribosomal RNA and the proteins of ribosomal subunits.

The eukaryotic cell contains a variety of structures in the cytoplasm. Ribosomes are the site of protein synthesis. They may exist freely or be attached to the endoplasmic reticulum. Several structures are part of the endomembrane system in the cell. The endoplasmic reticulum provides channels that transport substances through the cell. Substances are processed and packaged by the Golgi apparatus. Lysosomes contain enzymes that promote the breakdown of cell substances.

Some organelles are specialized to handle energy in the cell. Chloroplasts are the site of photosynthesis, whereas the mitochondria are regions involved in cellular respiration. These organelles may be remnants of prokaryotes that inhabited eukaryotic cells over evolutionary time.

The cytoskeleton contains microtubules, intermediate filaments, and actin filaments. They maintain cell shape and assist movement of cell parts.

## STUDY QUESTIONS

Study the text section by section. Answer the study questions so that you can fulfill the learning objectives for each section.

## 3.1 THE CELLULAR LEVEL OF ORGANIZATION (PP. 46—48)

The learning objectives for this section are:
- State the tenets of the cell theory.
- Describe the surface/volume ratio of cells.

1. Check the two statements that are tenets of the cell theory.
   a. _____ All organisms are made up of cells.
   b. _____ Cork cells are living.
   c. _____ Multicellular organisms are living.
   d. _____ Cells come only from preexisting cells.

2. As the volume of a cell a._____, the proportionate amount of cell surface area b._____.

3. A large cell requires more a._____ and produces more b._____ than a small cell. Materials are exchanged at the cell's c._____. Because the surface area of a large cell actually d._____, cell size stays e._____.

## 3.2 EUKARYOTIC CELLS (PP. 49—61)

The learning objectives for this section are:
- Describe parts of animal cells and plant cells.
- Describe the function of the plasma membrane.
- Discuss the specific functions carried out by the membranous organelles.

4. Label this diagram of an animal cell, using the following alphabetized list of terms.

centriole
Golgi apparatus
lysosome
microtubule
mitochondrion
nucleolus
nucleus
ribosome
rough ER
smooth ER
vacuole

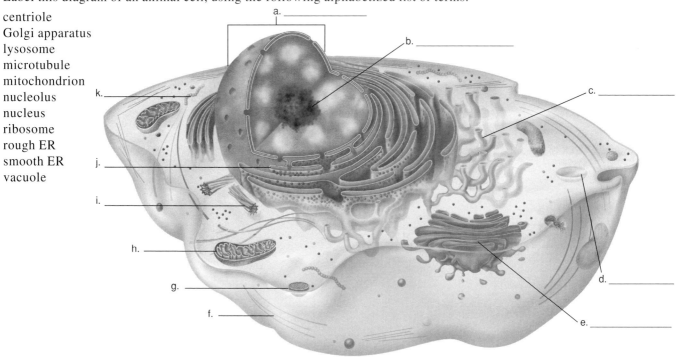

a. _____

b. _____

c. _____

d. _____

e. _____

f. _____

g. _____

h. _____

i. _____

j. _____

k. _____

5. Label this diagram of a plant cell, using the following alphabetized list of terms.

actin filament
cell wall
central vacuole
chloroplast
chromatin
Golgi apparatus
intracellular space
microtubule
middle lamella
mitochondrion
nuclear envelope
nuclear pore
nucleolus
plasma membrane
ribosome
rough ER
smooth ER

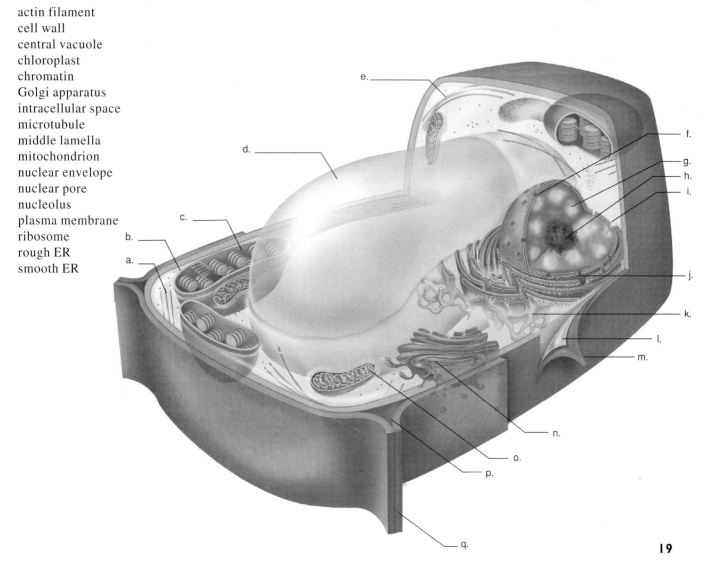

a. _____
b. _____
c. _____
d. _____
e. _____
f. _____
g. _____
h. _____
i. _____
j. _____
k. _____
l. _____
m. _____
n. _____
o. _____
p. _____
q. _____

6. Place the following terms in the appropriate column to compare plant and animal cell structures (some terms are used in both columns).
cell wall     centrioles     chloroplasts     large central vacuole     mitochondria
plasma membrane     small vacuoles only

| Animal | Plant |
|--------|-------|
|        |       |
|        |       |
|        |       |
|        |       |
|        |       |
|        |       |
|        |       |
|        |       |

7. The nucleus is enclosed by the a._____, which contains b._____ that open into the cytoplasm. At the time of cell division, chromatin c._____ to form chromosomes. The nucleus has a region called the d._____ , where e._____ is produced.

8. Explain how these organelles work together.
   a. ribosomes and endoplasmic reticulum _____

   _____

   b. endoplasmic reticulum and Golgi apparatus _____

   _____

   c. lysosomes and vacuoles _____

   _____

   d. chloroplasts and mitochondria _____

   _____

9. Label this diagram, using the following alphabetized list of terms:
   ATP
   carbohydrate
   $CO_2$ and $H_2O$
   chloroplast
   mitochondrion

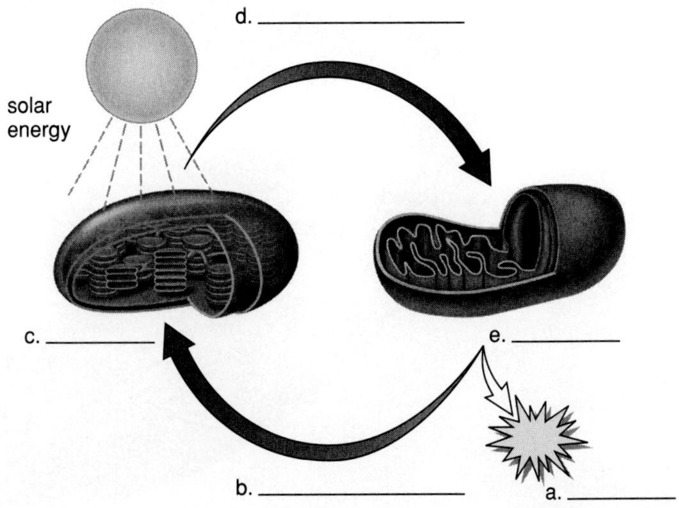

   solar energy

   d. _____

   c. _____

   e. _____

   b. _____

   a. _____

Use this diagram to answer the following: Chloroplasts build up f._____, and mitochondria break it down. Chloroplasts use the molecules that mitochondria give off, namely g._____ and h._____, as raw material for i._____, which utilizes the energy of the j._____. Mitochondria use k._____ produced by chloroplasts to build up a supply of l._____, the energy currency of cells. m._____ is the cellular breakdown of carbohydrate to acquire energy.

10. Label this diagram of a chloroplast, using the following alphabetized list of terms:
    granum
    inner membrane
    outer membrane
    stroma
    thylakoid space

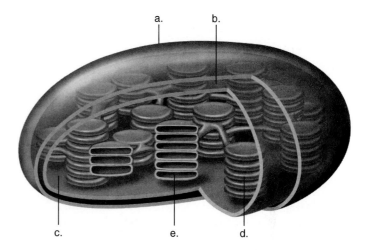

11. Label this diagram of a mitochondrion, using the following alphabetized list of terms:
    cristae
    inner membrane
    matrix
    outer membrane

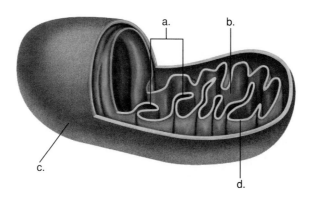

12. Match the definitions to these terms:
    actin filament    intermediate filament    microtubule
    a. _____ small cylinder made of the protein tubulin
    b. _____ long, extremely thin fiber that often interacts with myosin
    c. _____ fibrous polypeptide that varies according to the tissue

13. Microtubules, like actin filaments and intermediate filaments, are able to assemble and a._____. Microtubules radiate out from the centrosome, the main b._____ center in a cell. In animal cells, this center contains two c._____, which have a 9 + 0 pattern of microtubules. Centrosomes have long been associated with the formation of the d._____ during cell division. Centrioles are believed to give rise to e._____, which organize cilia and flagella. Cilia and flagella have a(n) f._____ pattern of microtubules.

## 3.3 PROKARYOTIC CELLS (P. 62)

The learning objective for this section is:
• Name the parts of a prokaryotic cell.

14. Label these diagrams of prokaryotic cells, using the following alphabetized list of terms:

capsule
cell wall
cytoplasm
nucleoid
plasma membrane
ribosome
slime layer
thylakoid

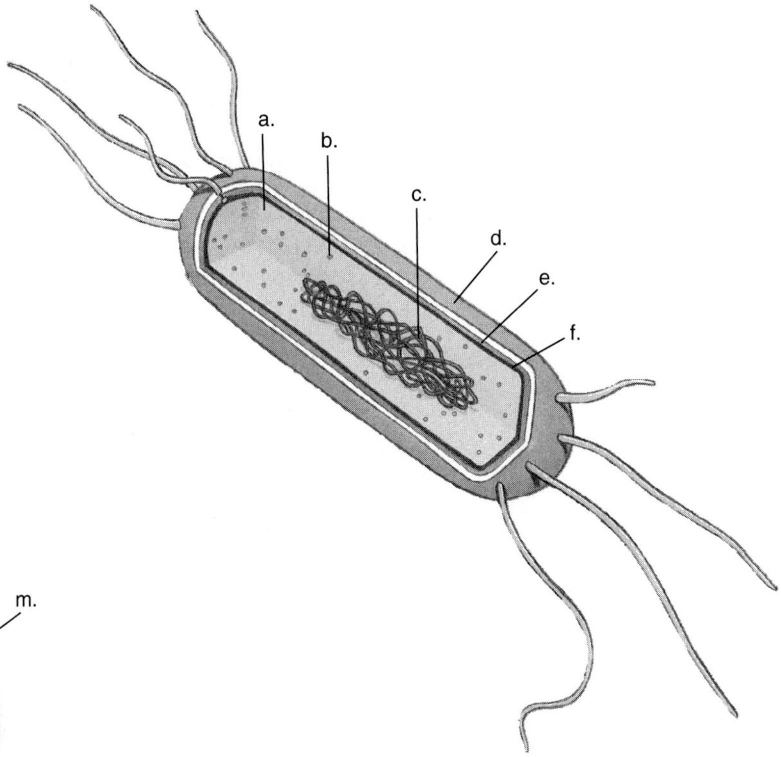

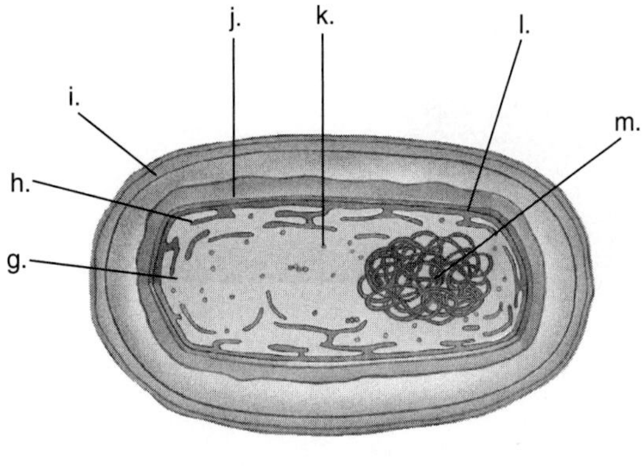

Which of these cells is a cyanobacterium? n._____

For questions 15–20, match each of the following prokaryotic cell parts to its description:
15._____ capsule
16._____ cell wall
17._____ flagellum
18._____ plasmid
19._____ ribosome
20._____ thylakoid

a. long, thin appendage from the cell
b. cytoplasmic granule
c. structure surrounding the cell wall
d. flattened membranous disks
e. external covering around plasma membrane
f. small accessory ring of DNA

21. Complete this table by writing *yes* (the structure is present) or *no* (it is not present) on the lines provided.

| | Prokaryotic | Eukaryotic (animal) |
|---|---|---|
| Plasma membrane | | |
| Cell wall | | |
| Nuclear envelope | | |
| Mitochondria | | |
| Endoplasmic reticulum | | |
| Ribosomes | | |
| Centrioles | | |

## 3.4 EVOLUTION OF THE EUKARYOTIC CELL (P. 63)

The learning objective for this section is:
- Discuss the endosymbiotic theory to explain the evolution of the eukaryotic cell.

22. Match the definitions to these terms:
    chloroplast    mitochondria    flagellum    eukaryote
    a. _____ arose from a photosynthetic cyanobacterial endosymbiont
    b. _____ host cell taking up prokaryotes by endosymbiosis
    c. _____ arose from heterotrophic bacterial endosymbionts
    d. _____ derived from a spirochete prokaryote

Review key terms by using the following alphabetized list of terms to fill in the blanks below. Then complete the wordsearch.

```
S E E T H E C A T R U N T T H I M O
E M O S O M O R H C B R P F A M I P
T H E R I M O T R Y C Y L T T O C K
T H E R I O T B G T D O A G D H R E
G I N N Y O C Y T O S K E L E T O N
R E N T O F R D C P T Y I O S H T I
A Y L I L L K A Q L W Z E U P P U P
F U N I D A R Y T A Y V L K C I B L
R E T N U G G U N S L O O H G F U M
L I T C O E N Z Y M E U I U O D L J
K I L Y T L N U T L R U R O M M E E
F L A G G L T M C I Y O T R E E S I
G O L G A U B U S O K J N O P L Y G
C I L I U M N R T E L L E N A G R O
N U C L A I U B F I T O C J U M P G
```

*centriole*
*chromosome*
*cilium*
*cytoplasm*
*cytoskeleton*
*flagellum*
*microtubule*
*nucleolus*
*organelle*

a. _____ Cylinder containing microtubules in a 9 + 0 arrangement and associated with the formation of basal bodies.

b. _____ Rodlike structure in the nucleus; DNA.

c. _____ Short, hairlike extension from the cell.

d. _____ Slender, long extension from the cell.

e. _____ Structure within the cell, surrounded by a membrane, with a specific function.

f. _____ A region inside the nucleus where rRNA is produced for the manufacture of ribosomes.

g. _____ Cytoskeleton component made up of the protein tubulin.

h. _____ A network of protein filaments found within the cell. Its function is to move organelles around, among other things.

i. _____ Semifluid medium outside the nucleus in the cell.

# CHAPTER TEST

## OBJECTIVE QUESTIONS

Do not refer to the text when taking this test. In questions 1–8, match each cell part with these descriptions.

    a. regulates passage of substances into the cell
    b. processing and transport channel
    c. contains enzymes for digestion
    d. site of protein synthesis
    e. location of the nucleolus
    f. site of cellular respiration
    g. found in plants, not animals
    h. maintains cell shape

_____ 1. chloroplast
_____ 2. cytoskeleton
_____ 3. endoplasmic reticulum
_____ 4. lysosome
_____ 5. mitochondrion
_____ 6. nucleus
_____ 7. plasma membrane
_____ 8. ribosome

_____ 9. Which of the following structures is part of the cell's endomembrane system?
    a. chloroplast
    b. endoplasmic reticulum
    c. mitochondrion
    d. nucleolus

_____10. How are mitochondria and chloroplasts similar to bacteria?
    a. They are bounded by a single membrane.
    b. They have a limited amount of genetic material.
    c. They lack ribosomes.
    d. They are larger than normal cells.

_____11. Plant cells
    a. have a cell wall but not a plasma membrane.
    b. have chloroplasts but no mitochondria.
    c. do not have any centrioles and yet divide.
    d. have a large central vacuole but do not have endoplasmic reticulum.

_____12. Which of these does NOT contain nucleic acid?
   a. chromosomes
   b. ribosomes
   c. chromatin
   d. centrioles
   e. genes

_____13. How are mitochondria like chloroplasts?
   a. They have the same structure.
   b. They both absorb the energy of the sun.
   c. They both are concerned with energy.
   d. They are both in animal cells.

_____14. Which of the following cell structures within the cytoplasm is connected to the nuclear envelope?
   a. nucleolus
   b. chromatin
   c. endoplasmic reticulum
   d. vacuoles
   e. lysosomes

_____15. Which organelle is used to produce steroid hormones and to detoxify drugs?
   a. lysosomes
   b. Golgi apparatus
   c. mitochondria
   d. rough endoplasmic reticulum
   e. smooth endoplasmic reticulum

_____16. Select the structure found in eukaryotic cells but not in prokaryotic cells.
   a. plasma membrane
   b. cell wall
   c. mitochondrion
   d. ribosome

_____17. Select the incorrect pair.
   a. capsule—covering
   b. cell wall—covering
   c. flagellum—movement
   d. plasmid—movement

_____18. The cytoplasmic structure in a bacterial cell is the
   a. cell wall.
   b. nucleoid.
   c. plasma membrane.
   d. ribosome.

_____19. Which organelle has a 9 + 0 arrangement of microtubules and is associated with the formation of basal bodies?
   a. actin filament
   b. spindle fiber
   c. flagellum
   d. centriole

_____20. The cytoskeleton of the cell is composed of
   a. protein.
   b. actin filaments.
   c. microtubules.
   d. All of these are correct.

## THOUGHT QUESTIONS

Answer in complete sentences.

21. What would be the effect on a cell if it were suddenly to lose its mitochondria?

22. How would the destruction of the Golgi apparatus affect a cell?

**Test Results:** _____     number correct ÷ 22 = _____ × 100 = _____ %

# ANSWER KEY

## STUDY QUESTIONS

**1.** a, d  **2. a.** increases **b.** decreases  **3 a.** nutrients **b.** wastes **c.** surface **d.** decreases proportionately **e.** small **4. a.** nucleus **b.** nucleolus **c.** smooth ER **d.** vacuole **e.** Golgi apparatus  **f.** microtubule **g.** lysosome **h.** mitochondrion **i.** centriole **j.** rough ER **k.** polyribosome **5. a.** actin filament **b.** ribosome **c.** chloroplast **d.** central vacuole **e.** microtubule **f.** nuclear pore **g.** chromatin **h.** nucleolus **i.** nuclear envelope **j.** rough ER **k.** smooth ER **l.** plasma membrane **m.** cell wall **n.** Golgi apparatus **o.** mitochondrion **p.** intracellular space **q.** middle lamella
**6.**

| Animal | Plant |
|---|---|
| centrioles | cell wall |
| mitochondria | mitochondria |
| small vacuoles only | large central vacuole |
| plasma membrane | plasma membrane |
|  | chloroplasts |

**7. a.** nuclear envelope **b.** nuclear pores **c.** condenses **d.** nucleolus **e.** rRNA  **8. a.** Proteins are made at the ribosomes located on the endoplasmic reticulum. **b.** Products made at the endoplasmic reticulum are sent to the Golgi apparatus for final processing and packaging. **c.** Vacuoles may contain a substance that can be digested after fusion with lysosomes. **d.** Chloroplasts synthesize carbohydrates, which mitochondria break down to produce ATP molecules.  **9. a.** ATP **b.** $CO_2$ and $H_2O$ **c.** chloroplast **d.** carbohydrate **e.** mitochondrion **f.** carbohydrate **g.** carbon dioxide **h.** water **i.** photosynthesis **j.** sun **k.** carbohydrate **l.** ATP **m.** Cellular respiration  **10. a.** outer membrane **b.** inner membrane **c.** stroma **d.** granum **e.** thylakoid space  **11. a.** cristae **b.** matrix **c.** outer membrane **d.** inner membrane  **12. a.** microtubule **b.** actin filament **c.** intermediate filament  **13. a.** disassemble **b.** microtubule organizing center **c.** centrioles **d.** spindle **e.** basal bodies **f.** 9 + 2  **14. a.** cytoplasm **b.** ribosome **c.** nucleoid region **d.** capsule **e.** cell wall **f.** plasma membrane **g.** cytoplasm **h.** thylakoid **i.** slime layer **j.** cell wall **k.** ribosome **l.** plasma membrane **m.** nucleoid **n.** the one on the right  **15.** c  **16.** e **17.** a  **18.** f  **19.** b  **20.** d

**21.**

| Prokaryotic | Eukaryotic (animal) |
|---|---|
| yes | yes |
| yes | no |
| no | yes |
| no | yes |
| no | yes |
| yes | yes |
| no | yes |

**22. a.** chloroplast **b.** eukaryote **c.** mitochondria **d.** flagellum

## DEFINITIONS WORDSEARCH

**a.** centriole **b.** chromosome **c.** cilium **d.** flagellum **e.** organelle **f.** nucleolus **g.** microtubule **h.** cytoskeleton **i.** cytoplasm

```
                                            M
    E M O S O M O R H C               I
                    Y                 C
                    T                 R
            C Y T O S K E L E T O N
        F           P         S   T
        L           L     E U     U
        A           A     L O     B
        G           S O   O       U
        E       M E   I   R       L
        L       L     L R   T     E
        L   C             T
        U   U             N
      C I L I U M N     E L L E N A G R O
                                C
```

## CHAPTER TEST

**1.** g  **2.** h  **3.** b  **4.** c  **5.** f  **6.** e  **7.** a  **8.** d  **9.** b **10.** b  **11.** c  **12.** d  **13.** c  **14.** c  **15.** e  **16.** c **17.** d  **18.** d  **19.** d  **20.** d  **21.** The cell would be unable to extract energy from carbohydrates. The ATP harvested by this process would be unavailable for cell functions. Therefore, the cell would die.  **22.** The smooth ER packages substances in vesicles. A large portion of these go to the Golgi apparatus for further processing. These vesicles would most likely accumulate in the cell to the point that the cell would be unable to function properly.

# 4

# MEMBRANE STRUCTURE AND FUNCTION

## CHAPTER REVIEW

According to the **fluid-mosaic model** of membrane structure, phospholipid molecules provide a fluid, lipid bilayer with their polar heads at the membrane surfaces. Hydrophobic portions of protein molecules are in the lipid bilayer; their hydrophilic portions are at the surfaces. The proteins form channels and function as receptors, enzymes, and carrier molecules.

Molecules move across a membrane in several ways. By **diffusion,** molecules move down their concentration gradient. **Osmosis** is the diffusion of water through a differentially permeable membrane. When cells are in a **hy-**potonic solution, they gain water; when they are in a **hypertonic solution,** they lose water. In **isotonic solutions,** cells neither gain nor lose water. Both diffusion and osmosis are passive processes that do not require energy. Facilitated transport is also passive and involves carrier molecules moving substances from higher to lower concentrations. **Active transport** moves substances in the opposite direction, with the function of a carrier molecule and energy. Larger substances pass through cells by **endocytosis** and **exocytosis.**

## STUDY QUESTIONS

Study the text section by section. Answer the study questions so that you can fulfill the learning objectives for each section.

### 4.1 PLASMA MEMBRANE STRUCTURE AND FUNCTION (PP. 68–69)

The learning objectives for this section are:
- Explain the function of the plasma membrane.
- Describe the structure of the plasma membrane, including the phospholipid bilayer and the various types of proteins.

1. Label this diagram of the plasma membrane, using the following alphabetized list of terms:
   carbohydrate chain
   cholesterol
   cytoskeleton filaments
   glycolipid
   glycoprotein
   phospholipid bilayer
   protein molecule

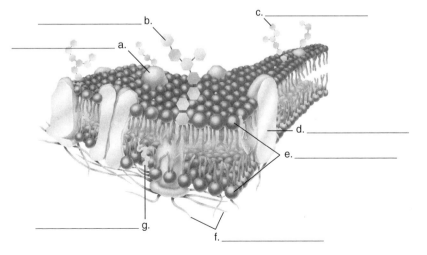

b. _____

c. _____

a. _____

d. _____

e. _____

g. _____

f. _____

In questions 2–5, fill in the blanks.

2. The two components of the fluid-mosaic model of membrane structure are a._____ and b._____.

3. a._____ form a bilayer, in which the b._____ heads are at the surfaces of the membranes, and the c._____ tails face each other, making up the interior of the membrane. The lipid d._____, which is also in the membrane, e._____ the membrane's permeability.

4. Complete the following sentences, using the terms *hydrophilic* and/or *hydrophobic:* Transmembrane proteins are found within the plasma membrane. a._____ regions are embedded within the membrane, and b._____ regions project from both surfaces of the bilayer.

5. Both glycolipids and glycoproteins have a(n) a._____ chain and are active in cell-to- b._____ recognition.

6. Label the following diagrams of proteins found in the membrane, and state a function on the lines provided:

carrier protein _____

cell recognition protein _____

channel protein _____

enzymatic protein _____

receptor protein _____

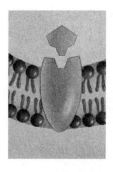

a. _____
_____

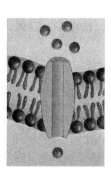

b. _____
_____

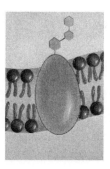

c. _____
_____

d. _____
_____

e. _____
_____

## 4.2 THE PERMEABILITY OF THE PLASMA MEMBRANE (P. 70)

The learning objective for this section is:
- Describe the passage of molecules into and out of the plasma membrane.

7. In the following diagram, assume that glucose and water can cross the membrane and that protein cannot.

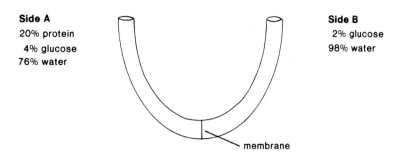

**Side A**
20% protein
4% glucose
76% water

**Side B**
2% glucose
98% water

membrane

a. Will the amount of water on side A stay the same, or increase or decrease with time? _____

b. Will the amount of protein on side A stay the same, or increase or decrease with time? _____

c. Will glucose cross the membrane toward side A or toward side B? _____

d. What will happen to the level of solution on each side of the membrane? _____

_____

## 4.3 DIFFUSION AND OSMOSIS (PP. 71–73)

The learning objectives for this section are:
- Describe the process of diffusion.
- Explain how water diffuses across the plasma membrane.

8. Place the appropriate letter next to each statement.

D—diffusion    O—osmosis

a. _____ Algae in a pond become dehydrated.
b. _____ A hypertonic solution draws water out of a cell.
c. _____ A red blood cell bursts in a test tube.
d. _____ Dye crystals spread out in a beaker of water.
e. _____ Gases move across the plasma membrane.
f. _____ Perfume is sensed from the other side of a room.

9. If a solution is 8% solute, it is [a.]_____% solvent. If a solution is 99.5% solvent, it is [b.]_____% solute. If solution A is 2% solute and solution B is 3% solute, then solution A is [c.]_____ to solution B, which is [d.]_____ to solution A. Compared to solution A, a solution with 2% solute is [e.]_____.

10. Complete this diagram to describe the effect of tonicity on red blood cells.

| Tonicity | Before | After |
|---|---|---|
| Isotonic Solution | (cell diagram) | a. |
| b. | (cell diagram) | (cell diagram) |
| Hypotonic Solution | (cell diagram) | c. |

11. Complete this diagram to describe the effect of tonicity on plant cells.

| Tonicity | Before | After |
|---|---|---|
| a. | cell wall / water vacuole / plasma membrane (cell diagram) | (cell diagram) |
| b. | (cell diagram) | vacuole (cell diagram) |
| Hypertonic Solution | (cell diagram) | c. |

## 4.4 TRANSPORT BY CARRIER PROTEINS (PP. 74–75)

The learning objectives for this section are:
- Explain the role of carrier proteins.
- Describe facilitated transport and active transport.

12. Place the appropriate letter(s) next to each statement.

F—facilitated transport    A—active transport    FA—both

a. _____ Uses a carrier molecule.
b. _____ Substances travel down a concentration gradient.
c. _____ Substances travel against a concentration gradient.
d. _____ Sodium-potassium pump.
e. _____ Energy is not required.

The learning objective for this section is:
- Define exocytosis and endocytosis.

13. Place the appropriate letters next to each statement.

    Ex—exocytosis    En—endocytosis

    a. _____ Vesicles formed by the Golgi apparatus fuse with the plasma membrane.
    b. _____ Materials leave the cell.
    c. _____ Phagocytosis is an example.
    d. _____ Pinocytosis is an example.
    e. _____ Occurs after molecules bind to receptor proteins.

## DEFINITIONS WORDSEARCH

Review key terms by using the following alphabetized list of terms to fill in the blanks below. Then complete the wordsearch.

```
S E E T H E C A T R U N T T H I M O O O
E M O S O M O R H C B R P F N M I P D P
L S I S O T Y C O D N E I M O P J C F C
T H O R I O T B G T D O A G I H R E Q E
G I G L Y C O P R O T E I N S T O N Z N
R E N T V F R D C P T Y I O U H T I M E
A Y L I L E K A Q L W Z E U F P P P L X
F U N I D A N Y T A Y V L K F I I L L O
R E T N U G G T N S L O O H I F N M M C
L I T C O E N Z Y M O U I U D D O J J Y
K I L Y T L N U T L S U R O M M C E E T
F L A G G L T M C I M O T R E E Y I I O
G O L S I S O T Y C O G A H P L T G G S
C I L I U M N R T E S L E N A G O O O I
N U C L A I U B F I I O S J U M S G G S
Q M B O Y B U P L A S M O L Y S I S R G
G I N R W X M P N K U W L J R O S X N T
C V Z M O I M H A N I B U R E E H G U G
O C H C G O I N F Y V O T S U F F T W S
D U C L A I U B N I T O E J U W J K I G
```

*diffusion*
*endocytosis*
*exocytosis*
*glycoproteins*
*osmosis*
*phagocytosis*
*pinocytosis*
*plasmolysis*
*solute*
*solvent*

a. _____ Proteins in plasma membranes that bear a carbohydrate chain.
b. _____ Movement of molecules or ions from a region of higher to lower concentration; it requires no energy and stops when the distribution is equal.
c. _____ Fluid, such as water, that dissolves solutes.
d. _____ Diffusion of water through a differentially permeable membrane.
e. _____ Contraction of the cell contents due to the loss of water.
f. _____ Process in which an intracellular vesicle fuses with the plasma membrane so that the vesicle's contents are released outside the cell.
g. _____ Process by which amoeboid-type cells engulf large substances, forming an intracellular vacuole.
h. _____ Process by which substances are moved into the cell from the environment by phagocytosis or pinocytosis.
i. _____ Process by which vesicle formation brings macromolecules into the cell.
j. _____ Substance that is dissolved in a solvent, forming a solution.

## OBJECTIVE QUESTIONS

Do not refer to the text when taking this test. In questions 1–7, match the descriptions to each transport process:

a. active transport
b. diffusion
c. exocytosis
d. facilitated transport
e. osmosis
f. phagocytosis
g. pinocytosis

_____ 1. Small particle or liquid intake into a cell

_____ 2. Requires vacuole formation

_____ 3. Carrier molecule, no energy

_____ 4. Carrier molecule, energy required

_____ 5. Water enters a hypertonic solution from a cell

_____ 6. Secretion from the cell

_____ 7. Dye molecules spread through water

_____ 8. Proteins form the nonactive matrix of the plasma membrane.
   a. true
   b. false

_____ 9. Which type of molecule forms a bilayer within the plasma membrane?
   a. carbohydrate
   b. protein
   c. phospholipid
   d. nucleic acid

_____ 10. Hydrophilic ends of proteins are oriented toward membrane surfaces.
   a. true
   b. false

_____ 11. Lipid-soluble molecules pass through the plasma membrane by
   a. active transport.
   b. diffusing through it.
   c. facilitated transport.
   d. use of the sodium-potassium pump.

_____ 12. A 2% salt solution is _____ to a 4% salt solution.
   a. hypertonic
   b. hypotonic
   c. isometric
   d. isotonic

_____ 13. Which molecule is directly required for operation of the sodium-potassium pump?
   a. ATP
   b. NAD$^+$
   c. DNA
   d. water

_____ 14. In cells, which process moves materials opposite to the direction of the other three?
   a. endocytosis
   b. exocytosis
   c. phagocytosis
   d. pinocytosis

_____ 15. Which of the following statements is NOT true about osmosis?
   a. A differentially permeable membrane must separate two solutions.
   b. One side of the membrane must have more water than the other side of the membrane.
   c. The membrane must permit water to pass.
   d. Over time, the side that initially had the larger concentration of solute will become even more concentrated.

_____ 16. A small lipid-soluble molecule passes easily through the plasma membrane. Which of these statements is the most likely explanation?
   a. A carrier protein must be at work.
   b. The plasma membrane is partially composed of lipid molecules.
   c. The cell is expending energy to do this.
   d. Phagocytosis has enclosed this molecule in a vacuole.

_____ 17. Which of these does NOT require an expenditure of energy?
   a. diffusion
   b. osmosis
   c. facilitated transport
   d. None of these require energy.

_____ 18. Which term refers to the bursting of an animal cell?
   a. plasmolysis
   b. crenation
   c. lysis
   d. turgor pressure

_____ 19. An animal cell always takes in water when placed in a(n) _____ solution.
   a. hypertonic
   b. osmotic
   c. isotonic
   d. hypotonic

_____ 20. Which of the following is actively transported across plasma membranes?
   a. carbon dioxide
   b. oxygen
   c. water
   d. sodium ions

## THOUGHT QUESTIONS

Answer in complete sentences.

21. Why is the plasma membrane considered so important to the cell?

22. What osmotic problem would plant cells experience if they lost their cell walls?

**Test Results:** _____ number correct ÷ 22 = _____ × 100 = _____ %

## ANSWER KEY

### STUDY QUESTIONS

**1. a.** glycoprotein **b.** carbohydrate chain **c.** glycolipid **d.** protein molecule **e.** phospholipid bilayer **f.** cytoskeleton filaments **g.** cholesterol **2. a.** lipids **b.** proteins **3. a.** Phospholipids **b.** hydrophilic (polar) **c.** hydrophobic **d.** cholesterol **e.** reduces **4. a.** hydrophobic **b.** hydrophilic **5. a.** carbohydrate **b.** cell **6. a.** receptor protein, shaped in such a way that a specific molecule can bind to it. **b.** channel protein, allows molecules to pass across the plasma membrane. **c.** cell recognition protein, functions in cell-to-cell recognition. **d.** enzymatic protein, catalyzes a specific reaction. **e.** carrier protein, allows selective passage of molecules across the plasma membrane. **7. a** increase **b.** stay the same **c.** toward side B **d.** Side A will rise, and side B will fall. **8. a.** O **b.** O **c.** O **d.** D **e.** D **f.** D **9. a.** 92 **b.** 0.5 **c.** hypotonic **d.** hypertonic **e.** isotonic **10. a.** cell is same size and shape **b.** Hypertonic Solution **c.** cell is bursting **11. a.** Isotonic Solution **b.** Hypotonic Solution **c.** vacuole is much smaller **12. a.** F, A **b.** F **c.** A **d.** A **e.** F **13. a.** Ex **b.** Ex **c.** En **d.** En **e.** En

### DEFINITIONS WORDSEARCH

```
                            N
   S I S O T Y C O D N E    O
   O                        I
   G L Y C O P R O T E I N S
      V                     U        E
       E                    F   P    X
        N                   F   I    O
         T                  I   N    C
              O             D   O    Y
              S                 C    T
              M                 Y    O
   S I S O T Y C O G A H P       T    S
              S                 O    I
              I     S           S    S
         P L A S M O L Y S I S
              L             S
              U
              T
              E
```

### CHAPTER TEST

**1.** g **2.** f **3.** d **4.** a **5.** e **6.** c **7.** b **8.** b **9.** c **10.** a **11.** b **12.** b **13.** a **14.** b **15.** d **16.** b **17.** d **18.** c **19.** d **20.** d **21.** The plasma membrane is very important because it is the outer living boundary of the cell; it provides for and regulates the passage of molecules into and out of the cell. **22.** In a hypotonic environment, plant cells would continue to gain water until they burst.

# 5

# CELL DIVISION

The **cell cycle** has four stages. During the G₁ stage, the organelles increase in number; during the S stage, DNA replication occurs; during the G₂ stage, various proteins are synthesized; and during the M stage, mitosis occurs.

Each species has a characteristic number of chromosomes. The total number is the **diploid** number, and half this number is the **haploid** number. Among eukaryotes, cell division involves nuclear division and division of the cytoplasm (**cytokinesis**).

Replication of DNA precedes cell division. The duplicated chromosome is composed of two sister chromatids held together at a **centromere.** During **mitosis,** the centromeres split and daughter chromosomes go into each new cell.

Mitosis has the following phases: **prophase,** when the chromosomes have no particular arrangement; **metaphase,** when the chromosomes are aligned at the metaphase plate; **anaphase,** when daughter chromosomes move toward the poles; and **telophase,** when new nuclear envelopes form around the daughter chromosomes and cytokinesis begins.

**Meiosis** is found in any life cycle that involves sexual reproduction. During meiosis, **homologues** (homologous chromosomes) separate, and this leads to gametes with half the number of chromosomes, ensuring that offspring will have the same chromosome number as their parents.

Meiosis utilizes two nuclear divisions. During meiosis I, homologues undergo **synapsis.** The nonsister chromatids of the resulting tetrad exchange chromosome pieces; this is called crossing-over. When the homologues separate during meiosis I, each daughter cell receives one from each pair of chromosomes. Separation of daughter chromosomes derived from sister chromatids during meiosis II then leads to a total of four new cells, each with the haploid number of chromosomes.

**Spermatogenesis** in males produces four viable sperm, while **oogenesis** in females produces one egg and two **polar bodies.** Each **gamete** is specialized for the job it does; the sperm is a tiny, flagellated cell that swims to the cytoplasm-laden egg.

Among sexually reproducing organisms, meiosis produces variation by independent assortment of homologues and crossing-over. Fertilization also contributes to variation.

## STUDY QUESTIONS

Study the text section by section. Answer the study questions so that you can fulfill the learning objectives for each section.

### 5.1 CELL INCREASE AND DECREASE (PP. 82–83)

The learning objectives for this section are:
- Identify the causes of cell increase and cell decrease.
- Identify the stages of the cell cycle.
- Discuss control of the cell cycle.
- Identify the significance of apoptosis.

1. Complete this diagram of the cell cycle by labeling the phases as $G_1$, S, $G_2$, or M. What do these letters stand for?

   a. $G_1$ _____

   b. S _____

   c. $G_2$ _____

   d. M _____

2. Match the letters in the following diagram to these phrases.

   _____ Mitosis and cytokinesis occur.

   _____ DNA replication occurs as chromosomes duplicate.

   _____ Growth occurs as cell prepares to divide.

   _____ Growth occurs as organelles double.

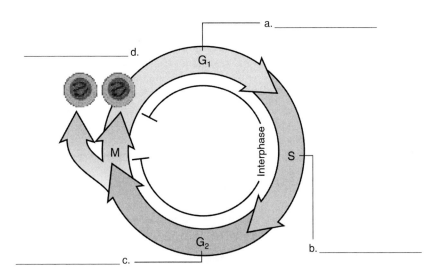

3. Match the checkpoints in the following diagram to these phrases.

   _____ Apoptosis can occur if DNA is damaged.

   _____ Mitosis will not occur if DNA is damaged or not replicated.

   _____ Mitosis stops if chromosomes are not properly aligned.

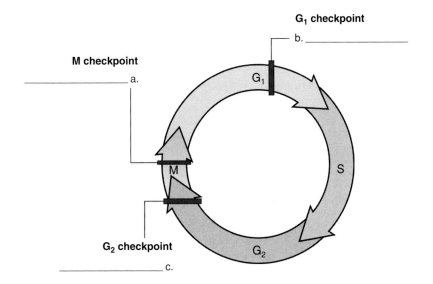

4. Place an X on the line beside each statement that is true of apoptosis.
   _____ a. Some cells in the body are always undergoing apoptosis.
   _____ b. Apoptosis is abnormal and occurs rarely.
   _____ c. Cells are destroyed during apoptosis by external enzymes.
   _____ d. Cells are destroyed during apoptosis by internal enzymes.
   _____ e. At the end of apoptosis, the cell looks like it did before.
   _____ f. At the end of apoptosis, the cell is fragmented.

## 5.2 MAINTAINING THE CHROMOSOME NUMBER (PP. 85–89)

The learning objective for this section is:
- Explain the functions of mitosis and its different phases.

5. Complete each of the following statements with the correct number:

   In corn, the haploid chromosome number is 10. Its body cells normally have a._____ chromosomes.

   The diploid chromosome number in the domestic cat is 38. Normally, its sex cells have b._____

   chromosomes. The horse has a haploid chromosome number of 32. In this animal, 2n = c._____.

   The sex cells of a dog normally have 39 chromosomes. In this animal, n = d._____.

6. Label the following diagram:

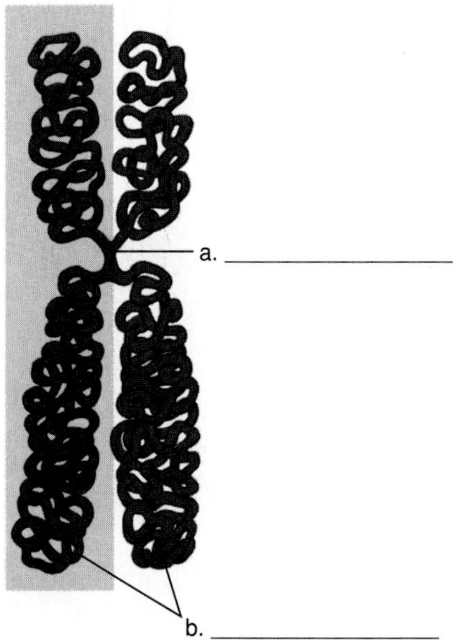

   a. _____

   b. _____

7. Label this diagram, using the following alphabetized list of terms.

   aster
   centriole
   centromere
   centrosome
   chromatid
   kinetochore
   nuclear membrane fragment

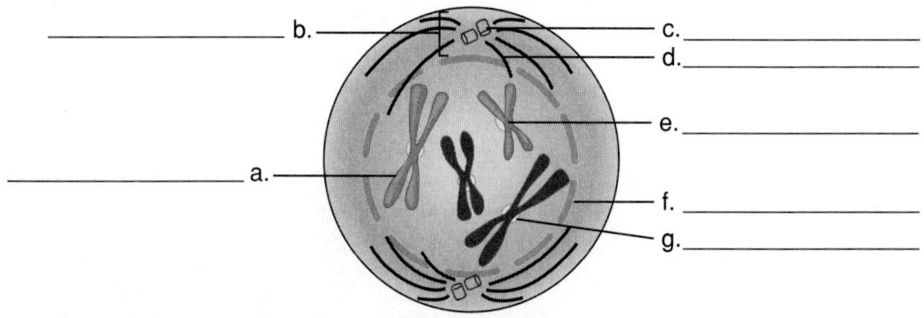

   b. _____
   a. _____
   c. _____
   d. _____
   e. _____
   f. _____
   g. _____

8. Complete the following diagrams to show the arrangement and movement of chromosomes during animal cell mitosis. Briefly describe the events of each phase on the lines provided below.

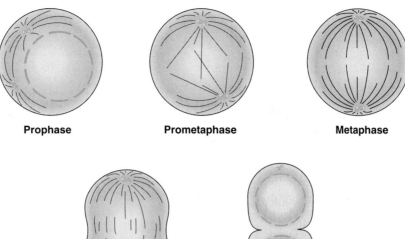

**Prophase**          **Prometaphase**          **Metaphase**

**Anaphase**          **Telophase**

   a. Prophase: _____

   b. Metaphase: _____

   c. Anaphase: _____

   d. Telophase: _____

9. a. How does an animal cell undergo cytokinesis? _____

   b. Which phases show the process of cytokinesis? _____

10. To show the difference between plant and animal mitosis, complete the following table by writing *yes* or *no*.

| Mitosis | Plant Cell | Animal Cell |
| --- | --- | --- |
| Same phases | | |
| Spindle fibers | | |
| Aster | | |
| Cell plate | | |
| Furrowing | | |

11. a. Do plant cells have a centrosome? _____

   _____

   b. Do plant cells have centrioles? _____

   _____

   c. Are centrioles necessary to spindle formation? Explain. _____

   _____

The learning objectives for this section are:
- Describe how meiosis reduces the chromosome number.
- Explain the phases of meiosis I and meiosis II.

In questions 12–14, fill in the blanks.

12. The nuclear division that reduces the chromosome number from the a._____ number to the
    b._____ number is called meiosis.

13. In a life cycle, the zygote always has the _____ number of chromosomes.

14. A pair of chromosomes having the same length and centromere position are called a. _____.
    The b._____, a product of fertilization, is always diploid.

15. Indicate whether these statements regarding the role of meiosis are true (T) or false (F):
    a. _____ In animals, meiosis forms gametes that fuse to form a zygote.
    b. _____ Meiosis forms haploid cells in the life cycle of animals.
    c. _____ Meiosis produces four diploid cells over two divisions.

16. Label this summary diagram of meiosis, using the following alphabetized list of terms.

    centromere
    DNA replication
    meiosis I
    meiosis II
    nucleolus
    sister chromatids
    synapsis

    Solve 2n = h._____

    Solve n = i._____

    Why is it correct to symbolize meiosis as 2n → n? j. _____
    _____

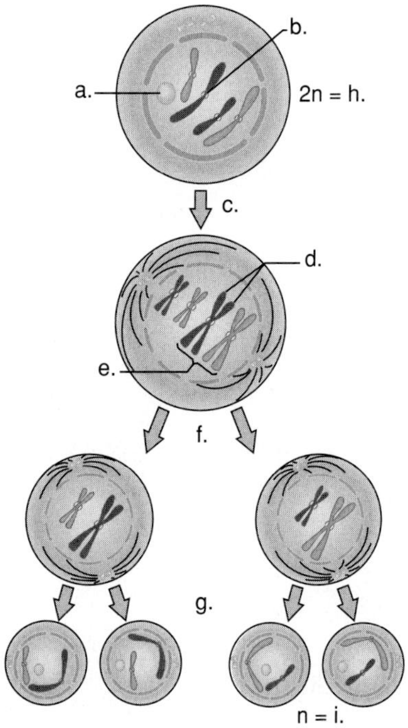

17. a. What is the diploid number of chromosomes in the diagram in question 16? _____
    b. Which structures separate during meiosis I? _____
    c. Which structures separate during meiosis II? _____
    d. Does chromosome duplication occur between meiosis I and meiosis II? _____
    e. Why or why not? _____

18. Match the following terms with the appropriate description:
    crossing-over     genetic variation     synapsis

    a. _____ Homologues line up side by side.
    b. _____ Nonsister chromatids exchange genetic material.
    c. _____ The arrangement of genetic material is new due to crossing-over.

19. Using ink for one duplicated chromosome and pencil for the other, color these homologues before and after crossing-over has occurred.

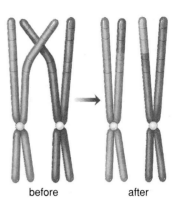

before          after

20. Label and complete these diagrams to show the arrangement and movement of chromosomes during meiosis I and meiosis II. (The diagram for meiosis II pertains to only one daughter cell from meiosis I.)

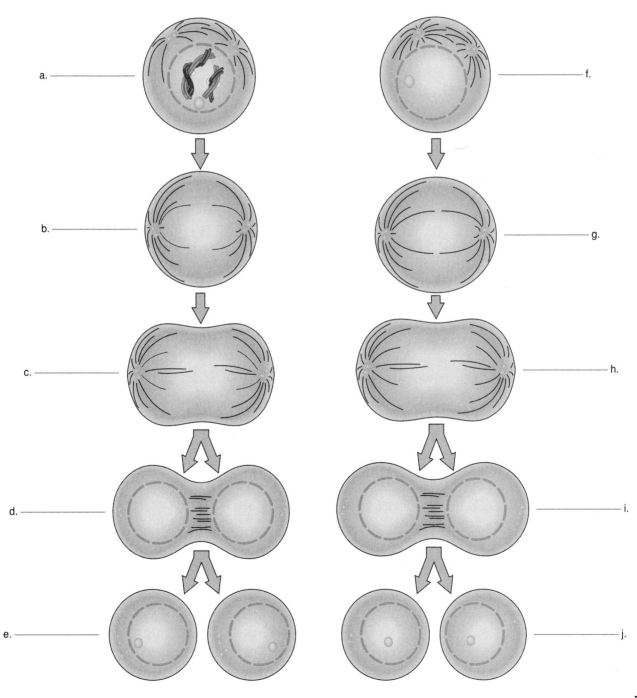

a.

b.

c.

d.

e.

f.

g.

h.

i.

j.

21. Show how genetic recombination occurs as a result of sexual reproduction by matching the following statements:
    1. Zygote carries a unique combination of chromosomes and genes.
    2. Gametes carry different combinations of chromosomes.
    3. Daughter chromosomes carry different combinations of genes.
    a. _____ Homologues separate independently.
    b. _____ Crossing-over occurs.
    c. _____ Gametes fuse during fertilization.

## 5.4 COMPARISON OF MEIOSIS WITH MITOSIS (PP. 94—95)

The learning objective for this section is:
• Understand the differences between the phases of meiosis and mitosis.

22. Complete the table by writing *yes* or *no* to distinguish meiosis from mitosis.

| | Meiosis | Mitosis |
|---|---|---|
| Complete after one division | | |
| Requires two successive divisions | | |
| During anaphase, daughter chromosomes separate | | |
| During anaphase I, homologous chromosomes separate | | |
| Results in daughter cells with the diploid number of chromosomes | | |
| Results in daughter cells with the haploid number of chromosomes | | |
| In animals, is unique to the somatic (body) cells | | |
| In animals, is unique to formation of gametes | | |

## 5.5 THE HUMAN LIFE CYCLE (PP. 96—97)

The learning objectives for this section are:
• Describe the roles of mitosis and meiosis in the human life cycle.
• Define spermatogenesis and oogenesis.

23. Indicate whether these statements are true (T) or false (F):
    a. _____ Meiosis in human males is a part of spermatogenesis.
    b. _____ Mitosis in human females is a part of oogenesis.
    c. _____ Oogenesis occurs in the testis.
    d. _____ A zygote undergoes mitosis during the development of the embryo.
    e. _____ Oogenesis produces four functional egg cells from one cell.
24. State whether the following processes occur in males (M), females (F), or both (B):
    a. _____ gamete formation
    b. _____ spermatogenesis
    c. _____ oogenesis
    d. _____ polar body formation

I. Make playing cards with these words or phrases on them:

- *nucleolus*
- *chromosomes (chromatids held by centromere)*
- *centrosome with centrioles*
- *nucleus (nuclear envelope)*
- *spindle fibers*
- *cell plate*
- *aster*
- *furrowing*

   a. From these cards, select those that have to do with the formation and structure of the spindle.

   b. From the cards remaining, select those that name structures that disappear during mitosis.

   c. Select the cards that have to do with cytokinesis.

II. Make cards with these phrases on them:

- *daughter chromosomes separate*
- *distinct chromosomes within daughter nuclei*
- *chromosomes arranged at the metaphase plate*
- *chromosomes are distinct; spindle appears; nucleolus disappears and nuclear envelope fragments*
- *homologues separate*

   a. Arrange the cards to describe the events of mitosis or meiosis II.

   b. Arrange the cards to describe the events of meiosis I.

III. Make three cards that are marked as follows:

   *16*

   *8 (diploid number)*

   *4 (haploid number)*

   a. Pick the card that tells how many chromosomes are in the parent nucleus before division.

   b. Pick the card that tells how many chromosomes each daughter nucleus will have after mitosis.

   c. Pick the card that tells how many chromosomes are in the parent nucleus during prophase of mitosis or prophase I of meiosis.

   d. Pick the card that tells how many chromatids are in the parent nucleus during prophase of mitosis or prophase I of meiosis. (For purposes of the game, assume the chromosomes are duplicated.)

   e. Pick the card that tells how many chromosomes are at *both* poles during anaphase of mitosis.

   f. Pick the card that tells how many chromosomes are at *both* poles during anaphase I of meiosis.

   g. Pick the card that tells how many chromosomes are in each daughter nucleus after meiosis I.

   h. Pick the card that tells how many chromosomes are in each daughter nucleus after meiosis II.

# DEFINITIONS WORDSEARCH

Review key terms by using the following alphabetized list of terms to fill in the blanks below. Then complete the wordsearch.

```
S J E T R E C P T V U I T T H I M K N E W F L M
N A F D Z C Y T O K I N E S I S B Y E H C I G X
L O I S O T Y C O D N T I M O P Y C F C L X Y K
T H I R I O T B G T D E A G I H R N Q E J W K U
G I G T Y C O P R O T R I N S T O N A N U D L P
R E N T A F R D C P T P I O U H T E M P R Z N A
A Y L I L Z K A Q L C H R O M A T I N X S V W O
F U I I D A I Y T A Y A L K E I I L L O N I N N
R E N N U G G L N S L S O H T F N M M C G G S G
L I T P O E N Z I M O E I U A D O J J Y N N S N
K I E Y R L N U T T S U R O P M C E E T N P I N
F L R G G O T M C I R O T R H E Y I I O I T S T
G O K S I S P T Y C O E A H A L T G G N Y Q O R
C I I I U M N H T E S L F N S G O O D I N N T N
N U N L A I U B A I I O S J E M S L G S G U I W
Q M E O Y B U P L S S M O L Y S E S R G U U M U
G I S R W X M P N C E N T R O M E R E T F M Z M
C V I M O I M H A N I B U R E E H G U G M S Q M
O C S P E R M A T O G E N E S I S T W S I Q F I
D U C L A I U B N I T O E J U W J K I G U F X U
```

*centromere*
*chromatin*
*cytokinesis*
*fertilization*
*interkinesis*
*interphase*
*metaphase*
*mitosis*
*prophase*
*spermatogenesis*
*spindle*
*synapsis*

a. _____ Type of cell division in which daughter cells receive the exact chromosome and genetic makeup of the parent cell; occurs during growth and repair.

b. _____ Division of the cytoplasm following mitosis and meiosis.

c. _____ Stages of the cell cycle ($G_1$, S, $G_2$) during which growth and DNA synthesis occur when the nucleus is not actively dividing.

d. _____ Network of fibrils consisting of DNA and associated proteins observed within a nucleus that is not dividing.

e. _____ Microtubule structure that brings about chromosomal movement during nuclear division.

f. _____ Mitosis phase during which chromatin condenses so that chromosomes appear.

g. _____ Constricted region of a chromosome where sister chromatids are attached to one another and where the chromosome attaches to a spindle fiber.

h. _____ Mitosis phase during which chromosomes are aligned at the metaphase plate of the mitotic spindle.

i. _____ Junction between neurons consisting of the presynaptic (axon) membrane, the synaptic cleft, and the postsynaptic (usually dendrite) membrane.

j. _____ Union of a sperm nucleus and an egg nucleus, which creates the zygote with the diploid number of chromosomes.

k. _____ Period of time between meiosis I and meiosis II during which no DNA replication takes place.

l. _____ Production of sperm in males by the process of meiosis and maturation.

Do not refer to the text when taking this test. In questions 1–4, match the description to the phase.

    a. anaphase
    b. metaphase
    c. prophase
    d. telophase

_____ 1. Chromosomes first become visible.

_____ 2. Chromatids separate at centromere.

_____ 3. Chromosomes are aligned at the metaphase plate.

_____ 4. Last phase of nuclear division.

_____ 5. The diploid chromosome number in an organism is 42. The number of chromosomes in its sex cells is normally
    a. 21.
    b. 42.
    c. 63.
    d. 84.

_____ 6. Which statement about mitosis is NOT correct?
    a. does not affect the nuclear envelope
    b. forms four daughter cells
    c. makes diploid nuclei
    d. prophase is the first active phase

_____ 7. How does mitosis in plant cells differ from that in animal cells?
    a. Animal cells do not form a spindle.
    b. Animal cells lack cytokinesis.
    c. Plant cells lack a cell plate.
    d. Plant cells lack centrioles.

_____ 8. Select the incorrect association.
    a. $G_1$—cell grows in size
    b. $G_2$—protein synthesis occurs
    c. mitosis—nuclear division
    d. S—DNA fails to duplicate

_____ 9. The phase of cell division in which the nuclear envelope and nucleolus are disappearing as the spindle fibers are appearing is called
    a. anaphase.
    b. prophase.
    c. telophase.
    d. metaphase.

_____ 10. In animal cells, cytokinesis takes place by
    a. membrane fusion.
    b. a furrowing process.
    c. formation of a cell plate.
    d. cytoplasmic contraction.

_____ 11. Cyclin
    a. is a molecule that regulates the cell cycle.
    b. combines with a kinase.
    c. occurs in two different forms.
    d. All of these are correct.

_____ 12. If a cell is to divide, DNA replication must occur during
    a. prophase.
    b. metaphase.
    c. anaphase.
    d. telophase.
    e. interphase.

_____ 13. If a cell had 18 chromosomes, how many chromosomes would each daughter cell have after mitosis?
    a. 9
    b. 36
    c. 18
    d. cannot be determined

_____ 14. Normal growth and repair of the human body requires
    a. mitosis.
    b. binary fission.
    c. both *a* and *b*.
    d. neither *a* nor *b*.

_____ 15. The cell cycle
    a. includes mitosis as an event.
    b. includes only the stages $G_1$, S, and $G_2$.
    c. is under cellular but not under genetic control.
    d. involves proteins but not chromosomes.

_____ 16. When do chromosomes move to opposite poles?
    a. prophase
    b. metaphase
    c. anaphase
    d. telophase

In questions 17–23, label each statement with one of the following choices:
    a. meiosis I
    b. meiosis II

_____ 17. Synapsis of homologues occurs.

_____ 18. Separation of homologues occurs.

_____ 19. Results in one oocyte and/or two polar bodies in human females.

_____ 20. Results in four sperm cells in human males.

_____ 21. Daughter cells have double-stranded chromosomes.

_____ 22. Daughter nuclei produced have single-stranded chromosomes.

_____ 23. Crossing-over occurs.

_____24. Which of the following is NOT a valid contrast between mitosis and meiosis?

| Mitosis | Meiosis |
|---|---|
| a. Requires one set of phases | Requires two sets of phases |
| b. Occurs when somatic (body) cells divide | Occurs during gamete production |
| c. Results in four daughter nuclei | Results in two daughter nuclei |
| d. Results in daughter nuclei with diploid number of chromosomes | Results in daughter nuclei with haploid number of chromosomes |

_____25. Polar bodies are formed during
    a. meiosis.
    b. mitosis.
    c. oogenesis.
    d. spermatogenesis.
    e. Both *a* and *c* are correct.

_____26. During anaphase of meiosis II,
    a. homologues separate.
    b. sister chromatids separate.
    c. daughter centrioles separate.
    d. duplicated chromosomes separate.

_____27. During interkinesis,
    a. chromosome duplication occurs.
    b. chromosomes consist of two chromatids.
    c. meiosis I is complete.
    d. Both *b* and *c* are correct.

_____28. By the end of meiosis I,
    a. crossing-over has occurred.
    b. daughter chromosomes have separated.
    c. synapsis of homologues has occurred.
    d. each daughter nucleus is genetically identical to the original cell.
    e. Both *a* and *c* are correct.

## THOUGHT QUESTIONS

Answer in complete sentences.
29. Compare how mitosis in animal cells and plant cells differs.

30. How does sexual reproduction increase genetic variations?

**Test Results:** _____ number correct ÷ 30 = _____ × 100 = _____ %

## ANSWER KEY

### STUDY QUESTIONS

**1. a.** Growth **b.** Synthesis **c.** Growth **d.** Mitosis **2. a.** Growth occurs as organelles . . . **b.** DNA replication . . . **c.** Growth occurs as cell prepares . . . **d.** Mitosis . . . **3. a.** Mitosis stops . . . **b.** Apoptosis can occur . . . **c.** Mitosis will not occur . . . **4.** a, d, f are true. **5. a.** 20 **b.** 19 **c.** 64 **d.** 39 **6. a.** centromere **b.** sister chromatids **7. a.** chromatid **b.** centrosome **c.** centriole **d.** aster **e.** centromere **f.** nuclear membrane fragment **g.** kinetochore

**Prophase**

**Prometaphase**

**Metaphase**

**Anaphase**

**Telophase**

**8. a.** Chromosomes are now distinct; nucleolus is disappearing; centrosomes begin moving apart, and nuclear envelope is fragmenting. **b.** Chromosomes are at the metaphase plate. **c.** Daughter chromosomes are at the poles of the spindle. **d.** Daughter cells are forming as nuclear envelopes and nucleoli appear. **9. a.** cleavage furrowing **b.** anaphase and telophase
**10.**

| Plant Cell | Animal Cell |
|---|---|
| yes | yes |
| yes | yes |
| no | yes |
| yes | no |
| no | yes |

**11. a.** Yes and they form a spindle. **b.** no **c.** It seems not, because plant cells have no centrioles, yet they have a spindle. **12. a.** diploid (2n) **b.** haploid (n) **13.** diploid (2n) (or full) **14. a.** homologues or homologous chromosomes **b.** zygote **15. a.** T **b.** T **c.** F **16. a.** nucleolus **b.** centromere **c.** DNA replication **d.** sister chromatids **e.** synapsis **f.** meiosis I **g.** meiosis II **h.** 4 **i.** 2 **j.** A diploid cell becomes haploid. The parent cell is diploid and undergoes meiosis, which results in four daughter cells, each of which is haploid. **17. a.** 4 **b.** homologues or homologous chromosomes **c.** daughter chromosomes **d.** no **e.** The chromosomes are already duplicated. **18. a.** synapsis **b.** crossing-over **c.** genetic variation **19.** See Fig. 5.10, page 91, in text **20. a.** prophase I **b.** metaphase I **c.** anaphase I **d.** telophase I **e.** interkinesis **f.** prophase II **g.** metaphase II **h.** anaphase II **i.** telophase II **j.** daughter cells. See Figs. 5.12 and 5.13 in text. **21. a.** 2 **b.** 3 **c.** 1
**22.**

| Meiosis | Mitosis |
|---|---|
| no | yes |
| yes | no |
| no | yes |
| yes | no |
| no | yes |
| yes | no |
| no | yes |
| yes | no |

**23. a.** T **b.** F **c.** F **d.** T **e.** F **24. a.** B **b.** M **c.** F **d.** F

**I. a.** 3, 5, 7 **b.** 1, 4 **c.** 6, 8 **II. a.** 4, 3, 1, 2 **b.** 4, 3, 5, 2 **III. a.** 8 **b.** 8 **c.** 8 **d.** 16 **e.** 16 **f.** 8 **g.** 4 **h.** 4

**DEFINITIONS WORDSEARCH**

**a.** mitosis **b.** cytokinesis **c.** interphase **d.** chromatin **e.** spindle **f.** prophase **g.** centromere **h.** metaphase **i.** synapsis **j.** fertilization **k.** interkinesis **l.** spermatogenesis

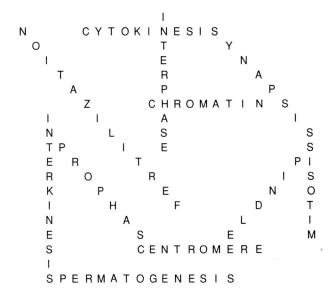

**CHAPTER TEST**

**1.** c **2.** a **3.** b **4.** d **5.** a **6.** b **7.** d **8.** d **9.** b **10.** b **11.** d **12.** e **13.** c **14.** a **15.** a **16.** c **17.** a **18.** a **19.** b **20.** b **21.** a **22.** b **23.** a **24.** c **25.** e **26.** b **27.** d **28.** c **29.** Mitosis in an animal cell utilizes centriole and asters, and the plasma membrane invaginates to form a cleavage furrow. In plant cells, mitosis does not utilize centriole or asters, and the cell plate forms to separate the two daughter cells. **30.** Sexual reproduction results in genetic recombinations among offspring due to (1) crossing-over of nonsister chromatids, (2) independent separation of homologues, and (3) fertilization.

# 6

# METABOLISM: ENERGY AND ENZYMES

Living things can't exhibit any of the characteristics of life without a supply of energy. There are two energy laws that are basic to understanding energy-use patterns in organisms at the cellular level. The first law says that energy cannot be created or destroyed, but can only be transferred or transformed. The second law states that a usable form of energy cannot be converted completely into another usable form. As a result of these laws, we know that the **entropy** (disorder) of the universe is increasing and that only a constant input of energy maintains the organization of living things.

**Metabolism** is all the reactions that occur in a cell. Only those reactions that result in a negative free energy difference—that is, the products have less usable energy than the reactants—occur spontaneously. Such reactions, called **exergonic reactions,** release energy. **Endergonic reactions,** which require an input of energy, occur because it is possible to **couple** an exergonic process with an endergonic process. For example, glucose breakdown is an exergonic metabolic pathway that drives the buildup of many **ATP** molecules. These ATP molecules then supply energy for cellular work. Thus, ATP goes through a cycle in which it is constantly being built up from, and then broken down to, **ADP +**(P).

A **metabolic pathway** is a series of reactions that proceed in an orderly, step-by-step manner. Each reaction has a specific **enzyme** that speeds the reaction by forming a complex with its **substrates.** Formation of the enzyme-substrate complex lowers the energy of activation, the amount of energy required to activate the reactants. Any environmental factor that affects the shape of a protein also affects the ability of an enzyme to do its job. (Many enzymes have **cofactors** or **coenzymes** that help them carry out a reaction.)

**Photosynthesis,** which transforms solar energy to chemical energy within carbohydrates, is a metabolic pathway that occurs in chloroplasts. Reduction is the gain of hydrogen atoms ($H^+ + e^-$). During photosynthesis, carbon dioxide is reduced to glucose, a carbohydrate. **Cellular respiration,** which is completed in mitochondria, is a metabolic pathway that transforms the energy of glucose (usually) into that of ATP molecules. Oxidation is the loss of hydrogen atoms. During cellular respiration, carbohydrate is oxidized to carbon dioxide and water. There is a cycling of molecules between chloroplasts and mitochondria, but energy flows one way. Eventually, all the solar energy captured by plants is lost as heat as ATP is utilized by cells.

Study the text section by section. Answer the study questions so that you can fulfill the learning objectives for each section.

## 6.1 CELLS AND THE FLOW OF ENERGY (PP. 102–3)

The learning objective for this section is:
• Explain the two laws of thermodynamics.

1. Indicate whether these statements, related to the energy laws, are true (T) or false (F), and if the statements are false, change them to true statements:
   a. _____ The chemical energy of ATP cannot be transformed into any other type of energy such as kinetic energy. Rewrite: _____

   b. _____ A cell produces ATP, and therefore cells do not obey the first law of thermodynamics. Rewrite: _____

c. _____ Because energy transformations always result in a loss of usable energy, the entropy of the universe is increasing. Rewrite: _____

d. _____ Because our society uses coal as an energy source, it is helping to decrease the entropy of the universe. Rewrite: _____

## 6.2 METABOLIC REACTIONS AND ENERGY TRANSFORMATIONS (PP. 104–5)

The learning objectives for this section are:
- Compare exergonic and endergonic reactions.
- Explain the cycle of ATP buildup and ATP breakdown.

2. Place the appropriate letters next to each statement.

    En—endergonic    Ex—exergonic

    a. _____ Energy is released as the reaction occurs.
    b. _____ Energy is required to make the reaction go.
    c. _____ Reaction used by the body for muscle contraction and nerve conduction.
    d. _____ ATP → ADP + $\textcircled{P}$.
    e. _____ ADP + $\textcircled{P}$ → ATP.

3. Label this diagram, using these terms:
    ATP
    ADP
    –P
    +P

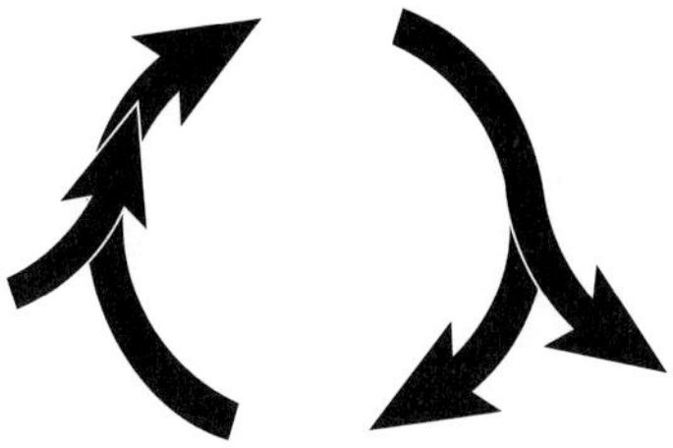

b. _____

a. _____

c. _____

d. _____

4. Label each of the following as pertaining to the left (L) or right (R) side of the diagram in question 3. Explain your choice.
    a. _____ cellular respiration. Explain: _____

    b. _____ muscle contraction. Explain: _____

    c. _____ active transport. Explain: _____

5. ATP is the common a._____ of cells; when cells require energy, they "spend" ATP. ATP breakdown provides energy for b._____ work, such as synthesizing macromolecules; c._____ work, such as pumping substances across plasma membranes; and d._____ work, such as the beating of flagella. Because ATP breakdown is e._____ to endergonic reactions, energy transformation occurs with minimal loss to the cell.

The learning objectives for this section are:
* Define metabolic pathway.
* Explain why enzymes speed reactions.
* Explain how environmental factors affect the activity of enzymes, and give examples.
* Describe the functions of cofactors and coenzymes.

6. Consider the following diagram of a metabolic pathway:

$$E_1 \quad\quad E_2 \quad\quad E_3 \quad\quad E_4 \quad\quad E_5 \quad\quad E_6$$
$$A \rightarrow B \rightarrow C \rightarrow D \rightarrow E \rightarrow F \rightarrow G$$

A–F are a._____, and B–G are b._____. $E_1$–$E_6$ are c._____. A is a

d._____ for the first enzyme, and B is the product.

7. Label this diagram using the following alphabetized list of phrases. You will use each phrase twice.

　　energy of activation
　　energy of product
　　energy of reactant

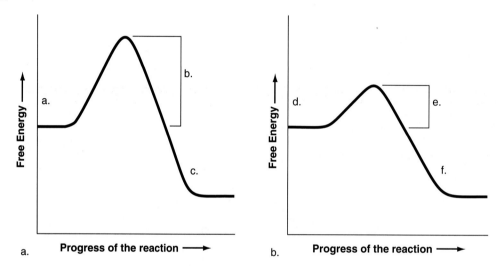

8. Label this diagram, using the following alphabetized list of terms.

　　active site
　　enzyme (used more than once)
　　enzyme-substrate complex
　　products
　　substrate

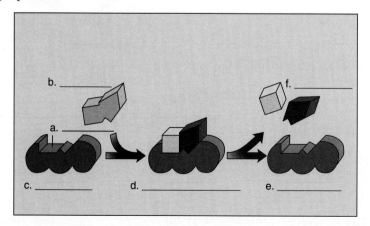

9. Which portion of the diagram in question 8 pertains to enzymes lowering the energy of
activation? a._____

Why? b._____

10. Express the reaction in question 8 in equation form, using E (for enzyme), S (for substrate), and P (for product). a._____

Is the reaction shown in question 8 a synthetic reaction or a degradative reaction? b._____

How do you know? c._____

The enzyme-substrate complex and the reaction occur at the d._____ site of the enzyme. What is the significance of using the label "enzyme" twice in the diagram in question 8? It shows that e._____
_____

Why are enzymes named for their substrates (e.g., maltase speeds the breakdown of maltose)? f._____
_____

11. Complete each statement with the term *increases* or *decreases*.

Raising the temperature generally a._____ the rate of an enzymatic reaction.

Boiling an enzyme drastically b._____ the rate of the reaction.

Changing the pH toward the optimum pH for an enzyme c._____ the rate of the reaction.

Introducing a competitive inhibitor d._____ the availability of an enzyme for its normal substrate.

Due to feedback inhibition, the affinity of the active site for the substrate e._____.

12. Enzymes have helpers called a._____, which b._____

## 6.4 OXIDATION-REDUCTION AND THE FLOW OF ENERGY (P. 110)

The learning objective for this section is:
- Compare the equations for photosynthesis and cellular respiration.

13. Oxidation is defined as the a._____ (*gain/loss*) of electrons, and reduction is defined as the b._____ (*gain/loss*) of electrons. In living things, hydrogen ions often accompany electrons. Therefore, in living things, oxidation is the c._____ of d._____ atoms, and reduction is the e._____ of f._____ atoms.

14. An overall equation for photosynthesis, a metabolic pathway in plant cells, is

$$6\ CO_2 + 6\ H_2O + energy\ \rightarrow\ C_6H_{12}O_6 + 6\ O_2$$

carbon    water              glucose    oxygen
dioxide

In this equation, a._____ molecules are oxidized, releasing b._____ molecules to the air. c._____ is reduced and becomes d._____. e._____ drives this reaction.

15. An overall equation for cellular respiration is

$$C_6H_{12}O_6 + 6\ O_2\ \rightarrow\ 6\ CO_2 + 6\ H_2O + energy$$

glucose    oxygen       carbon    water
                          dioxide

In this equation, a._____ is oxidized and b._____ is released to the air. c._____ is reduced to d._____, a low-energy molecule. The oxidation of glucose makes energy available for the production of e._____ molecules.

16. Show that there is a cycling of matter by completing this diagram using these terms:

$H_2O$ and $CO_2$ (used twice)    $O_2$ and $C_6H_{12}O_6$

a. _____ → chloroplasts ⟶ b. _____ → mitochondria ⟶ c. _____

cycling

17. Show that there is a flow of energy through living things by completing this diagram using these terms:

heat    ATP    solar energy    glucose

a. _____ chloroplasts produce    b. _____ mitochondria produce    c. _____    d. _____
(in plants)                              (in all organisms)

## DEFINITIONS WORDSEARCH

Review key terms by using the following alphabetized list of terms to fill in the blanks below. Then complete the wordsearch.

```
S J E T R E C P T V U I T T H I M K N E W F
N A F D Z C Y T O K I M O D E L B Y E H C I
L O V S O T Y C O D N E I M O P Y C F C L X
T H I R I O T B G T D T N G I H R N Q E J W
G I T T Y C D E R U T A N E D T O N A N E D
R E A T A F R D C P T B C O R H T E M P N Z
A Y M I L Z K A N L C O R O M G T I N X T V
F U I I D A I O T A F L L K E I Y L L O R I
R E N N U G I L N A L I O H T F N M M C O G
L I S C O T N Z C M O S I U A D O J E Y P N
K I E Y A L N T T T S M R O P M C E T T Y P
F L R D G L O M C I R O T R H E Y I A O I T
G O I S I R O T Y C O E A H A L T G R N Y Q
C X I I S M N R T E S L E N S G O O T I N N
O U N L A I R E A C T A N T E M S L S S G U
Q M E O Y B U P L A S M Z L Y S E S B G U U
G I S R W X M P N C E N Y R O M E R U T F M
C V I M O I M H A N I B M R E E H G S G M S
O C S P E R M A T O G E E E S I S T W S I Q
D U C L A I U B N I T O E J U W J K I G U F
```

*cofactors*
*denatured*
*energy*
*entropy*
*enzyme*
*metabolism*
*model*
*oxidation*
*reactant*
*substrate*
*vitamins*

a. _____ Capacity to do work and bring about change; occurs in a variety of forms.

b. _____ Measure of disorder or randomness.

c. _____ All of the chemical changes that occur within a cell.

d. _____ Substance that participates in a reaction.

e. _____ Organic catalyst, usually a protein, that speeds up a reaction in cells due to its particular shape.

f. _____ Reactant in a reaction controlled by an enzyme.

g. _____ Simulation of a process that aids conceptual understanding until the process can be studied firsthand; a hypothesis that describes how a particular process could possibly be carried out.

h. _____ Loss of an enzyme's normal shape so that it no longer functions; caused by a less than optimal pH and temperature.

i. _____ Nonprotein adjunct required by an enzyme in order to function; many are metal ions, while others are coenzymes.

j. _____ Essential requirement in the diet, needed in small amounts. They are often part of coenzymes.

k. _____ Loss of one or more electrons from an atom or molecule; in biological systems, generally the loss of hydrogen atoms.

Do not refer to the text when taking this test.

_____ 1. The useful energy conversion in photosynthesis is
  a. chemical to solar.
  b. heat to mechanical.
  c. mechanical to heat.
  d. solar to chemical.

_____ 2. Any energy transformation involves the loss of some energy as
  a. electricity.
  b. heat.
  c. light.
  d. motion.

_____ 3. In the enzymatically controlled chemical reaction A → B + C, A is the
  a. cofactor.
  b. enzyme.
  c. product.
  d. substrate.

_____ 4. An enzyme, functioning best at a pH of 3, is in a neutral solution at a temperature of 40°C. Its activity will increase by
  a. decreasing the amount of substrate.
  b. denaturing the enzyme.
  c. increasing the temperature 10 more degrees.
  d. making the pH more acidic.

_____ 5. In the reaction A + B → C, the reaction rate may slow down through feedback inhibition by
  a. increasing the concentration of A.
  b. increasing the concentration of B.
  c. increasing the concentration of C.
  d. decreasing the concentration of B.

_____ 6. The energy laws
  a. account for why energy does not cycle.
  b. say that some loss of energy always accompanies transformation.
  c. say that energy can be made available to living things.
  d. All of these are correct.

_____ 7. In a metabolic pathway A → B → C → D → E,
  a. A, B, C, and D are substrates.
  b. B, C, D, and E are products.
  c. each reaction requires its own enzyme.
  d. All of these are correct.

_____ 8. The enzyme-substrate complex
  a. indicates that an enzyme has denatured.
  b. accounts for why enzymes lower the energy of activation.
  c. is nonspecific.
  d. All of these are correct.

_____ 9. The tendency for an ordered system to become spontaneously disordered is called
  a. thermodynamics.
  b. entropy.
  c. activation.
  d. energy conversion.

_____10. A coupled reaction occurs when energy released from a(n) _____ reaction is used to drive a(n) _____ reaction.
  a. endergonic; exergonic
  b. breakdown; exergonic
  c. exergonic; endergonic
  d. chemical; mechanical

_____11. NAD$^+$ and FAD are
  a. dehydrogenases.
  b. proteins.
  c. coenzymes.
  d. Both _a_ and _c_ are correct.

_____12. Reduction has occurred
  a. when electrons are lost.
  b. when $C_6H_{12}O_6$ becomes $CO_2$.
  c. when $O_2$ becomes $H_2O$.
  d. when heat is given off.
  e. when ADP becomes ATP.

_____13. Chloroplasts
  a. take in $CO_2$.
  b. give off $H_2O$.
  c. pass on solar energy.
  d. occur in all living things.
  e. Both _a_ and _c_ are correct.

_____14. ATP is used for
  a. chemical work.
  b. transport work.
  c. mechanical work.
  d. All of these are correct.

_____15. NAD and NADP
  a. are found only in plants.
  b. do not participate in metabolic reactions.
  c. are coenzymes of oxidation-reduction.
  d. carry hydrogen atoms.
  e. Both _c_ and _d_ are correct.

_____16. Which statement is NOT correct about enzymes?
  a. They usually end in the suffix "-ase."
  b. They catalyze only one reaction.
  c. They increase the energy of activation.
  d. They bind temporarily with the substrate.

_____17. Which of these is NOT expected to increase the rate of an enzymatic reaction?
  a. add more enzyme
  b. remove inhibitions
  c. boil rapidly
  d. adjust the pH to optimum level

_____18. Which of these accurately represents a flow of energy from the sun?
  a. Plants take in solar energy and use it to oxidize glucose, which is used by mitochondria to produce ATP.
  b. Mitochondria break down glucose to ATP, which is returned to plants to produce glucose.
  c. Plants take in solar energy and use it to transport water up stems so that water is available to all animals.
  d. Plants take in solar energy and use it to reduce carbon dioxide so that glucose is made available to animals.
  e. Both plants and animals make and use ATP.

_____19. Since energy does not cycle, animal cells
  a. require a continuing source of glucose.
  b. are dependent on plant cells.
  c. must produce ATP nonstop.
  d. All of these are correct.

_____20. Synthetic reactions
  a. require the participation of ATP.
  b. do not require enzymes.
  c. are represented by S + E → ES → P.
  d. are coupled directly to glucose breakdown.

## THOUGHT QUESTIONS

Answer in complete sentences.

21. Why couldn't life exist without a continual supply of solar energy?

22. Why are enzymes absolutely necessary to the continued existence of a cell?

**Test Results:** _____ number correct ÷ 22 = _____ × 100 = _____ %

## ANSWER KEY

### STUDY QUESTIONS

**1. a.** F, The chemical energy of ATP can be transformed into other types of energy such as kinetic energy (muscle contraction). **b.** F, Cells transform the energy of glucose breakdown into ATP molecules, and they do obey the first law of thermodynamics. **c.** T **d.** F, Because our society uses coal as an energy source, it is increasing the entropy of the universe. **2. a.** Ex **b.** En **c.** Ex **d.** Ex **e.** En **3. a.** +P **b.** ATP **c.** –P **d.** ADP **4. a.** L, because during cellular respiration, the chemical energy within a glucose molecule is converted to the chemical energy within ATP. **b.** R, because when muscles contract, the chemical energy within ATP is converted to the kinetic energy of muscle contraction. **c.** R, because when active transport occurs, the energy released by ATP breakdown is used to pump a molecule across the plasma membrane. **5. a.** energy currency **b.** chemical **c.** transport **d.** mechanical **e.** coupled **6. a.** reactants **b.** products **c.** enzymes **d.** substrate **7. a.** energy of reactant **b.** energy of activation **c.** energy of product **d.** energy of reactant **e.** energy of activation **f.** energy of product **8. a.** active site **b.** substrate **c.** enzyme **d.** enzyme-substrate complex **e.** enzyme **f.** products **9. a.** enzyme-substrate complex **b.** Reactants come together when the enzyme-substrate complex forms. **10. a.** E + S → ES → E + P **b.** degradative **c.** The reactant is broken down. **d.** active **e.** The enzyme is not broken down and can be used over and over again. **f.** Enzymes are specific to their substrates. **11. a.** increases **b.** decreases **c.** increases **d.** decreases **e.** decreases **12. a.** coenzymes **b.** help enzymes function. **13. a.** loss **b.** gain **c.** loss **d.** hydrogen **e.** gain **f.** hydrogen **14. a.** water **b.** oxygen **c.** carbon dioxide **d.** glucose **e.** sun **15. a.** glucose **b.** carbon dioxide **c.** oxygen **d.** water **e.** ATP **16. a.** $H_2O$ and $CO_2$ **b.** $O_2$ and $C_6H_{12}O_6$ **c.** $H_2O$ and $CO_2$ **17. a.** solar energy **b.** glucose **c.** ATP **d.** heat

a. energy  b. entropy  c. metabolism  d. reactant  e. enzyme  f. substrate  g. model  h. denatured  i. cofactors  j. vitamins  k. oxidation

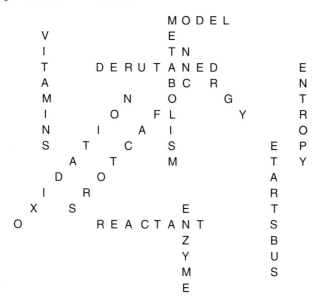

**1.** d  **2.** b  **3.** d  **4.** d  **5.** c  **6.** d  **7.** d  **8.** b  **9.** b  **10.** c  **11.** d  **12.** c  **13.** a  **14.** d  **15.** e  **16.** c  **17.** c  **18.** d  **19.** d  **20.** a  **21.** When chloroplasts carry on photosynthesis, solar energy is converted to the energy of carbohydrates, and when mitochondria complete cellular respiration, the energy stored in carbohydrates is converted to energy temporarily held by ATP. The energy released by ATP breakdown is used by the cell to do various types of work, and eventually it becomes nonusable heat.  **22.** Enzymes are absolutely essential for the existence of a cell because they lower the energy of activation of a reaction, thereby requiring less heat to bring about the reaction. At high temperatures, proteins would denature.

# 7

# CELLULAR RESPIRATION

**Cellular respiration,** the oxidation of glucose to carbon dioxide and water, is an exergonic reaction that drives ATP synthesis, an endergonic reaction. Oxidation involves the removal of hydrogen atoms ($H^+ + e^-$) from substrate molecules, usually by the coenzyme **$NAD^+$**, but in one case by **FAD.** Four phases are required: glycolysis, the transition reaction, the citric acid cycle, and the electron transport system.

During **glycolysis,** glucose is converted to **pyruvate** in the cytosol. Glycolysis produces two ATP by **substrate-level phosphorylation.** When oxygen is available, pyruvate from glycolysis enters **mitochondria.** In mitochondria, the **transition reaction** and the **citric acid cycle** are located in the matrix, and the electron transport system is located on the cristae. Both the transition reaction and the citric acid cycle release carbon dioxide as a result of the oxidation of carbohydrate breakdown products.

The electrons carried by NADH and $FADH_2$ enter the **electron transport system.** The electrons pass down a chain of carriers until they are finally received by oxygen, which combines with $H^+$, forming water. As electrons pass down the electron transport system, energy is released and stored for ATP production. The term **oxidative phosphorylation** is sometimes used for 34 ATPs produced as a result of the electron transport system. The protein complexes of the electron transport system pump $H^+$ received from NADH and $FADH_2$ into the intermembrane space, setting up an electrochemical gradient. When $H^+$ flows down this gradient through the ATP synthase complex, energy is released and used to form ATP. This process of producing ATP is called **chemiosmosis.**

**Fermentation,** which occurs when oxygen is not available for cellular respiration, involves glycolysis followed by the reduction of pyruvate by NADH. The end product can be lactate or alcohol and carbon dioxide. Fermentation produces a net yield of 2 ATP molecules per glucose molecule. Fermentation provides a quick, immediate source of ATP, but lactate buildup is toxic to the cell and creates an **oxygen debt** by the organism.

Other carbohydrates, as well as protein and fat, can also generate ATP by entering various steps in the degradative paths of glycolysis and the citric acid cycle. These pathways also provide metabolites needed for the synthesis of various important cellular substances.

## STUDY QUESTIONS

Study the text section by section. Answer the study questions so that you can fulfill the learning objectives for each section.

## 7.1 OVERVIEW OF CELLULAR RESPIRATION (PP. 116–17)

The learning objectives for this section are:
- Explain the overall equation of cellular respiration as an oxidation-reduction reaction.
- Identify and explain the behavior of $NAD^+$ and FAD as coenzymes of oxidation-reduction.
- Identify the four phases of complete glucose breakdown.

1. Consider the following equation:

$$C_6H_{12}O_6 + 6\,O_2 \rightarrow 6\,H_2O + 6\,CO_2 + energy$$

The molecule glucose is [a.]_____ (oxidized or reduced) while oxygen is [b.]_____ (oxidized or reduced). This is an [c.]_____ (endergonic or exergonic) reaction and therefore is used by cells to build up ATP.

2. Complete the adjacent diagram by labeling a–d, and answer these questions:

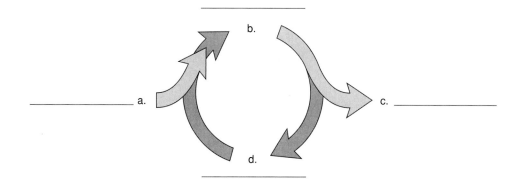

The (left or right) <sup>e.</sup>_____ side of the diagram represents oxidation, and the (left or

right) <sup>f.</sup>_____ side of the diagram represents reduction of NAD. Why is NAD$^+$ called a coenzyme

of oxidation-reduction? <sup>g.</sup> _____

3. Place the appropriate letters next to each statement.

GL—glycolysis    TR—transition reaction    CA—citric acid cycle    ETS—electron transport system

a. _____ Series of carriers that pass electrons from one to the other.
b. _____ Cyclical series of oxidation reactions that release $CO_2$.
c. _____ Pyruvate is oxidized to an acetyl group.
d. _____ Breakdown of glucose to two molecules of pyruvate.
e. _____ Energy is released and stored for ATP production.
f. _____ Occurs inside the mitochondria.
g. _____ Occurs outside the mitochondria in the cytoplasm.
h. _____ Results in only 2 ATP.

## 7.2 OUTSIDE THE MITOCHONDRIA: GLYCOLYSIS (PP. 118–19)

The learning objective for this section is:
• Describe the process of glycolysis

4. Where does glycolysis occur? <sup>a.</sup>_____

Does it require oxygen? <sup>b.</sup>_____

Glycolysis begins with <sup>c.</sup>_____.

Glycolysis ends with <sup>d.</sup>_____.

How many ATP are produced per glucose molecule as a direct result of glycolysis? <sup>e.</sup>_____

What type of phosphorylation occurs during glycolysis? <sup>f.</sup>_____

What coenzyme carries out the oxidation of substrates during glycolysis? <sup>g.</sup>_____

Considering your answers to these questions, what is the output of glycolysis? <sup>h.</sup>_____,

_____, and _____

When glycolysis is a part of fermentation, what is the end product in humans? <sup>i.</sup>_____.

## 7.3 INSIDE THE MITOCHONDRIA (PP. 120–24)

The learning objectives for this section are:
• Describe the reactions that take place inside the mitochondria: transition reaction, citric acid cycle, and electron transport system.
• Calculate the energy yield from glucose metabolism.

5. Label this diagram of a mitochondrion, using the following alphabetized list of terms:
   cristae
   cytoplasm
   inner membrane
   intermembrane space
   matrix
   outer membrane

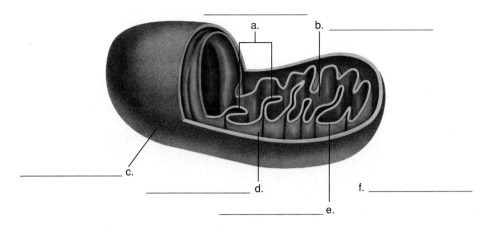

a. _____

b. _____

c. _____

d. _____

e. _____

f. _____

6. Using your labels from question 5, where does each of the following processes occur?

   glycolysis  a._____

   citric acid cycle  b._____

   electron transport system  c._____

7. The citric acid cycle begins and ends with what molecule?  a._____

   A two-carbon molecule acetyl group enters the citric acid cycle. What carbon molecules leave the
   cycle?  b._____

   How many ATP are produced per glucose molecule as a direct result of the citric acid
   cycle?  c._____

   What coenzymes carry out the oxidation of substrates in the citric acid cycle?  d._____

   Considering your answers to these questions, what are the outputs of the citric acid cycle?  e._____,
   _____, _____, and _____

8. What coenzymes bring hydrogen atoms ($H^+ + e^-$) to the electron transport system?  a._____

   What happens to the electrons?  b._____

   What happens to the hydrogen ions?  c._____

   What molecule is the final acceptor of electrons from the electron transport system?  d._____

   Each pair of electrons carried by NADH from the citric acid cycle that passes down the electron transport
   system accounts for the buildup of how many ATP?  e._____

   What type of phosphorylation is associated with the electron transport system?  f._____

9. During chemiosmotic ATP synthesis in mitochondria, $H^+$ build up in the  a._____ space.

   When these $H^+$ flow  b._____ their concentration gradient into the matrix,  c._____ is
   produced from ADP + Ⓟ.

10. Match the complexes to these functions:

   a. _____ NADH dehydrogenase complex
   b. _____ cytochrome *b-c* complex
   c. _____ cytochrome oxidase complex
   d. _____ ATP synthase complex

   1 passes on electrons and pumps $H^+$ into intermembrane space
   2 carries out ATP synthesis
   3 receives electrons and passes them on to oxygen
   4 oxidizes NADH and pumps $H^+$ into intermembrane space

11. In the following diagram, fill in the blanks with the correct numbers and with the terms NADH, FADH₂, and ATP:

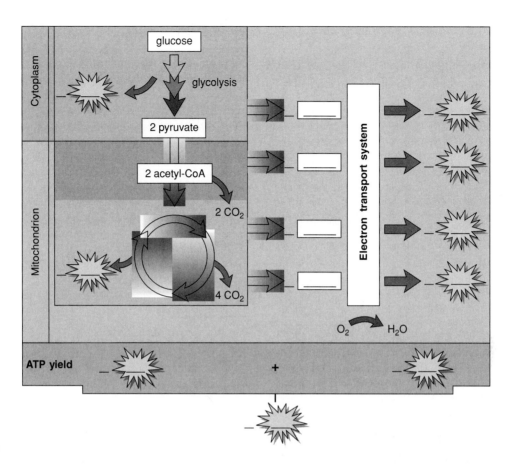

## 7.4 FERMENTATION (P. 125)

The learning objective for this section is:
• Explain the process of fermentation.

12. What happens to pyruvate during fermentation? In humans? ª._____ In yeast? ᵇ._____

What is fermentation wasteful? ᶜ._____

What is its advantage? ᵈ. _____

What is oxygen debt in humans? ᵉ. _____

13. Label the following processes I, II, and/or III, based on this pyruvate diagram:
    a. _____ occurs under anaerobic conditions
    b. _____ fermentation
    c. _____ glycolysis
    d. _____ transition reaction

14. Consider III in the diagram for question 13.

    Which has more hydrogen atoms, pyruvate or lactate? ª._____

    In yeast, the product of this reaction is ᵇ._____.

    What happens to NAD⁺ produced by the reaction? ᶜ._____

15. Consider II in the diagram for question 13.

    What happens to NADH? ª._____

    What happens to the acetyl group? ᵇ._____

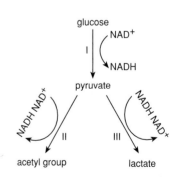

The learning objective for this section is:
• Explain how molecules from a metabolic pool can be used for catabolism or for anabolism.

16. The carbon skeleton of amino acids can be respired if the amino acid first undergoes a._____.

When fats are respired, glycerol is converted to b._____, fatty acids are converted to the two-carbon

molecule c._____, and the acetyl group enters the citric acid cycle. Excess acetyl groups from

glucose metabolism can be used to build up fat. Explain why the consumption of carbohydrate makes us

fat. d._____

_____

 *Cellular Respiration Roulette*

On the table before you are three locations to place a bet. These locations are labeled:

| G | C | E |
|---|---|---|
| **Glycolysis** | **Citric Acid Cycle** | **Electron Transport System** |

You can place a chip on more than one of these locations. For every chip properly placed, you would hypothetically win $5.00. Place your bets! Where should you place chips for each of the following:

1. _____ occurs in cytoplasm
2. _____ occurs in mitochondrion
3. _____ glucose
4. _____ oxygen
5. _____ carbon dioxide
6. _____ NADH produced
7. _____ NADH received
8. _____ water produced
9. _____ 2 ATP
10. _____ 32 ATP
11. _____ Product from transition reaction enters here.
12. _____ How much money did you win?

# DEFINITIONS WORDSEARCH

Review key terms by using the following alphabetized list of terms to fill in the blanks below. Then complete the wordsearch.

```
S J E T R E C M I T O C H O N D R I O N W F
N A C D Z C Y T O K I M O D E L B Y E H C I
L O Y S O C H E M I O S M O S I S C F C D X
T H T R I O T B G T D T N G I H R N Q E E W
G I O T Y C D E R U T A S E D T O N A N A D
R E C T A F R D C P T B I O R H T E M P M Z
A Y H I L Z K A N L C O S O M G T I N X I V
F U R I D F I O T A F L Y K E I Y L L O N I
R E O N U G E L N A L I L H T F N M M C A G
L I M C O T N R C M O S O U A D O J E Y T N
K I E Y A L N T M T S M C O P M C E T T I P
F L R D G L O M C E R O Y R H E C I A O O T
G O I S I R O T Y C N E L H A L I G R N N Q
C X I A E R O B I C S T G N S G B O T I N N
O U N L A I R E A C T A A T E M O L S S G U
Q M E O Y B U P L A S M Z T Y S R S B G U U
G I S R W X M P N C E N Y R I M E R U T F M
C V I M O I M H A N I B M R E O A G S G M S
O C S C A T A B O L I S M E S I N T W S I Q
D U C L A I U B M S I L O B A N A K I G U F
```

*aerobic*
*anabolism*
*anaerobic*
*catabolism*
*chemiosmosis*
*cytochrome*
*deamination*
*fermentation*
*glycolysis*
*mitochondrion*

a. _____ Growing or metabolizing only in the presence of oxygen, as in cellular respiration.

b. _____ Metabolic pathway found in the cytoplasm that participates in cellular respiration and fermentation; it converts glucose to two molecules of pyruvate.

c. _____ Anaerobic breakdown of glucose that results in a gain of two ATP and end products such as alcohol and lactate.

d. _____ Growing or metabolizing in the absence of oxygen.

e. _____ Membrane-bounded organelle in which ATP molecules are produced during the process of cellular respiration.

f. _____ Protein within the inner mitochondrial membrane that is an electron carrier in cellular respiration (electron transport system).

g. _____ Ability of certain membranes to use a hydrogen ion gradient to drive ATP formation.

h. _____ Removal of an amino group (—$NH_2$) from an amino acid or other organic compound.

i. _____ Metabolic process that breaks down large molecules into smaller ones.

j. _____ Metabolic process by which larger molecules are synthesized from smaller ones.

# CHAPTER TEST

## OBJECTIVE QUESTIONS

Do not refer to the text when taking this test.

____ 1. Fermentation is
    a. glycolysis and the citric acid cycle.
    b. glycolysis and the reduction of pyruvate.
    c. glycolysis only.
    d. the reduction of pyruvate only.

____ 2. Each of the following is a product of cellular respiration EXCEPT
    a. ATP.
    b. carbon dioxide.
    c. oxygen.
    d. water.

_____ 3. Per glucose molecule, the net gain of ATP molecules from glycolysis is
   a. two.
   b. four.
   c. six.
   d. eight.
_____ 4. Fermentation supplies
   a. glycolysis with free $NAD^+$.
   b. hydrogen to the transition reaction.
   c. oxygen as an electron acceptor.
   d. the citric acid cycle with oxygen.
_____ 5. The process that occurs first is
   a. chemiosmosis.
   b. glycolysis.
   c. the electron transport system.
   d. the citric acid cycle.
_____ 6. Which is NOT an event of the transition reaction?
   a. breaks down pyruvate
   b. converts a citrate molecule
   c. oxidizes pyruvate
   d. transfers an acetyl group
_____ 7. Select the incorrect pair.
   a. electron transport system—cristae
   b. fermentation—plasma membrane
   c. glycolysis—cytoplasm
   d. citric acid cycle—matrix
_____ 8. Select the process which results in the greatest number of ATP per glucose molecule.
   a. glycolysis
   b. citric acid cycle
   c. substrate-level phosphorylation
   d. transition reaction
_____ 9. The energy yield by ATP molecules per glucose molecule is closest to
   a. 25%.
   b. 40%.
   c. 50%.
   d. 60%.
_____10. Inside a cell, glycerol is broken down into
   a. amino acids.
   b. acetyl-CoA.
   c. fatty acids.
   d. PGAL.
_____11. Which of the following reactions is NOT a part of cellular respiration?
   a. glycolysis
   b. citric acid cycle
   c. electron transport system
   d. transition reaction
   e. fermentation
_____12. The coenzyme used in the transition reaction of cellular respiration is
   a. ATP.
   b. $NAD^+$.
   c. NADH.
   d. coenzyme A.
   e. RuBP.

_____13. The carbon dioxide given off by cellular respiration is produced by
   a. glycolysis.
   b. the transition reaction.
   c. the citric acid cycle.
   d. the electron transport system.
   e. Both _b_ and _c_ are correct.
_____14. The final acceptor for electrons in cellular respiration is
   a. ATP.
   b. $NAD^+$.
   c. FAD.
   d. oxygen.
   e. carbon dioxide.
_____15. Which of the following reactions occurs on the inner membrane of mitochondria?
   a. the citric acid cycle
   b. the transition reaction
   c. the electron transport system
   d. glycolysis
   e. the Calvin cycle
_____16. The coenzymes $NAD^+$ and FAD carry hydrogen atoms ($H^+ + e^-$) to the
   a. glycolysis reactions.
   b. transition reaction.
   c. citric acid cycle.
   d. Calvin cycle.
   e. electron transport system.
_____17. Which of the following statements is NOT true about fermentation?
   a. It is an anaerobic process.
   b. The end products are toxic to cells.
   c. It results in two ATPs per glucose molecule.
   d. In the absence of $O_2$, muscle cells form $CO_2$ and alcohol.
   e. It can be used to make bread rise.
_____18. Which of the following statements is NOT true regarding fats?
   a. Fatty acids are converted to acetyl-CoA.
   b. Eighteen-carbon fatty acids are converted to nine acetyl-CoA molecules.
   c. Glycerol is converted to PGAL.
   d. Fats are the least efficient form of stored energy.
   e. Carbohydrates can be converted to fats.
_____19. The process directly responsible for most of the ATP formed during cellular respiration is
   a. the citric acid cycle.
   b. the transition reaction.
   c. the electron transport system.
   d. chemiosmosis.
_____20. A pathway that begins with glucose and ends with pyruvate is
   a. glycolysis.
   b. the citric acid cycle.
   c. the electron transport system.
   d. the transition reaction.

Answer in complete sentences.

21. Explain how the human body obtains the reactants for cellular respiration and what happens to the products.

22. Why would you expect the citric acid cycle to be located in the matrix of the mitochondrion and the electron transport system to be located on the cristae?

**Test Results:** _____ number correct ÷ 22 = _____ × 100 = _____ %

## ANSWER KEY

### STUDY QUESTIONS

**1. a.** oxidized **b.** reduced **c.** exergonic  **2. a.** 2H **b.** NADH + H$^+$ **c.** 2H **d.** NAD$^+$ **e.** right **f.** left **g.** It becomes reduced when it accepts electrons from a substrate and becomes oxidized when it passes electrons on to another carrier.  **3. a.** ETS **b.** CA **c.** TR **d.** GL **e.** ETS **f.** TR, CA, ETS **g.** GL **h.** GL, CA (two turns per glucose molecule)  **4. a.** cytoplasm **b.** no **c.** glucose **d.** pyruvate **e.** two ATP **f.** substrate-level **g.** NAD$^+$ **h.** NADH, ATP, pyruvate **i.** lactate  **5. a.** cristae **b.** matrix **c.** outer membrane **d.** intermembrane space **e.** inner membrane **f.** cytoplasm  **6. a.** cytoplasm **b.** matrix **c.** cristae  **7. a.** citrate **b.** CO$_2$ **c.** 2 ATP **d.** NAD$^+$ and FAD **e.** NADH, FADH$_2$, ATP, and CO$_2$  **8. a.** NADH and FADH$_2$ **b.** pass down the system **c.** pumped into intermembrane space **d.** O$_2$ **e.** three ATP **f.** oxidative  **9. a.** intermembrane **b.** down **c.** ATP  **10. a.** 4 **b.** 1 **c.** 3 **d.** 2  **11.** see Figure 7.9, page 124, in text. **12. a.** reduced to lactate **b.** reduced to alcohol and CO$_2$ **c.** produces only two ATP **d.** does not require oxygen **e.** O$_2$ needed to metabolize lactate  **13. a.** I, III **b.** III **c.** I **d.** II  **14. a.** lactate **b.** alcohol and CO$_2$ **c.** returns to glycolysis  **15. a.** goes to the electron transport system **b.** enters the citric acid cycle  **16. a.** deamination **b.** PGAL **c.** acetyl-CoA **d.** The acetyl groups, which result from carbohydrate breakdown, can be used to make fat.

### CELLULAR RESPIRATION ROULETTE

**1.** G  **2.** C, E  **3.** G  **4.** E  **5.** C  **6.** G, C  **7.** E  **8.** E **9.** G, C  **10.** E  **11.** C  **12.** Calculate your winnings.

### DEFINITIONS WORDSEARCH

**a.** aerobic **b.** glycolysis **c.** fermentation **d.** anaerobic **e.** mitochondrion **f.** cytochrome **g.** chemiosmosis **h.** deamination **i.** catabolism **j.** anabolism

```
        C                 M I T O C H O N D R I O N
        Y       C H E M I O S M O S I S         D
        T                       S               E
        O                       I               A
        C                       S               M
        H       F               Y               I
        R       E               L               N
        O       R               O               A
        M       M               C               T
        E       E               Y       C       I
                    N           L       I       O
        A E R O B I C       T G         B       N
                        A               O
                    T                   R
                I                       E
                        O               A
        C A T A B O L I S M             N
            M S I L O B A N A
```

### CHAPTER TEST

**1.** b **2.** c **3.** a **4.** a **5.** b **6.** b **7.** b **8.** a **9.** b **10.** d **11.** e **12.** d **13.** c **14.** d **15.** c **16.** e **17.** d **18.** d **19.** d **20.** a **21.** Glucose enters the body at the digestive tract, and oxygen enters at the lungs. Glucose and oxygen are delivered to cells by the cardiovascular system. Water from cellular respiration enters the blood and is utilized by the body or excreted; we breathe out the carbon dioxide.  **22.** The citric acid cycle is a series of enzymatic reactions, and one reaction is tied to the next because the product of one reaction is the substrate for the next. The electron transport system is the passage of electrons from one carrier to the next. These carriers are stationed in their correct order on the cristae; otherwise, the passage of electrons would not release energy in the most efficient manner.

# 8

# PHOTOSYNTHESIS

## CHAPTER REVIEW

**Photosynthesis** provides food for living organisms and replenishes oxygen in the atmosphere. Photosynthesis utilizes the portion of the electromagnetic spectrum known as visible light. **Chlorophyll**—both chlorophyll *a* and chlorophyll *b*—absorbs violet, blue, and red light better than light of other colors. These pigments are present in the **thylakoid** membrane within the grana or **chloroplasts.** They participate in the so-called **light-dependent reactions.**

The light-dependent reactions involve a cyclic electron pathway and a noncyclic electron pathway. Only **photosystem I** is required for the **cyclic electron pathway,** in which electrons energized by the sun leave the reaction-center chlorophyll *a* and then pass down an **electron transport system** with the concomitant buildup of ATP before returning to chlorophyll *a*. In the **noncyclic electron pathway,** energized electrons leave chlorophyll *a* of **photosystem II,** pass down an electron transport system, and enter photosystem I, where they are energized once more before being accepted by NADPH. The overall result from the noncyclic pathway is the production of NADPH and ATP. Oxygen is also liberated when water is split and electrons enter chlorophyll *a* (photosystem II) to replace those lost. The NADPH and ATP from the light-dependent reactions of photosynthesis are used to build a carbohydrate in the **light-independent reactions,** which occur in the **stroma** of chloroplasts.

ATP production during the light-dependent reactions requires chemiosmosis. Hydrogen ions are concentrated in the thylakoid space; when water splits, it releases hydrogen ions, and carriers within the cytochrome complex of the electron transport system pump the hydrogen ions to the thylakoid space. The hydrogen ions flow down their concentration gradient through a channel in a protein having an ATP synthase, which forms ATP from ADP and (P).

The light-independent reactions occur during the **Calvin cycle:** Carbon dioxide is fixed by **RuBP,** and reduced to PGAL (this requires the ATP and NADPH from the light-dependent reactions), and RuBP is regenerated. One PGAL out of every six joins with another PGAL to form glucose-6-phosphate.

**C$_3$** photosynthesis (the first molecule after fixation is a C$_3$ molecule) occurs when a plant uses the Calvin cycle directly. Plants have also evolved two other types of photosynthesis: **C$_4$** photosynthesis and CAM photosynthesis. C$_4$ plants fix carbon dioxide in mesophyll cells (which results in a C$_4$ molecule) and then transport it to bundle sheath cells, where it enters the Calvin cycle. CAM plants fix carbon dioxide at night and then release it during the day to the Calvin cycle. C$_4$ and CAM photosynthesis are adaptations to hot, dry environments, since these processes allow the stomata to close to conserve water.

## STUDY QUESTIONS

Study the text section by section. Answer the study questions so that you can fulfill the learning objectives for each section.

## 8.1 RADIANT ENERGY (PP. 132–33)

The learning objective for this section is:
• Define photosynthesis and explain its importance.

1. Indicate whether these statements about the importance of photosynthesis are true (T) or false (F).
   a. _____ It makes food for animals.
   b. _____ It promotes the breakdown of biodegradable wastes.
   c. _____ It returns carbon dioxide to the atmosphere.
   d. _____ It returns oxygen to the atmosphere.

2. Indicate whether these statements about solar energy are true (T) or false (F).
   a. _____ Chlorophylls *a* and *b* absorb violet, blue, and red light best.
   b. _____ Photons of visible light energize electrons without harming cells.
   c. _____ Photosynthesis uses infrared light efficiently in the daytime.
   d. _____ Photosynthesis uses ultraviolet light efficiently at night.

## 8.2 STRUCTURE AND FUNCTION OF CHLOROPLASTS (PP. 134–35)

The learning objectives for this section are:
- Identify the two main portions of chloroplasts.
- Describe the reactions involved in photosynthesis.

In questions 3–5, fill in the blanks.

3. Photosynthesis refers to the ability of plants, algae, and a few kinds of bacteria to make their own a._____ in the presence of b._____. In plants, photosynthesis is carried on in c._____.

4. The green pigment a._____ is found within the membrane of the b._____, and it is here that c._____ energy is captured.

5. A chloroplast contains flattened, membranous sacs called a._____ that are stacked like poker chips into b._____. The fluid-filled space surrounding the grana is called the c._____.

6. Label the adjacent diagram of a chloroplast using the following list of terms.

   granum
   stroma
   thylakoid

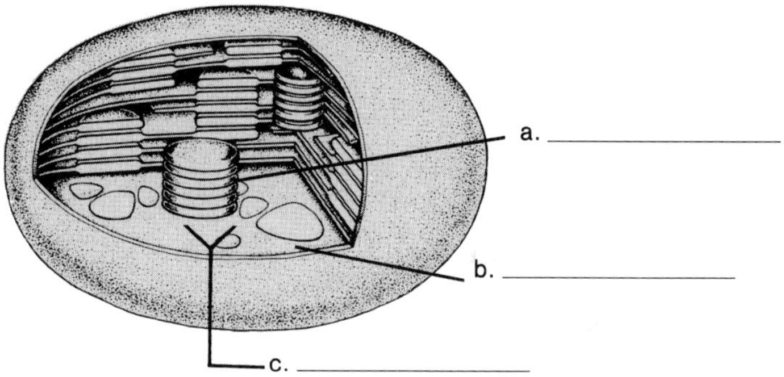

7. From the diagram in question 6, the light-dependent reactions would be associated with the a._____, and the light-independent reactions would be associated with the b._____.

a._____

b._____

c._____

8. The light-dependent reactions drive the light-independent reactions. The light-independent reactions use the NADPH and a._____ from the light-dependent reactions to reduce b._____ to a c._____.

9. Place the appropriate letters next to each reaction.
   LD—light-dependent    LI—light-independent

   a. _____ energy-capturing reactions
   b. _____ synthesis reactions
   c. _____ Carbon dioxide becomes carbohydrate.
   d. _____ Water gives off oxygen.
   e. _____ NADPH and ATP are made.
   f. _____ NADPH and ATP are used.

## 8.3 SOLAR ENERGY CAPTURE (PP. 136–38)

The learning objective for this section is:
- Describe how solar energy energizes electrons and permits a buildup of ATP.

10. The thylakoid membrane has two light-gathering units, called a._____ and b._____. Within each unit are green pigments called c._____ and yellow-orange pigments called d._____.

11. Label this diagram of the cyclic electron pathway, using the following alphabetized list of terms.

ADP
ATP
electron acceptor
light-dependent reactions
photosystem I
pigment complex

12. In the diagram below, label the noncyclic electron pathway, using the following alphabetized list of terms. (One term is used more than once.)

electron acceptor
electron transport system
light-dependent reactions
light-independent reactions
photosystem I
photosystem II

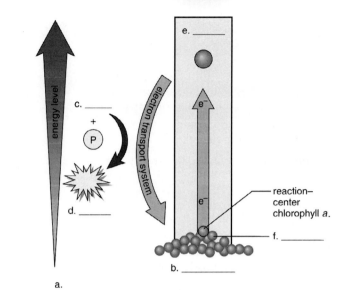

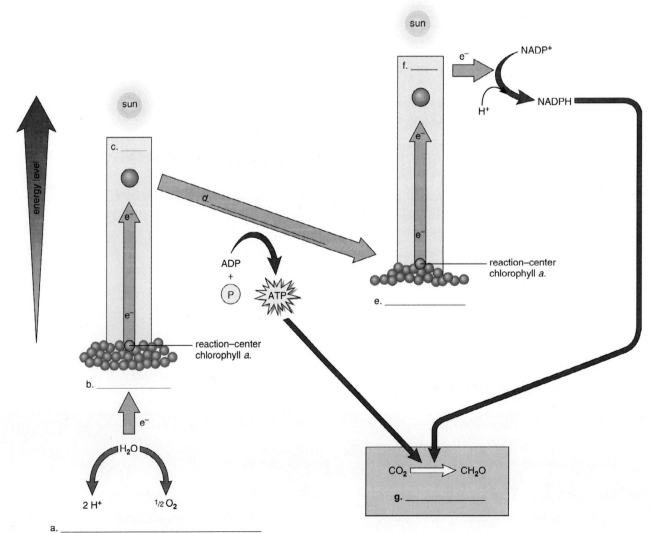

13. What is the role of each of these in the noncyclic electron pathway?

a. reaction-center chlorophyll *a* _____

_____

b. electron acceptors _____

_____

c. NADP$^+$ _____

_____

d. electron transport system _____

_____

e. water _____

_____

_____

_____

14. The concentration of H$^+$ in the $^{a.}$_____ space is $^{b.}$_____, compared to the lower H$^+$ concentration in the $^{c.}$_____. The flow of H$^+$ down its concentration gradient provides the energy for an enzyme called $^{d.}$_____ to produce $^{e.}$_____ from ADP + Ⓟ.

15. Match the functions to the complexes in the thylakoid membrane by writing one of the following terms on lines *a–d*.

photosystem II
cytochrome complex
photosystem I
ATP synthase complex

a. _____ produces ATP from ADP + Ⓟ
b. _____ transports electrons and stores H$^+$ in the thylakoid space
c. _____ captures solar energy; water is split, releasing electrons and oxygen
d. _____ captures solar energy; NADP$^+$ is reduced to NADPH

## 8.4 CARBOHYDRATE SYNTHESIS (PP. 139–41)

The learning objective for this section is:
• Explain the light-independent reactions in the second stage of photosynthesis.

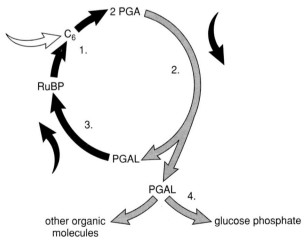

16. Match the numbers in the diagram to the following descriptions of events in the Calvin cycle. (Some numbers are used more than once.)
a. _____ ATP only required
b. _____ CO$_2$ reduction reaction
c. _____ NADPH and ATP required
d. _____ CO$_2$ taken up, CO$_2$ fixation
e. _____ glucose formed after six turns of cycle
f. _____ five PGAL required to form three molecules of product

17. Indicate the sequence in which molecules *a–e*
    appear during or as a result of the Calvin cycle.

    _____

    a. $CO_2$
    b. glucose phosphate
    c. PGA
    d. PGAL
    e. starch

## 8.5 OTHER TYPES OF PHOTOSYNTHESIS (P. 142)

The learning objective for this section is:
- Categorize different types of photosynthesis.

18. Label the following as describing $C_3$, $C_4$, and/or CAM plants:
    a. _____ predominate in spring and cooler summer weather
    b. _____ succulent plants, cacti; live in hot, arid regions
    c. _____ wheat, rice, oats
    d. _____ predominate in hot, dry summer weather
    e. _____ product of $CO_2$ fixation is PGA
    f. _____ $CO_2$ fixation occurs at night and $C_4$ molecules are stored until daylight
    g. _____ photorespiration occurs
    h. _____ stomata are closed *during the day* to conserve water
    i. _____ end product of $CO_2$ fixation is oxaloacetate
    j. _____ sugarcane, corn, Bermuda grass
    k. _____ chloroplasts only in mesophyll cells
    l. _____ chloroplasts in bundle sheath cells and mesophyll

## 8.6 PHOTOSYNTHESIS VERSUS CELLULAR RESPIRATION (P. 143)

The learning objective for this section is:
- Contrast photosynthesis and cellular respiration.

Indicate whether these statements are true (T) or false (F). If false, rewrite the statement as a true one.

19. Carbon dioxide is oxidized during photosynthesis, and oxygen is oxidized during cellular respiration.

    Answer: _____ Rewrite: _____

    _____

20. The coenzyme active during photosynthesis is $NADP^+$, and the coenzyme active during cellular respiration is $NAD^+$.

    Answer: _____ Rewrite: _____

    _____

21. PGAL becomes PGA during photosynthesis, and PGA becomes PGAL during cellular respiration.

    Answer: _____ Rewrite: _____

    _____

22. Oxygen is given off by photosynthesis, and carbon dioxide is given off by cellular respiration.

    Answer: _____ Rewrite: _____

    _____

Photosynthesis, which is now playing in chloroplasts, has two acts. The first act is the light-dependent reactions, and the second act is the light-independent reactions. The theater has a revolving stage of two parts, shown in the diagram below.

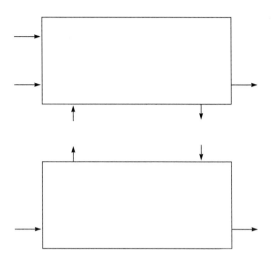

1. Place these "props" on the correct "stage": grana (thylakoids) and stroma.

2. Place these "props" on the correct "stage": Calvin cycle, energy-capturing reactions, and synthesis reactions.

3. Place these "props" on the correct "stage": chlorophyll and RuBP.

4. Certain actors/actresses walk on during either the first act or the second act (not both acts), and certain others walk off during either the first act or the second act. Entrances and exits are represented by the arrows. Show where the following "actresses/actors" either walk on or walk off: $CO_2$, $H_2O$, light, $O_2$, and PGAL.

5. Certain actors/actresses are in the first act, but then they change their costumes and move to the second act. Naturally, they have to get ready again for the first act. Movement of these actors/actresses is indicated by arrows between the acts. Show where the following "actresses/actors" belong: ADP, ATP, $NADP^+$, and NADPH.

Review key terms by using the following alphabetized list of terms to fill in the blanks below. Then complete the wordsearch.

```
S J P H O T O S Y N T H E S I S R I O
N A R D Z C Y T O K I M O D E T B Y E
P H O T O S Y S T E M S M O S R S C F
T H D R I O T B G T D T N G I O R N Q
G I U T Y C D E R U T A S E D M O N A
R E C T A F R D C P T D I O R A T E M
A Y E I L Z K A N L I O S O M G T R N
F U R I D F I O T O F L Y K E I Y E L
R E O N U G E L N A L I L H T F N M M
L I M C O T N E C M O S O C A D O U E
K I E Y A L T T M T S M C H P T C S T
F L R D G O O M C E R O Y L H H C N A
G O I S R R O T Y C N E L O A Y I O R
C X I A E R O L L Y H P O R O L H C T
O U C L A I R E A C T A A O E A O L S
Q M E O Y B U P L A S M Z P Y K R S B
G I S R W X M P N C E N Y L I O E R U
C V I M O I M H A N I B M A E I A G S
O C S C A T A B O L I S M S S D N T W
D U C L A I U B M S I L O T A N A K I
```

carotenoids
chlorophylls
chloroplast
consumer
photosynthesis
photosystem
producer
stroma
thylakoid

a. _____ Process by which plants and algae make their own food using the energy of the sun.

b. _____ Photosynthetic organism at the start of a grazing food chain that makes its own food (e.g., green plants on land and algae in water).

c. _____ Organism that feeds on another organism in the food chain; primary consumers eat plants, and secondary (or higher) consumers eat animals.

d. _____ Green pigment that captures solar energy during photosynthesis.

e. _____ Yellow or orange pigment that serves as an accessory to chlorophyll in photosynthesis.

f. _____ Fluid within a chloroplast that contains enzymes involved in the synthesis of carbohydrates during photosynthesis.

g. _____ Flattened sac within a granum whose membrane contains chlorophyll and where the light-dependent reactions of photosynthesis occur.

h. _____ Membranous organelle that contains chlorophyll and is the site of photosynthesis.

i. _____ Cluster of light-absorbing pigment molecules within thylakoid membranes.

## OBJECTIVE QUESTIONS

Do not refer to the text when taking this test. In questions 1–6, match the definitions to these terms.

   a. chlorophyll
   b. oxygen
   c. stroma
   d. sugar
   e. thylakoid membrane
   f. water

_____ 1. organic product of photosynthesis
_____ 2. released by photosynthesis
_____ 3. reactant of photosynthesis
_____ 4. site of light-dependent reactions
_____ 5. site of light-independent reactions
_____ 6. molecule-absorbing solar energy

_____ 7. Each of the following is a product of photosynthesis EXCEPT
   a. carbon dioxide.
   b. organic food.
   c. oxygen.
   d. carbohydrate.

_____ 8. Photosynthesis occurs best at wavelengths that are
   a. blue.
   b. gamma.
   c. infrared.
   d. ultraviolet.

_____ 9. Each is a product of light-dependent reactions EXCEPT
   a. ATP.
   b. NADPH.
   c. oxygen.
   d. sugar.

_____ 10. The cyclic pathways of photosynthesis produce
   a. ATP only.
   b. NADPH only.
   c. ATP and NADPH.
   d. organic sugars only.

_____ 11. Carbon dioxide fixation occurs when $CO_2$ combines with
   a. ATP.
   b. NADPH.
   c. PGAL.
   d. RuBP.

_____ 12. Which of the following pathways uses the enzyme RuBP carboxylase?
   a. $C_2$
   b. $C_3$
   c. CAM
   d. CAP

_____ 13. The enzyme that produces ATP from ADP + $\text{P}$ in the thylakoid is
   a. RuBP carboxylase.
   b. rubisco.
   c. ATPase.
   d. ATP synthase.
   e. coenzyme A.

_____ 14. Which statement is NOT true regarding chemiosmosis?
   a. $H^+$ concentration is higher in the stroma than in the thylakoid space.
   b. The electron transport system pumps $H^+$ from the stroma into the thylakoid space.
   c. The ATP synthase complex is present in the thylakoid membrane.
   d. All of these are true.

_____ 15. Which of the following is NOT a stage in the Calvin cycle?
   a. carbon dioxide fixation
   b. carbon dioxide oxidation
   c. carbon dioxide reduction
   d. ribulose bisphosphate regeneration

_____ 16. Which of these descriptions is NOT true of photosynthesis?
   a. not affected by temperature
   b. not affected by solar energy
   c. requires a supply of oxygen
   d. involves a reduction reaction
   e. more likely to occur during the day

_____ 17. Which of these descriptions is NOT true of chlorophyll?
   a. absorbs solar energy
   b. located in the grana
   c. located in thylakoid membranes
   d. passes electrons directly to $NADP^+$
   e. passes electrons to an acceptor molecule

_____ 18. The two major sets of reactions involved in photosynthesis are
   a. the cyclic and noncyclic electron pathways.
   b. glycolysis and the citric acid cycle.
   c. the Calvin and citric acid cycles.
   d. the Calvin cycle and the electron transport system.
   e. the light-dependent and light-independent reactions.

_____19. Which of the following statements is NOT true of the Calvin cycle?
 a. RuBP is regenerated with the use of ATP.
 b. Glucose phosphate is synthesized from PGAL.
 c. NADPH is used to reduce PGAL to PGA.
 d. Five molecules of PGAL are used to reform three molecules of RuBP.

_____20. Photosystem II gets replacement electrons from
 a. the sun.
 b. water molecules.
 c. ATP.
 d. photosystem I.
 e. NADPH.

## THOUGHT QUESTIONS

Answer in complete sentences.
21. How is life dependent on photosynthesis?

22. Why is photosynthesis dependent on the high degree of compartmentalization in chloroplasts?

**Test Results:** _____ number correct ÷ 22 = _____ × 100 = _____ %

## ANSWER KEY

### STUDY QUESTIONS

**1. a.** T **b.** F **c.** F **d.** T **2. a.** T **b.** T **c.** F **d.** F **3. a.** food **b.** sunlight **c.** chloroplasts **4. a.** chlorophyll **b.** thylakoids **c.** solar **5. a.** thylakoids **b.** grana **c.** stroma **6. a.** thylakoid **b.** stroma **c.** granum **7. a.** thylakoid and granum **b.** stroma **8. a.** ATP **b.** carbon dioxide **c.** carbohydrate **9. a.** LD **b.** LI **c.** LI **d.** LD **e.** LD **f.** LI **10. a.** photosystem I **b.** photosystem II **c.** chlorophyll **d.** carotenoids **11. a.** light-dependent reactions **b.** photosystem I **c.** ADP **d.** ATP **e.** electron acceptor **f.** pigment complex **12. a.** light-dependent reactions **b.** photosystem II **c.** electron acceptor **d.** electron transport system **e.** photosystem I **f.** electron acceptor **g.** light-independent reactions **13. a.** releases electrons that become excited from solar energy **b.** accepts energized electrons from the reaction-center chlorophyll *a* and sends them to the electron transport system **c.** accepts electrons and hydrogen and becomes NADPH **d.** stores energy as the electrons fall to a lower energy level **e.** splits, releasing oxygen and hydrogen ions **14. a.** thylakoid **b.** higher **c.** stroma **d.** ATP synthase **e.** ATP **15. a.** ATP synthase complex **b.** cytochrome complex **c.** photosystem II **d.** photosystem I **16. a.** 2, 3 **b.** 2 **c.** 2 **d.** 1 **e.** 4 **f.** 3 **17.** a, c, d, b, e **18. a.** $C_3$ **b.** CAM **c.** $C_3$ **d.** $C_4$ **e.** $C_3$ **f.** CAM **g.** $C_3$ **h.** CAM **i.** $C_4$ **j.** $C_4$ **k.** $C_3$ **l.** $C_4$ **19.** F, Carbon dioxide is reduced during photosynthesis, and oxygen is reduced during cellular respiration. **20.** T **21.** F, PGA becomes PGAL during photosynthesis, and PGAL becomes PGA during cellular respiration. **22.** T

### PHOTOSYNTHESIS: A PLAY IN TWO ACTS

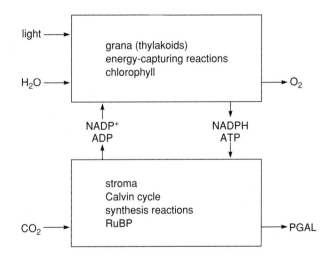

70

**a.** photosynthesis   **b.** producer   **c.** consumer   **d.** chlorophyll   **e.** carotenoid   **f.** stroma   **g.** thykaloid   **h.** chloroplast   **i.** photosystem

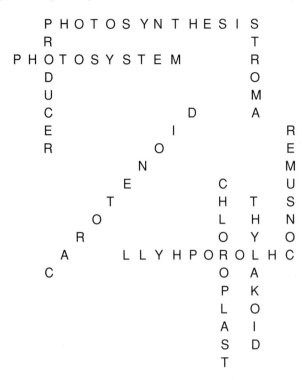

1. d  2. b  3. f  4. e  5. c  6. a  7. a  8. a  9. d  10. a  11. d  12. b  13. d  14. a  15. b  16. c  17. d  18. e  19. c  20. b  21. Through photosynthesis, plants (and algae) produce food for themselves and all other living things. These organisms are the producers at the start of food chains of all types. Animals feed directly on photosynthesizers or on other animals that have fed on photosynthesizers.  22. Chemiosmosis cannot occur without a membrane that maintains a difference in hydrogen ion concentration. In plant cells, hydrogen ions build up within the thylakoid space, and then they flow down their concentration gradient through an ATP synthase complex located in the thylakoid membrane.

# 9

# PLANT ORGANIZATION

A flowering plant has three vegetative organs: roots, stems, and leaves. A **root** anchors a plant, absorbs water and minerals, and stores the products of photosynthesis. **Stems** support leaves, conduct materials to and from roots and leaves, and help store plant products. **Leaves** carry on photosynthesis.

Three types of tissue are found in each organ: **dermal, ground,** and **vascular. Epidermis** is the dermal tissue in most parts of the plant. **Parenchyma, collenchyma,** and **sclerenchyma** are the types of ground tissue. **Xylem,** which transports water and minerals, and **phloem,** which transports organic solutes, are the two types of vascular tissue.

Root tissues can be studied in longitudinal and cross section. There are two main types of roots: **taproot** and **fibrous.** The layers and organization of vascular tissue differ in the roots of the **monocot,** the herbaceous **dicot,** and the woody dicot. The stems of woody, temperate plants experience pronounced secondary growth, developing **annual rings** of xylem.

The cross section of a leaf shows **mesophyll** between upper and lower layers of epidermis. The venation differs between monocots and dicots. Leaves can be simple or compound. Specializations of the leaf include **epidermis, stomata,** and mesophyll cells.

## STUDY QUESTIONS

Study the text section by section. Answer the study questions so that you can fulfill the learning objectives for each section.

### 9.1 PLANT ORGANS (PP. 148–49)

The learning objective for this section is:
• Describe the three vegetative organs of a flowering plant.

1. Flowering plants are adapted to living on land. The root system a._____ the plant in the soil and takes up b._____ and c._____, especially by means of d._____. Stems lift the leaves and e._____ water and minerals to the leaves. The leaves carry on f._____ after receiving water from the roots and g._____ from the air.

### 9.2 MONOCOT VERSUS DICOT PLANTS (P. 150)

The learning objective for this section is:
• Classify plants into two groups, the monocots and the dicots.

2. Complete the following table to indicate the differences between monocot and dicot plants.

| Plant Part | Monocot | Dicot |
|---|---|---|
| Leaf veins | | |
| Vascular bundles in stem | | |
| Cotyledons in seed | | |
| Parts in a flower | | |
| Vascular tissue in root | | |

## 9.3 PLANT TISSUES (PP. 151–53)

The learning objective for this section is:
• Describe the three types of plant tissues.

3. Match the cell types with the following tissue types:
   1 epidermal tissue
   2 ground tissue
   3 vascular tissue
   a. _____ sclerenchyma cells
   b. _____ tracheids
   c. _____ root hair cells
   d. _____ parenchyma cells
   e. _____ sieve-tube elements
   f. _____ vessel elements

4. Indicate whether these statements are true (T) or false (F).

**Epidermal Tissue:**

   a. _____ Epidermis covers the entire body of nonwoody plants.
   b. _____ Epidermis covers the entire body of young woody plants.
   c. _____ Epidermis replaces cork in the stems of older plants.
   d. _____ Guard cells regulate the entrance of water into the roots.

**Ground Tissue:**

   e. _____ Collenchyma cells are a major site of photosynthesis.
   f. _____ Collenchyma has thinner primary walls than parenchyma.
   g. _____ Parenchyma is the least specialized cell type.
   h. _____ Sclerenchyma secondary walls are impregnated with lignin.
   i. _____ Sclereids give pears their gritty texture.

**Vascular Tissue:**

   j. _____ Phloem transports organic nutrients from leaves to roots.
   k. _____ Sieve-tube elements are found in xylem.
   l. _____ Tracheids are a type of cell in phloem.
   m. _____ Xylem transports water from roots to leaves.

## 9.4 ORGANIZATION OF ROOTS (PP. 154–56)

The learning objectives for this section are:
• Name the three zones in a root tip.
• Compare dicot and monocot roots.

5. Label this diagram of a dicot root tip, using the following alphabetized list of terms.

cortex
endodermis
epidermis
pericycle
phloem
root cap
root hair
vascular cylinder
xylem
zone of cell division
zone of elongation
zone of maturation

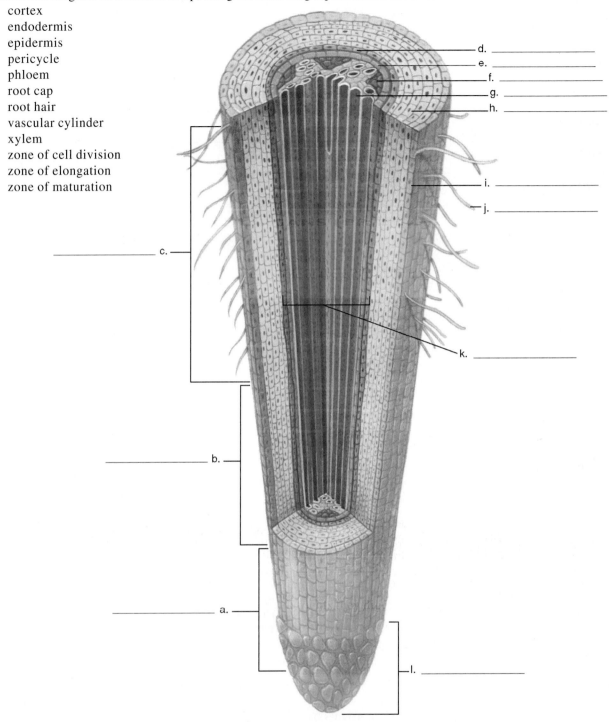

d. _____
e. _____
f. _____
g. _____
h. _____

i. _____
j. _____

k. _____

c. _____

b. _____

a. _____

l. _____

6. Explain what is happening in each of these zones of a root tip.

a. cell division _____

b. elongation _____

c. maturation _____

7. Complete the following table to describe and give a function for the specific tissues within a root.

| Tissue | Description | Function |
|---|---|---|
| Epidermis | | |
| Cortex | | |
| Endodermis | | |
| Vascular tissue | | |

8. Label each type of root shown, and describe its special function.

a.                                    b.

Type of root:       _____        _____

Function:           _____        _____

                    _____        _____

## 9.5 ORGANIZATION OF STEMS (PP. 158–62)

The learning objectives for this section are:
- Describe the process of primary and secondary stem growth.
- Compare the organization of dicot and monocot herbaceous stems.
- Describe the main parts of a woody stem.

9. During primary growth, a._____ at the shoot tip produces new cells that
   b._____ and thereby increase the length of the stem. c._____ primordia are
   produced at the nodes, and they form the terminal bud. Specialized types of meristem give rise to specialized
   tissues such as d._____ tissue, which transports water in plants. During secondary growth,
   lateral meristem gives rise to new e._____ and f._____ every year. The xylem
   builds up and becomes g._____. Actually, a woody stem has three main portions:
   h._____, i._____, and j._____. Phloem is in the
   k._____.

10. Study the diagram and write the term *herbaceous* or *woody* on the lines provided.

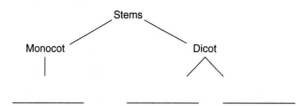

Stems

Monocot          Dicot

_____     _____     _____

11. Label these diagrams of stems, using the following alphabetized list of terms. (Some are used more than once.)

bark
cortex
epidermis
phloem
pith
vascular bundle
wood
xylem

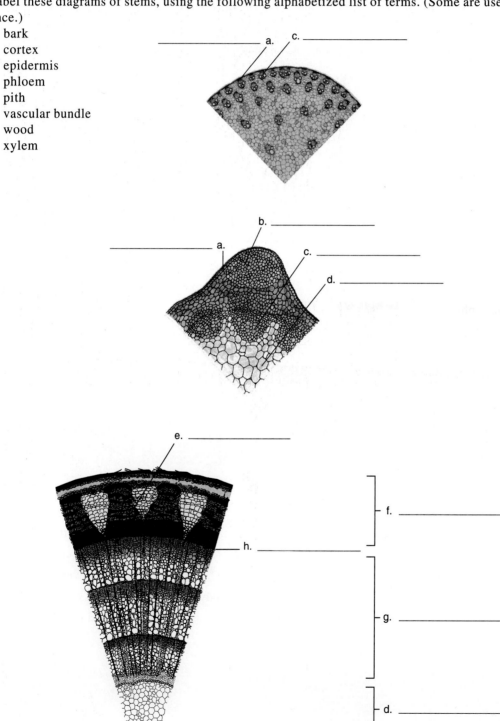

_____ a.     c. _____

b. _____

_____ a.     c. _____

d. _____

e. _____

h. _____

f. _____

g. _____

d. _____

12. a. How would you describe the arrangement of vascular bundles in a herbaceous monocot stem?

   _____

   b. In a herbaceous dicot stem? _____

   c. Annual rings appear in which type of stem? _____

   d. What causes annual rings? _____

13. Match the descriptions to these types of stems.
      1 corm
      2 rhizome
      3 stolon
      4 tuber
   a. _____ bulbous underground stem
   b. _____ enlarged area of an underground stem that serves as storage area
   c. _____ underground stem that survives the winter
   d. _____ runner; new plants start from nodes

## 9.6 ORGANIZATION OF LEAVES (PP. 164–65)

The learning objectives for this section are:
- Describe the main parts of a leaf.
- Provide examples of different types of leaves.

14. Label this diagram of a leaf, using the following alphabetized list of terms, and use the same terms to fill in the blanks h–k.
      epidermis
      guard cell
      leaf vein
      mesophyll
      palisade mesophyll
      spongy mesophyll
      stoma

   An extension of vascular bundles of stem is the h._____. The i._____ allows gas exchange. Longitudinal cells that photosynthesize are the j._____. Loosely arranged cells that photosynthesize and exchange gases are the k._____.

15. Place a check in front of the correct descriptions of leaves.
   a. _____ spines of a cactus
   b. _____ tendrils of a cucumber
   c. _____ trap of a Venus's-flytrap
   d. _____ stolon of a strawberry plant

# DEFINITIONS WORDSEARCH

Review key terms by using the following alphabetized list of terms to fill in the blanks below. Then complete the wordsearch.

```
S J E T R E C M I T O C H A N D R I O
N O C D Z C Y E P I D E R M I S B Y C
L O R S O C H E M I O S M Y S I S C O
T H T G I O T B G T D T N H I H I N L
G I O T A C D E R U T A S C D T N N L
R E C T A N R D C P T B I N R H T E E
A Y H I L Z K A N L C O S E M G E I N
F U R I D F I O L A F L Y R E I R L C
R E O N U G E L I A L I L A T F N M H
L I M C O T N R X M O S O P A D O J Y
K P E R E N N I A L S M C O P M D E M
F E R D G L O M C E R O Y R H E E I A
G T I S I R O T M C N M M H A L I G R
C I I A E R O B O C S T E N S G B O T
C O T Y L E D O N C T A T R E M O L S
Q L E O Y B U P O A S M S T I S R S B
G E S R W X M P C C E N Y R I S E R U
C V I M O I M H O N I B M R E O T G S
O C S C A T A B T L I S M E S I N E W
D U C L A I U B S S I L O B A N A K M
```

*axil*
*collenchyma*
*cotyledon*
*epidermis*
*internode*
*meristem*
*monocots*
*organ*
*parenchyma*
*perennial*
*petiole*
*stem*

a. _____ Combination of two or more different tissues performing a common function.

b. _____ Flowering plant that lives more than one growing season because the underground parts regrow each season.

c. _____ In vascular plants, the region of a stem between two successive nodes.

d. _____ Usually the upright, vertical portion of a plant, which transports substances to and from the leaves.

e. _____ Part of the plant leaf that connects the blade to the stem.

f. _____ Angle between petiole and stem.

g. _____ Seed leaf for embryo of a flowering plant; provides nutrient molecules for the developing plant before photosynthesis begins.

h. _____ Flowering plant group; members have one embryonic leaf (cotyledon), parallel-veined leaves, scattered vascular bundles, and other characteristics.

i. _____ Undifferentiated embryonic tissue in the active growth regions of plants.

j. _____ Tissue that covers the roots, leaves, and stems of nonwoody plants.

k. _____ Thin-walled, minimally differentiated cell that photosynthesizes or stores the products of photosynthesis.

l. _____ Plant tissue composed of cells with unevenly thickened walls; supports growth of stems and petioles.

Do not refer to the text when taking this test.

_____ 1. Each is an adult tissue in plants EXCEPT
    a. epidermal.
    b. ground.
    c. meristem.
    d. vascular.

_____ 2. Select the incorrect association.
    a. collenchyma—flexible support
    b. parenchyma—unspecialized
    c. sclerenchyma—tough and hard
    d. sieve tube—mechanical strength

_____ 3. Select the correct association.
    a. phloem—minerals
    b. phloem—photosynthesis
    c. xylem—sugar
    d. xylem—water

_____ 4. The zone farthest from the root cap is the zone of
    a. cell division.
    b. elongation.
    c. maturation.
    d. primary growth.

_____ 5. The cortex of the root mainly functions for
    a. entrance of substances into the root.
    b. photosynthesis.
    c. protection.
    d. starch storage.

_____ 6. The root type in the carrot is the
    a. adventitious root.
    b. fibrous root.
    c. taproot.
    d. prop root.

_____ 7. In the vascular bundles of a herbaceous stem,
    a. phloem develops to the inside.
    b. xylem develops to the outside.
    c. Both _a_ and _b_ are correct.
    d. Neither _a_ nor _b_ is correct.

_____ 8. A stolon is a(n)
    a. aboveground stem.
    b. cork covering on a stem.
    c. specialized type of leaf.
    d. underground stem.

_____ 9. The palisade mesophyll
    a. stores food in the form of starch for the rest of the plant.
    b. absorbs water and minerals.
    c. contains elongated cells where photosynthesis takes place.
    d. opens to allow gases to move in and out.

_____ 10. The wood portion of a woody stem is composed of
    a. pith.
    b. cambium.
    c. bark.
    d. secondary xylem.
    e. secondary phloem.

_____ 11. Trace the path of water from the roots to the leaves.
    a. root hairs→cortex→vascular cylinder→ vascular bundles→leaf veins
    b. root hairs→pith→xylem→phloem→ wood→leaf veins
    c. cortex→endodermis→xylem→tracheids→ parenchyma→leaf veins
    d. root hairs→sclerenchyma→vascular cylinder→cortex→leaf veins

_____ 12. Stomata are
    a. a type of transport tissue.
    b. openings in leaf epidermis.
    c. found in woody trees only.
    d. a universal type of cell.
    e. All of these are correct.

_____ 13. Which of these comparisons of monocots and dicots is NOT correct?

| **Monocots** | **Dicots** |
| --- | --- |
| a. net veined | parallel veined |
| b. one cotyledon | two cotyledons |
| c. scattered vascular bundles | circular pattern |
| d. flower parts in threes | flower parts in fours/fives |

_____ 14. The cells in the root cap are produced from the zone of
    a. cell division.
    b. elongation.
    c. maturation.
    d. pericycle.
    e. vascular bundle.

_____ 15. A cross section through the zone of maturation of a dicot root would show
    a. cells of the root cap.
    b. the apical meristem.
    c. transport tissues.
    d. greatly elongated, undifferentiated cells.
    e. All of these are correct.

_____ 16. Annual rings in woody stems are caused by an increase in rings of the
    a. primary phloem.
    b. secondary phloem.
    c. primary xylem.
    d. secondary xylem.

____17. The point of a stem at which leaves or buds are attached is termed the
   a. node.
   b. internode.
   c. lenticel.
   d. endodermis.
   e. plasmodesmata.

____18. Which of these is mismatched?
   a. monocot stem—vascular bundles scattered
   b. dicot herbaceous stem—vascular bundles in a ring
   c. dicot woody stem—no vascular tissue
   d. All of these are properly matched.

____19. Water taken up by root hairs has to enter the vascular cylinder because
   a. osmosis draws it in.
   b. guard cells regulate the opening of stomata.
   c. the Casparian strip prevents anything else.
   d. Both *a* and *c* are correct.

____20. A leaf is like a root because they both
   a. photosynthesize.
   b. store the products of photosynthesis in bad times.
   c. have vascular tissue.
   d. have a double layer of epidermis.

## THOUGHT QUESTIONS

Answer in complete sentences.

21. Do plants have a true organ structure, just as animals do? What characteristics of their anatomy support your answer?

22. Why is the evolution of xylem such an important adaptation for the success of plants on land?

**Test Results:** _____ number correct ÷ 22 = _____ × 100 = _____%

## STUDY QUESTIONS

**1. a.** anchors **b.** water **c.** minerals **d.** root hairs **e.** transport **f.** photosynthesis **g.** carbon dioxide
**2.**

| Monocot | Dicot |
|---|---|
| parallel pattern | net pattern |
| scattered | in a ring |
| one | two |
| in threes | in fours and fives |
| in a ring | in a vascular cylinder |

**3. a.** 2 **b.** 3 **c.** 1 **d.** 2 **e.** 3 **f.** 3 **4. a.** T **b.** T **c.** F **d.** F **e.** F **f.** F **g.** T **h.** T **i.** T **j.** T **k.** F **l.** F **m.** T **5.** See Figure 9.8, page 154, in text. **6. a.** New cells are appearing. **b.** Cells are getting longer. **c.** Cells are mature and specialized.
**7.**

| Description | Function |
|---|---|
| outer single layer of cells | root hairs, especially, absorb water and minerals |
| thin-walled parenchyma cells | food storage |
| single layer of rectangular cells | regulates entrance of minerals into vascular cylinder |
| xylem and phloem | transport water, minerals, and organic nutrients |

**8. a.** fibrous root; holds soil **b.** taproot; stores food **9. a.** apical meristem **b.** elongate **c.** leaf **d.** xylem **e.** xylem **f.** phloem **g.** wood **h.** bark **i.** wood **j.** pith **k.** bark **10.** herbaceous, herbaceous, woody **11. a.** epidermis **b.** cortex **c.** vascular bundle **d.** pith **e.** phloem **f.** bark **g.** xylem **h.** vascular cambium **12. a.** scattered **b.** ring **c.** woody stem **d.** summer xylem vessels are small, and spring vessels are large **13. a.** 1 **b.** 4 **c.** 2 **d.** 3 **14. a.** epidermis **b.** palisade mesophyll **c.** leaf vein **d.** spongy mesophyll **e.** epidermis **f.** stoma **g.** guard cell **h.** leaf veins **i.** stoma **j.** palisade mesophyll **k.** spongy mesophyll **15.** a, b, c, d

## DEFINITIONS WORDSEARCH

**a.** organ **b.** perennial **c.** internode **d.** stem **e.** petiole **f.** axil **g.** cotyledon **h.** monocots **i.** meristem **j.** epidermis **k.** parenchyma **l.** collenchyma

## CHAPTER TEST

**1.** c **2.** d **3.** d **4.** c **5.** d **6.** c **7.** d **8.** a **9.** c **10.** d **11.** a **12.** b **13.** a **14.** a **15.** c **16.** d **17.** a **18.** c **19.** c **20.** c **21.** Several specialized tissues (epidermal, ground, vascular) are integrated structurally in the leaf, stem, and root. All are organs, consisting of two or more tissues working together. **22.** Xylem allows plants to transport water against gravity, from roots to leaves. Without water transport, plants have to be low lying.

# 10

# PLANT REPRODUCTION, GROWTH, AND DEVELOPMENT

Flowering plants have an **alternation of generations** life cycle, which includes separate microgametophytes and megagametophytes. The **pollen grain,** the microgametophyte, is produced within the **stamens** of a flower. The megagametophyte is produced within the ovule of a flower. Following **pollination** and fertilization, the ovule matures to become the seed and the ovary becomes the **fruit.** The enclosed seeds contain the embryo (hypocotyl, epicotyl, **plumule,** radicle) and stored food (endosperm and/or cotyledons). When a seed **germinates,** the root appears below and the shoot appears above.

Many flowering plants reproduce asexually, as when the nodes of stems (either aboveground or underground) give rise to entire plants, or when an isolated root produces new shoots. Micropropagation, the production of clonal plants utilizing tissue culture, is now a commercial venture. In cell suspension cultures, plant cells produce chemicals of medical importance. The practice of plant tissue culture facilitates genetic engineering of plants. One aim of genetic engineering is to produce plants that have improved agricultural or food quality traits. Plants can also be engineered to produce chemicals of use to humans.

Plant **hormones** control plant responses to environmental stimuli. **Tropisms** are growth responses toward or away from unidirectional stimuli. When a plant is exposed to light, **auxin** moves laterally from the bright to the shady side of a stem. Thereafter, the cells on the shady side elongate, and the stem moves toward the light.

Plant hormones most likely control **photoperiodism.** Short-day plants flower when the days are shorter (nights are longer) than a critical length, and long-day plants flower when the days are longer (nights are shorter) than a critical length. Some plants are day-length neutral. **Phytochrome,** a plant pigment that responds to daylight, is believed to be a part of a biological clock system that in some unknown way brings about flowering.

Water transport in plants occurs within **xylem.** The **cohesion-tension** model of xylem transport states that **transpiration** (evaporation of water at stomata) creates tension, which pulls water upward in xylem. This method works only because water molecules are cohesive.

**Stomata** open when **guard cells** take up potassium ions ($K^+$), and water follows by osmosis. Stomata open because the entrance of water causes the guard cells to buckle out.

Transport of organic nutrients in plants occurs within phloem. The **pressure-flow model** of phloem transport states that sugar is actively transported into **phloem** at a source, and water follows by osmosis. The resulting increase in pressure creates a flow, which moves water and sucrose to a sink.

Study the text section by section. Answer the study questions so that you can fulfill the learning objectives for each section.

## 10.1 SEXUAL REPRODUCTION IN FLOWERING PLANTS (PP. 170–75)

The learning objectives for this section are:
- Name the parts of a flower.
- Define microgametophytes and megagametophytes.
- Explain the life cycle of flowering plants.

1. Label this diagram of a flower, using the following alphabetized list of terms.

anther
filament
ovary
ovule
petal
pistil
pollen grain
pollen tube
sepal
stamen
stigma
style

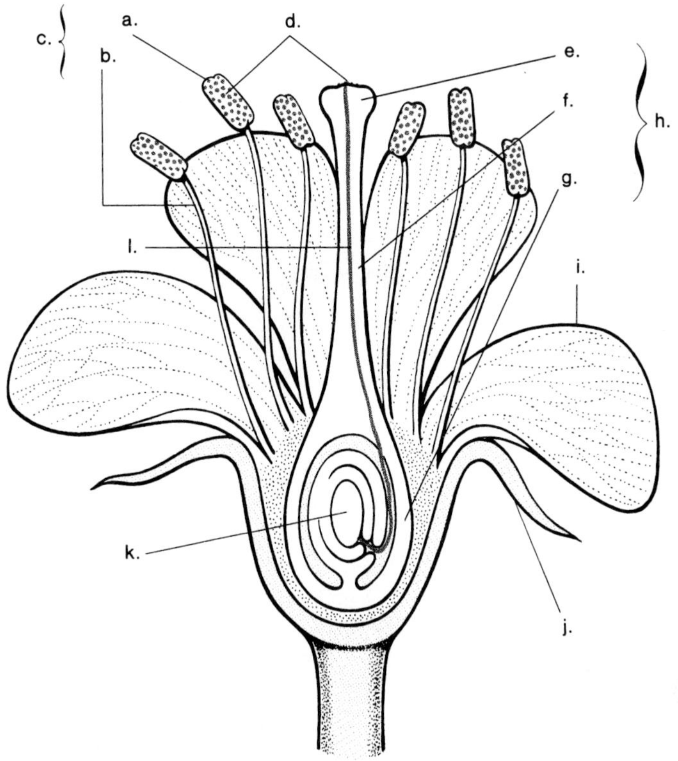

2. Name the following parts of a flower:

a. green leaves that form a whorl _____

b. colored leaves of a flower _____

c. a vaselike structure, _____, that consists of:

  d. a sticky enlarged knob _____

  e. a slender stalk _____

  f. an enlarged base containing ovules _____

Stamen components include:

  g. a portion that produces pollen grains _____

  h. a slender stalk _____

3. a. Label this diagram of the life cycle of a flowering plant, using the following alphabetized list of terms. Tell whether each structure is n or 2n.

anther
megagametophyte
megaspore
microgametophyte
microspore
ovary
sporophyte
zygote

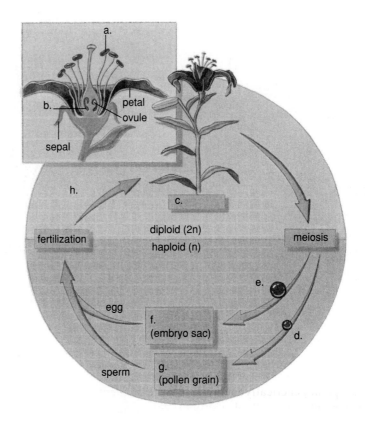

Fill in the blanks.

4. In the life cycle shown in question 3, the dominant sporophyte produces two types of spores, called the

a. _____ and b. _____.

The microspore develops into the mature c. _____, a pollen grain that contains a

d. _____.

The megaspore develops into the e. _____, an embryo sac that contains an

f. _____.

When the sperm fertilizes the egg, a g. _____ results. The zygote becomes an embryo

enclosed within a seed.

5. What is the pericarp? a. _____

Explain the term *simple fruit.* b. _____

Explain the term *dry fruit.* c. _____

Explain the term *aggregate fruit.* d. _____

A peach is a simple fruit with a(n) e. _____ pericarp, while a nut is a simple fruit with a(n)

f. _____, hard pericarp.

6. List four ways that seeds and fruits are dispersed.

a. _____

b. _____

c. _____

d. _____

7. Label each statement as describing the structure and germination of either the bean seed (B) or the corn kernel (C).
   a. _____ It is actually a fruit.
   b. _____ The plumule is enclosed in a sheath called the coleoptile.
   c. _____ Most of the food storage tissue is endosperm.
   d. _____ The shoot is hook shaped.

## 10.2 ASEXUAL REPRODUCTION IN FLOWERING PLANTS (PP. 176–77)

The learning objectives for this section are:
- Explain how flowering plants reproduce asexually.
- Describe how identical plants are produced commercially.

8. Place a check in front of the example(s) of vegetative propagation.
   a. _____ strawberry plants grown from the nodes of stolons
   b. _____ potato plants grown from the eyes of a potato
   c. _____ ornamental plants grown from stem cuttings
9. Place a check in front of the characteristics of vegetative propagation.
   a. _____ sexual reproduction
   b. _____ asexual reproduction
   c. _____ new plant genetically identical to original plant
   d. _____ new plant genetically dissimilar to original plant
10. Micropropagation is a commercial way to produce thousands of a._____

    (*identical/dissimilar*) seedlings utilizing b._____ culture. Somatic embryos can be grown

    from flower c._____, which is free of viruses. The process of culturing individual cells—

    derived from root, stem, or leaf—to produce desirable chemicals is called d._____.

    A protoplast is a(n) e._____ that can go on to develop into an entire plant.

## 10.3 CONTROL OF PLANT GROWTH AND DEVELOPMENT (PP. 177–81)

The learning objectives for this section are:
- Discuss how plants respond to outside stimuli by changing their growth patterns.
- Describe the functions of different plant hormones.

11. Explain this diagram.

_____

_____

_____

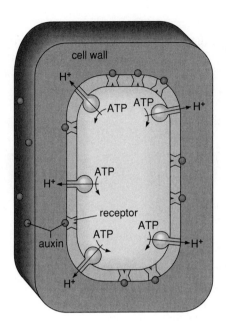

12. Complete the first column of the table below by filling in the names of plant hormones.

**Plant Hormones**

| Name | Primary Example | Notable Function |
|---|---|---|
| **_Growth Promoters_** | | |
| a. _____ | Indoleacetic acid (IAA) | Promotes cell elongation in stems; phototropism, gravitropism, apical dominance; formation of roots, development of fruit |
| b. _____ | Gibberellic acid (GA) | Promotes stem elongation; releases some buds and seeds from dormancy |
| c. _____ | Zeatin | Promotes cell division and embryo development; prevents leaf senescence and promotes bud activation |
| **_Growth Inhibitors_** | | |
| d. _____ | Abscisic acid (ABA) | Resistance to stress conditions; causes stomatal closure; maintains dormancy |
| e. _____ | Ethylene | Promotes fruit ripening; promotes abscission and fruit drop; inhibits growth |

13. State whether the following is _phototropism, gravitropism,_ or _thigmotropism:_
    a. _____ movement in response to touch
    b. _____ movement in response to light stimulus
    c. _____ movement in response to gravity

    d. Explain what is meant by "movement" of a plant. _____

    _____

    _____

14. State whether the following is an example of *positive tropism* or *negative tropism.*
    a. _____ Stems curve upward, opposite the direction of gravity.
    b. _____ Stems curve toward light.

15. Study the following diagram and then complete lines *a–h* below.

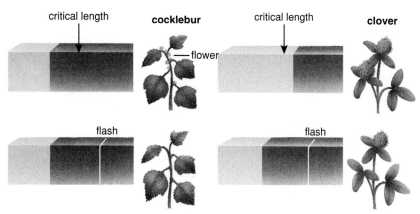

a. Short-day (long-night) plant          b. Long-day (short-night) plant

A long-day (short-night) plant flowers when the night is  a._____ than a critical length. A
short-day (long-night) plant will not flower when the night is  b._____ than a critical
length.

A long-day (short-night) plant will not flower when the night is  c._____ than a critical length.
A short-day (long-night) plant will flower when the night is  d._____ than a critical length.

A long-day (short-night) plant will flower if a flash of light interrupts a night that is  e._____
than a critical length. A short-day (long-night) plant will not flower if a flash of light interrupts a night that
is  f._____ than a critical length.

The conclusion is that it is the length of the  g._____, and not the  h._____,
that controls flowering.

16. a.  Label the two arrows in the diagram with these terms:

    absorbs red light      absorbs far-red light

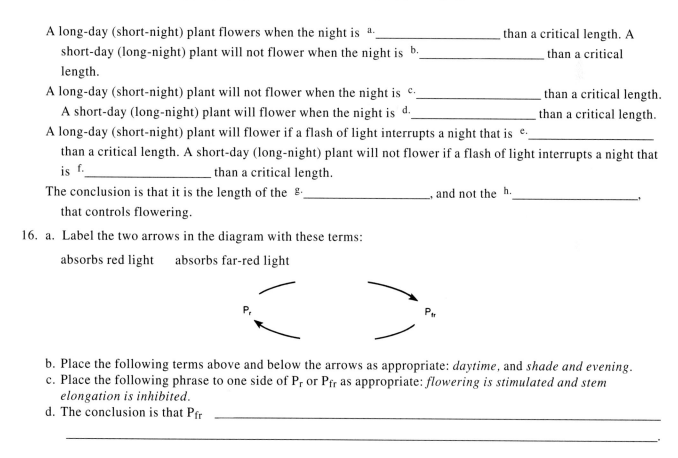

    b. Place the following terms above and below the arrows as appropriate: *daytime,* and *shade and evening.*
    c. Place the following phrase to one side of $P_r$ or $P_{fr}$ as appropriate: *flowering is stimulated and stem
       elongation is inhibited.*
    d. The conclusion is that $P_{fr}$  _____

    _____.

The learning objectives for this section are:
- Explain the transport of water and minerals in plants.
- Explain how the opening of stomata is regulated.
- Describe the transport of organic nutrients in plants.

17. With the help of the accompanying illustration and the following list of terms, fill in the blanks for lines *a–j* on the following page.

> cohesion and adhesion of water molecules
> positive pressure
> root hairs
> tension
> tracheids and vessel elements
> transpiration
> water column

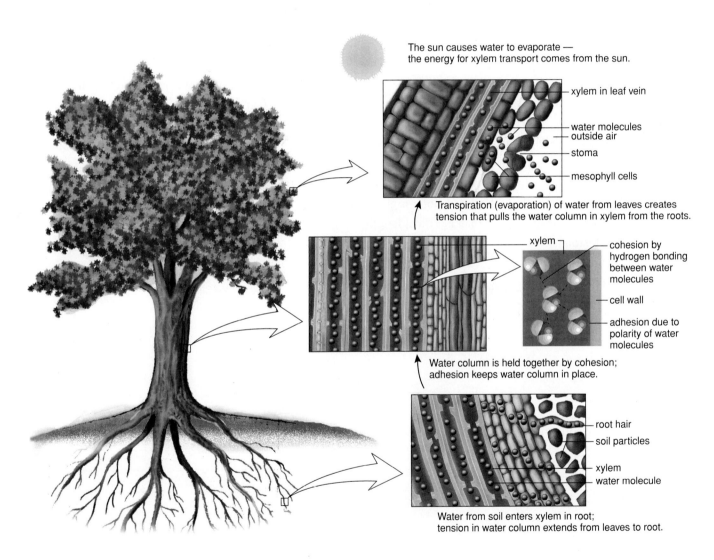

The sun causes water to evaporate —
the energy for xylem transport comes from the sun.

xylem in leaf vein
water molecules
outside air
stoma
mesophyll cells

Transpiration (evaporation) of water from leaves creates
tension that pulls the water column in xylem from the roots.

xylem
cohesion by
hydrogen bonding
between water
molecules
cell wall
adhesion due to
polarity of water
molecules

Water column is held together by cohesion;
adhesion keeps water column in place.

root hair
soil particles
xylem
water molecule

Water from soil enters xylem in root;
tension in water column extends from leaves to root.

a._____ are the conducting cells in xylem. Why is there an open pipeline from the roots to the leaves? b._____

c._____ absorb water from soil. When water enters the root and makes its way to xylem, what type of pressure results? d._____

e._____ is evaporation of water. Why does water evaporate? f._____

_____

g._____ is created by the evaporation of water. It pulls the h._____
from the roots to the leaves.

i._____ causes the water column to hold together. Why must the water column be continuous?

j._____

_____

18. Using the input from question 17, explain the cohesion-tension model of xylem transport. _____

_____

19. Stomata open during photosynthesis when a._____ ions are actively transported
    b._____ (*into*/*out of*) the guard cells. Water now enters the guard cells by
    c._____, causing the cells to buckle out due to their d._____ (*thick*/*thin*) inner
    walls.

20. What is the proper sequence for these steps to demonstrate how stomata proceed from closed to open
    positions? (Indicate by letters.) _____
    a. $CO_2$ decreases in leaf.
    b. $K^+$ enters guard cells.
    c. Stomata are closed.
    d. Stomata are open.
    e. Water enters guard cells.

21. What two important events are occurring when stomata are open?
    a._____

    b._____

22. With the help of the accompanying illustration and the following list of terms, fill in the blanks for *a–i* below.

active transport (used twice)
osmosis
positive pressure
sieve-tube elements
sink
source

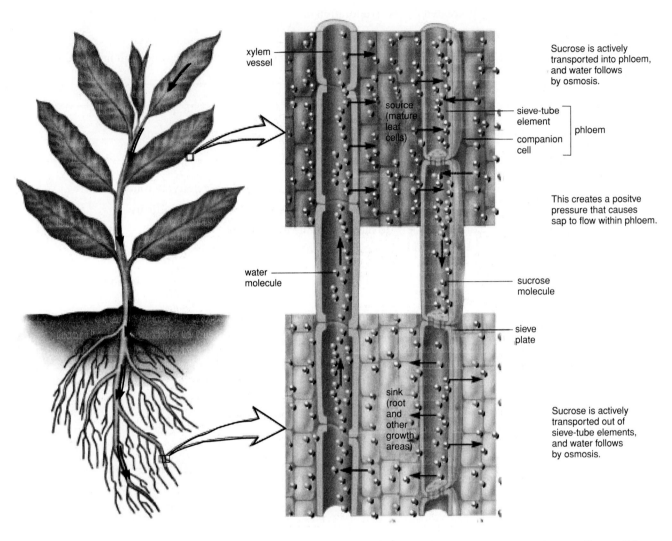

a._____ are the conducting cells of phloem. Why can substances move from cell to cell in

phloem? b. _____

c._____ means sugar is made in leaves.

d._____ causes sugar to enter phloem.

e._____ occurs after sugar enters phloem.

f._____ exists in phloem after sugar and water enter.

g._____ causes sugar to exit phloem. Why is the exit of sugar from phloem a necessary part

of phloem transport? h. _____

i._____ means sugar is removed in roots.

23. Using the input from question 22, explain the cohesion-tension model of phloem transport.

_____

Review key terms by using the following alphabetized list of terms to fill in the blanks below. Then complete the wordsearch.

```
S L E T R E C M I T O C H A N D R I O R
N E C D Z C Y E P I D E R O V A R Y C B
L D R S O C H E M I O S M Y S I S C O S
T O T G I O T B A T D T N H I H I N L I
G M O T A C D U R U T A S C D T N N L N
R W C T A N X D C P T B I N R H N E E T
C O H E S I O N T E N S I O N M O D E L
F L R I N F I O R A F L Y R E I I L T R
R F O N U G E L A A L E L A T F T M Y N
L E M C O T N R N M O N O P A D A J H O
K R E R E N N I S L S D C O P M N E P D
F U R D G L O M P E R O Y R H E I I O E
G S I S I R O H I C N S M H A L L G T I
C S I A E R L B R C S P P N S G L O E B
C E T Y L O D O A C T E T O E M O L M O
Q R E O E B U P T A S R S T R S P S A R
G P S M M X M P I C E M Y R I E E R G E
C V I M O A M H O V U L E R E O T G S T
O C S C A T T B N L I S M E S I N E W N
D U C L A I U A S S I L O B A N A K M A
```

*auxin*
*cohesion-tension model*
*endosperm*
*gametophyte*
*ovary*
*ovule*
*phloem*
*pollination*
*pressure-flow model*
*spore*
*stomata*
*transpiration*

a. _____ In angiosperms, the 3n tissue that nourishes the embryo and seedling and is formed as a result of a sperm joining with two polar nuclei.

b. _____ Plant hormone regulating growth, particularly cell elongation; most often indoleacetic acid.

c. _____ In seed plants, a structure in which the megaspore becomes an egg-producing megagametophyte develops into a seed following fertilization.

d. _____ In seed plants, the delivery of pollen to the vicinity of the egg, producing the megagametophyte.

e. _____ In the life cycle of a plant, the haploid generation that produces gametes.

f. _____ Microscopic opening bordered by guard cells in the leaves of plants through which gas exchange takes place.

g. _____ In flowering plants, the base of the pistil that protects ovules and, along with associated tissues, becomes a fruit.

h. _____ Evaporation of water from a leaf; pulls water from the roots through a stem to leaves.

i. _____ Explanation for upward transportation of water in xylem based upon transpiration-created tension and the cohesive properties of water molecules.

j. _____ Explanation for phloem transport; osmotic pressure following active transport of sugar into phloem brings about a flow of sap from a source to a sink.

k. _____ Haploid reproductive cell, sometimes resistant to unfavorable environmental conditions, which is capable of producing a new individual that is also haploid.

l. _____ Vascular tissue that conducts organic solutes in plants; contains sieve-tube elements and companion cells.

Do not refer to the text when taking this test.

_____ 1. What role does transpiration play in water transport?
   a. no role
   b. pushes the water
   c. pulls the water

_____ 2. For transpiration to occur in the leaves,
   a. water must exhibit cohesiveness.
   b. the stomata must be open.
   c. water must evaporate.
   d. All of these are correct.

_____ 3. Turgor pressure is important to
   a. opening and closing of a stoma.
   b. plant cell rigidity.
   c. water flow in xylem.
   d. Both *a* and *b* are correct.
   e. All of these are correct.

_____ 4. The Casparian strip causes
   a. water to enter the vascular cylinder.
   b. sucrose to enter phloem.
   c. water to stay within xylem during transport.
   d. All of these are correct.

_____ 5. During water transport, negative pressure is due to the
   a. cohesion of water molecules.
   b. active transport of sucrose into root cells.
   c. evaporation of water from the leaves.
   d. All of these are correct.

_____ 6. When stomata are open,
   a. carbon dioxide enters leaves.
   b. potassium and water have entered guard cells.
   c. negative pressure pulls water upward.
   d. All of these are correct.

_____ 7. During phloem transport, the sink has
   a. the higher solute concentration, accounting for why water flows to it.
   b. the lower solute concentration, due to the active transport of sucrose out of it.
   c. the higher solute concentration because that is where sucrose is needed.
   d. Both *a* and *c* are correct.

_____ 8. Select the incorrect association.
   a. positive gravitropism—response to gravity
   b. phototropism—response to light stimulus
   c. negative gravitropism—response to gravity
   d. thigmotropism—response to chemical stimulus

_____ 9. Interrupting the dark period with a flash of white light prevents flowering in a
   a. long-day plant.
   b. short-day plant.

_____ 10. Select the incorrect statement about auxins.
   a. They cause breakdown of polysaccharides in the cell wall.
   b. Their concentration in leaves and fruits prevents leaves and fruits from falling to the ground.
   c. They are produced in the shoot apex of the plant.
   d. They are transported to the side of the plant receiving light.

_____ 11. Oat seedlings will NOT bend
   a. when exposed to artificial plant hormones.
   b. when coleoptile tips are cut off.
   c. when $P_r$ has become $P_{fr}$.
   d. All of these are correct.

_____ 12. The reception of auxin leads to
   a. the removal of $H^+$ from the cell.
   b. $ATP \rightarrow ADP + P$.
   c. the weakening of plant cell walls.
   d. All of these are correct.

_____ 13. In a short-day plant, flowering is believed to depend on
   a. proper lighting.
   b. the presence of phytochrome.
   c. the proper hormonal balance.
   d. All of these are correct.

_____ 14. The ovule is to the pistil as the
   a. anther is to the stamen.
   b. anther is to the filament.
   c. filament is to the anther.
   d. All of these are correct.

_____ 15. The structure immediately preceding the megagametophyte is the
   a. microgametophyte.
   b. ovule.
   c. megaspore.
   d. pistil.

_____ 16. The pollen grain contains
   a. the sperm.
   b. the egg.
   c. the embryo.
   d. It depends on the photoperiod.

_____ 17. Insects
   a. routinely disperse seeds in angiosperms.
   b. are flower pollinators.
   c. are the only flower pollinators.
   d. All of these are correct.

_____18. Vegetative propagation
    a. is asexual propagation.
    b. can occur from stems.
    c. requires the use of leaves.
    d. Both *a* and *b* are correct.
    e. All of these are correct.

_____19. In the life cycle of flowering plants, the embryo sac
    a. is the equivalent of the pollen grain.
    b. has ten cells.
    c. contains an embryo.
    d. All of these are correct.

_____20. A(n) _____ produces four pollen grains.
    a. microspore
    b. microsporocyte
    c. anther
    d. microsporangium

_____21. Fertilization in flowering plants
    a. results in a 2n zygote and 3n endosperm.
    b. results in a 3n embryo and 2n endosperm.
    c. involves the polar nuclei but not the egg nucleus.
    d. occurs rarely.

_____22. Fruits are classified according to whether they are
    a. big or small.
    b. dry or fleshy.
    c. simple or compound.
    d. Both *b* and *c* are correct.
    e. All of these are correct.

_____23. A seed typically
    a. contains an embryo and stored food.
    b. germinates before it starts to grow.
    c. has cotyledons.
    d. All of these are correct.

## THOUGHT QUESTIONS

Answer in complete sentences.

24. Discuss how plants regulate the opening and closing of stomata.

25. What are one advantage and one disadvantage of asexual plant reproduction, compared to sexual plant reproduction?

**Test Results:** _____ number correct ÷ 25 = _____ × 100 = _____ %

## ANSWER KEY

### STUDY QUESTIONS

**1. a.** anther **b.** filament **c.** stamen **d.** pollen grain **e.** stigma **f.** style **g.** ovary **h.** pistil **i.** petals **j.** sepal **k.** ovule **l.** pollen tube **2. a.** sepals **b.** petals **c.** pistil **d.** stigma **e.** style **f.** ovary **g.** anther **h.** filament **3. a.** anther (2n) **b.** ovary (2n) **c.** sporophyte (2n) **d.** microspore (n) **e.** megaspore (n) **f.** megagametophyte (n) **g.** microgametophyte (n) **h.** seed (2n) **4. a.** microspore **b.** megaspore **c.** microgametophyte **d.** sperm **e.** megagametophyte **f.** egg **g.** zygote **5. a.** thickened ovary wall **b.** developed from an individual ovary **c.** pericarp is dry **d.** developed from a group of individual ovaries **e.** fleshy **f.** dry **6. a.** hooks and spines **b.** defecation by birds and mammals **c.** squirrel activity **d.** wind and ocean currents **7. a.** B **b.** C **c.** C **d.** B **8.** a, b, c **9.** b, c. **10. a.** identical **b.** tissue **c.** meristem **d.** cell suspension culture **e.** naked cell **11.** Auxin is attaching to receptors. Pump is breaking down ATP and pumping H$^+$ out of the cell. H$^+$ causes the cell wall to break down, water and then solutes enter, and the cell increases in size. **12. a.** auxin **b.** gibberellins **c.** cytokinins **d.** abscisic acid **e.** ethylene **13. a.** thigmotropism **b.** phototropism **c.** gravitropism **d.** Plants grow in a certain direction. **14. a.** negative tropism **b.** positive tropism **15. a.** shorter **b.** shorter **c.** longer **d.** longer **e.** longer **f.** longer **g.** night **h.** day **16.** a, b, c (see following diagram) **d.** is the metabolically active form.

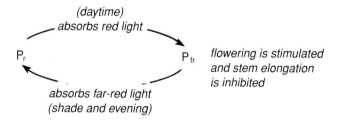

*(daytime)*
absorbs red light

$P_r$          $P_{fr}$   *flowering is stimulated and stem elongation is inhibited*

absorbs far-red light
*(shade and evening)*

**17. a.** tracheids and vessel elements **b.** Vessel elements are piled on top of one another. **c.** root hairs **d.** positive pressure **e.** transpiration **f.** Solar energy causes water to evaporate. **g.** tension **h.** water column **i.** cohesion and adhesion **j.** Tension can only work if the water column is continuous. **18.** The sun causes water to evaporate, and this creates a tension that pulls the water column, continuous due to the cohesion of water molecules that adhere to the sides of xylem. **19. a.** potassium **b.** into **c.** osmosis **d.** thick **20.** c, a, b, e, d **21. a.** Water is lost by transpiration **b.** Carbon dioxide is being absorbed **22. a.** sieve-tube elements **b.** Sieve-tube elements are connected by strands of cytoplasm (plasmodesmata). **c.** source **d.** active transport **e.** osmosis **f.** positive pressure **g.** active transport **h.** causes sugar to move from area of higher to lower concentration **i.** sink **23.** When sugar is actively transported into phloem, water follows by osmosis, creating a positive pressure. When sucrose is actively transported out of sieve-tube elements, a gradient is established that causes sugar to flow from source to sink.

**1.** c **2.** d **3.** d **4.** a **5.** c **6.** d **7.** b **8.** d **9.** b **10.** d **11.** b **12.** d **13.** d **14.** a **15.** c **16.** a **17.** b **18.** d **19.** a **20.** a **21.** a **22.** d **23.** d **24.** Each stoma has two guard cells with a pore between them. Their inner walls are thicker than the outer walls. When water enters the guard cells, their lengthwise expansion causes them to buckle out and the stoma opens. During photosynthesis, an ATP-driven proton pump actively transports $H^+$ out of the cell. New potassium ions enter the guard cells, which take up water by osmosis, causing the stoma to open. **25.** Asexual reproduction offers a quick, effective means of reproducing plants; however, it lacks the mechanism for genetic variability available through sexual reproduction. Through the latter, more adaptive forms of the organism may be produced.

## DEFINITIONS WORDSEARCH

**a.** endosperm **b.** auxin **c.** ovule **d.** pollination **e.** gametophyte **f.** stomata **g.** ovary **h.** transpiration **i.** cohesion-tension model **j.** pressure-flow model **k.** spore **l.** phloem

# 11

# HUMAN ORGANIZATION

## CHAPTER REVIEW

Human tissues are categorized into four groups. **Epithelial tissue** covers the body and lines its cavities. The different types of epithelial tissue (**squamous, cuboidal,** and **columnar**) can be simple or stratified and have cilia or microvilli. Also, columnar cells can be pseudostratified. Epithelial cells sometimes form glands that secrete either into ducts or into blood.

**Connective tissues,** in which cells are separated by a **matrix,** often bind body parts together. Loose connective tissue has both white and yellow fibers. Adipose tissue has fat cells. Dense fibrous connective tissue, such as in **tendons** and **ligaments,** contains closely packed white fibers. Both **cartilage** and **bone** have cells within lacunae, but the matrix for cartilage is more flexible than that for bone, which contains calcium salts. In bone, the lacunae lie in concentric circles within an osteon, or Haversian system, about a central canal. Blood is a connective tissue in which the matrix is a liquid called **plasma.**

**Muscular tissue** is of three types. Both **skeletal muscle** and **cardiac muscle** are **striated;** both cardiac and **smooth muscle** are involuntary. Skeletal muscle is found in muscles attached to bones, and smooth muscle is found in internal organs. Cardiac muscle makes up the heart.

**Nervous tissue** has one main type of cell, the **neuron.** Each neuron has dendrites, a cell body, and an axon. The brain and spinal cord contain complete neurons, while the nerves contain only fibers. Neurons and their fibers are specialized to conduct nerve impulses.

Tissues are joined together to form organs, each one having a specific function. **Skin** is a two-layered organ that waterproofs and protects the body. The **epidermis** contains a germinal layer that produces new epithelial cells that become keratinized as they move toward the surface. The **dermis,** a largely fibrous connective tissue, contains epidermally derived **glands** and **hair follicles,** nerve endings, and blood vessels. Receptors for touch, pressure, temperature, and pain are present. (For this reason, skin is sometimes considered the integumentary system.) Sweat glands and blood vessels help control body temperature. A **subcutaneous layer,** which is made up of loose connective tissue containing **adipose** cells, lies beneath the skin.

Organs are grouped into organ systems. In the dorsal cavity, the brain is in the cranial cavity, and the spinal cord is in the vertebral canal. Other internal organs are located in the ventral cavity, where the thoracic cavity is separated from the abdominal cavity by the diaphragm.

**Homeostasis** is the relative constancy of the internal environment. All organ systems contribute to the constancy of tissue fluid and blood. Special contributions are made by the liver, which keeps blood glucose constant, and the kidneys, which regulate the pH. The nervous and endocrine systems regulate the other systems. Both of these are controlled by a feedback mechanism, which results in fluctuation above and below the desired levels illustrated by body temperature.

## STUDY QUESTIONS

Study the text section by section. Answer the study questions so that you can fulfill the learning objectives for each section.

### 11.1 TYPES OF TISSUES (PP. 194–200)

The learning objectives for this section are:
- Describe the features and functions of epithelial tissues.
- Explain how junctions between cells aid communication between them.
- Discuss the different types of connective tissues and their functions.
- Describe the types of muscle tissue and how they operate.
- State the characteristics and functions of nervous tissue.

1. Draw a diagram of squamous epithelium.
   a.

In *b–d,* fill in the blanks.

    b. Name one place in the human body where simple squamous epithelium can be found. _____

    c. What is the function of this tissue? _____

    _____

    d. When this tissue is found in layers, it is called _____ squamous epithelium.

In question 2, fill in the blanks.

2. Epithelium can be classified by a._____. b._____ epithelium has cube-shaped cells, while
   c._____ epithelium has elongated, cylindrical cells.

3. The windpipe is lined by pseudostratified ciliated columnar epithelium. Use the space provided to describe this tissue in complete sentences.

    _____

    _____

    _____

In questions 4–8, fill in the blanks.

4. a._____ junctions allow for materials and information to be exchanged between cells. b._____ junctions are a means of reinforcement in which intercellular filaments are attached to buttonlike thickenings.
   c._____ junctions bind cells firmly together and form an impermeable barrier.

5. Loose fibrous connective tissue has cells called a._____, plus fibers made from b._____ and
   c._____. It binds the skin to underlying organs.

6. Tendons and ligaments are made up of a._____ tissue. Tendons join b._____ to bone, and ligaments join c._____ to bone.

7. a._____ cartilage can be found at the ends of bones and makes up the fetal skeleton. In it, cells lie within
   b._____, surrounded by a gel-like c._____.

8. The type of cartilage found in the ear, called a._____ cartilage, has many b._____ fibers to add flexibility. c._____, found in the intervertebral disks, aids in cushioning against jolts.

9. Draw a sketch of bone tissue. Label these parts: lacunae, matrix, and osteons (Haversian systems).

In question 10, fill in the blanks.

10. Blood is considered a connective tissue because its cells are separated by a fluid matrix called a._____.
    b._____ blood cells carry oxygen throughout the body in association with molecules of hemoglobin.
    c._____ blood cells are responsible for immunity, and cell fragments called d._____ aid in blood clotting.

11. Complete this table.

|  | Fiber Appearance | Location | Control |
|---|---|---|---|
| Skeletal | a. | | |
| Smooth | b. | | |
| Cardiac | c. | | |

In question 12, fill in the blanks.

12. The brain and spinal cord are made up of cells called a._____. Outside the central nervous system, connective tissue binds the long fibers of these cells to form b._____. The function of a neuron is to c._____. The other type of cell in nervous tissue are d._____ cells. This type of cell provides e._____ to neurons and keeps tissue free of debris. Neuroglia outnumber neurons f._____ to one and take up more than g._____ the volume of the brain.

## 11.2 BODY CAVITIES AND BODY MEMBRANES (P. 201)

The learning objective for this section is:
• Describe the membranes of the body cavities.

13. Specify the correct cavity for each organ, using these letters:

T—thoracic cavity    A—abdominal cavity    P—pelvic cavity

a. _____ small intestine          e. _____ lungs
b. _____ ovaries                  f. _____ stomach
c. _____ bladder                  g. _____ liver
d. _____ heart                    h. _____ kidneys

14. A large, ventral cavity called a(n) a._____ can be seen during embryological development. It later develops into the b._____ cavity of the chest and the c._____ cavity of the lower abdomen.

15. The a._____ are connective tissue membranes protecting the brain and spinal cord. b._____ lines the gut and secretes mucus. c._____ membrane lines joint cavities and secretes lubricating d._____.

16. Pleura and peritoneum are examples of a._____ membranes. These secrete b._____ to keep the membranes moist and prevent sticking.

## 11.3 ORGAN SYSTEMS (PP. 202–5)

The learning objectives for this section are:
• Give examples of the various organ systems of the body and their general functions.
• List the three regions of the skin, their functions, and the structures located within them.
• Describe how body temperature is regulated.

17. Place the appropriate letter next to each system.

M—maintenance of the body          S—support and movement
R—regulation of body systems       C—continuance of the species

a. _____ digestive system
b. _____ reproductive system
c. _____ skeletal system
d. _____ nervous system
e. _____ respiratory system

In questions 18–20, fill in the blanks.

18. The a._____ system is responsible for reducing food to molecules that can be used by the cells of the body. The b._____ system allows us to move from one place to another and generates heat. The c._____ system consists of the brain, spinal cord, nerves, and sensory receptors. The d._____ system sends out hormones to regulate bodily functions.

19. A(n) a._____ is made up of two or more tissues, functioning together. The skin is an example. It has an outer layer called the b._____. The cells of this layer become waterproof once they are filled with the protein c._____.

20. An accessory organ of the skin, the a._____, forms hair shafts. b._____ muscles cause goose bumps to appear. c._____ glands provide sebum to moisturize hairs.

21. Label this diagram of the skin, using the following alphabetized list of terms.
    adipose tissue
    arrector pili muscle
    blood vessels
    dermis
    epidermis
    fibrous connective tissue
    hair root
    hair shaft
    nerve
    oil gland
    sensory receptor
    subcutaneous layer
    sweat gland

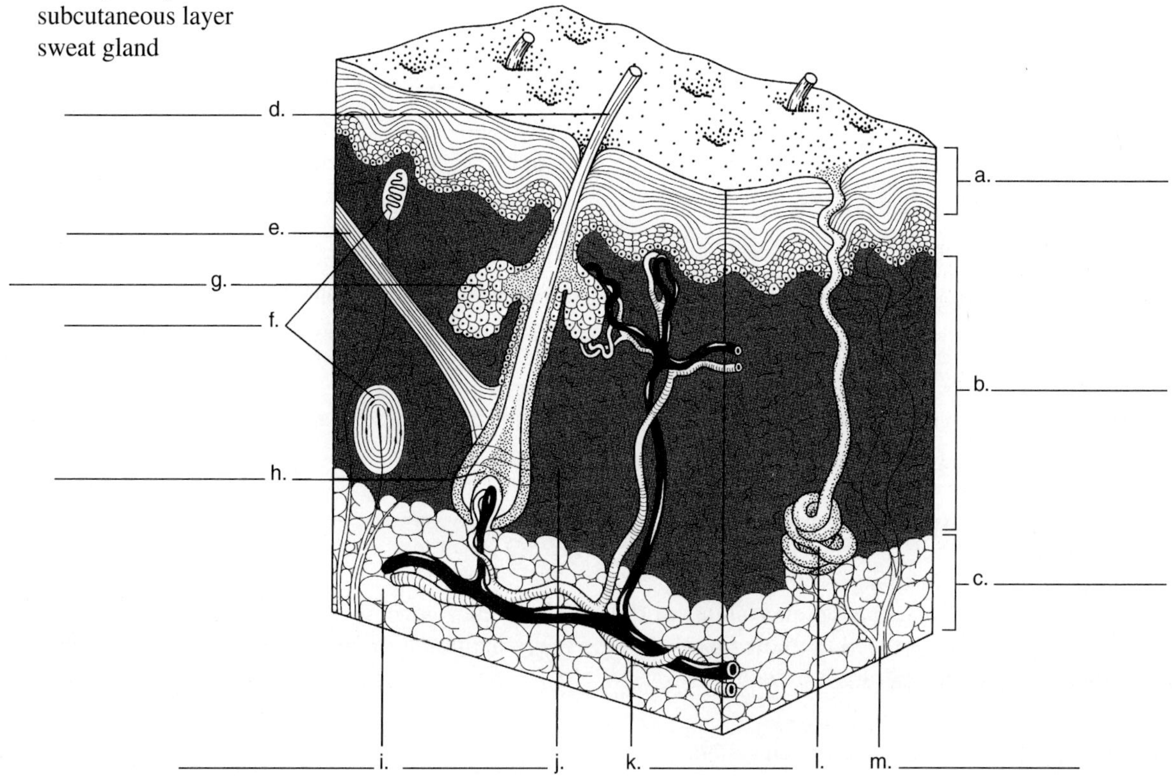

In question 22, fill in the blanks.

22. a. What events raise body temperature? _____

    _____

    b. What events lower body temperature? _____

    _____

The learning objectives for this section are:
- Define homeostasis, tell how it is controlled, and explain its importance.
- Understand how the nervous and endocrine systems coordinate to maintain homeostasis.
- Use the maintenance of room temperature as an example of negative feedback control.

23. a. What is homeostasis? _____

    b. Give an example of homeostasis. _____

24. Label this diagram of a negative feedback control, using these terms:

    effector    regulatory center    reversal    sensor

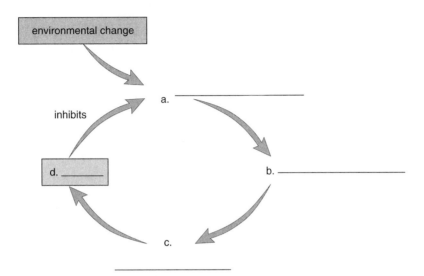

25. Give an example of negative feedback control in the body. _____

    _____

    _____

26. Indicate whether these statements about positive feedback are true (T) or false (F).

    a. _____ Like negative feedback, positive feedback maintains homeostasis.

    b. _____ Brings about change in the same direction.

    c. _____ Occurs during childbirth.

    d. _____ Occurs when a thermostat regulates room temperature.

In questions 27–28, fill in the blanks.

27. How does each of the following systems contribute to homeostasis?

    a. cardiovascular system _____

    b. digestive system _____

    c. respiratory system _____

    d. urinary system (i.e., the kidneys) _____

28. The two systems of the body that control homeostasis are a. _____

    and b. _____.

29. Figure 11.8 on pages 202–3 of your textbook illustrates the ways that the systems of the body contribute to homeostasis. Match the descriptions to the organ systems.

| integumentary system | respiratory system | lymphatic system | digestive system |
| skeletal system | nervous system | urinary system | reproductive system |
| endocrine system | muscular system | cardiovascular system | |

a. _____ Breakdown and absorption of food materials
b. _____ Gaseous exchange between external environment and blood
c. _____ Regulation of all body activities; learning and memory
d. _____ Body movement; production of body heat
e. _____ External support and protection of body
f. _____ Secretion of hormones for chemical regulation
g. _____ Production of sperm; transfer of sperm to female reproductive system
h. _____ Transport of nutrients to body cells; removal of wastes from cells
i. _____ Immunity; absorption of fats; drainage of tissue fluid
j. _____ Maintenance of volume and chemical composition of blood
k. _____ Internal support and protection; body movement; production of blood cells

## DEFINITIONS WORDSEARCH

Review key terms by using the following alphabetized list of terms to fill in the blanks below. Then complete the wordsearch.

```
E H J Y R D E R M I S Y T X
P T W F E D P P O P U L I C
I W E D I L I G A M E N T L
T A S R T W D I O P T L N J
H M I B O N E G U B V E R O
E J T L B C R N O R U E N L
L C O E L O M K R L D K I D
I O F E T I I H U G N U F O
A M S E F Y S T R I A T E D
L P M I N D L E S H L I F B
C A R T I L A G E O G I F T
D C P T J U N E T F D I L T
C T E N D O N J U N P T I F
```

*bone*
*cartilage*
*coelom*
*compact*
*dermis*
*epidermis*
*epithelial*
*gland*
*ligament*
*neuron*
*striated*
*tendon*

a. _____ Type of tissue that covers the body and lines its cavities.

b. _____ Inner layer of skin that lies beneath the epidermis.

c. _____ Fibrous connective tissue joining bone to bone.

d. _____ Rigid connective tissue containing mineral salts.

e. _____ Outer layer of skin, protective in nature.

f. _____ Specialized epithelial cell that secretes substances.

g. _____ Having alternating light and dark bands.

h. _____ Connective tissue found on ends of bones or in the ear.

i. _____ Fibrous connective tissue connecting muscle to bone.

j. _____ Type of bone with densely packed osteons.

k. _____ Embryonic body cavity that becomes the thoracic and abdominal cavities.

l. _____ A nerve cell.

## OBJECTIVE QUESTIONS

Do not refer to the text when taking this test.

_____1. Cells of a similar structure working together constitute a(n)
   a. tissue.
   b. organ.
   c. organism.
   d. organ system.

_____2. Epithelial tissues do which of the following?
   a. secrete
   b. line body cavities
   c. protect
   d. All of these are correct.

_____3. Which of these pairs is mismatched?
   a. fat–subcutaneous layer
   b. receptors–dermis
   c. keratinization–epidermis
   d. nerves and blood vessels–epidermis

_____4. If epithelial tissue is made up of many layers of cells, what would be true?
   a. It is called simple.
   b. It is called stratified.
   c. It is called striated.
   d. It is pseudostriated.

_____5. Which type of epithelial tissue is composed of flattened cells?
   a. glandular
   b. squamous
   c. columnar
   d. cuboidal

_____6. What type of tissue supports epithelium?
   a. fibrous connective tissue
   b. nervous tissue
   c. muscular tissue
   d. loose connective tissue

_____7. Which type of cell junction is characterized by intercellular filaments?
   a. tight junction
   b. gap junction
   c. adhesion junction
   d. intercalated disks

_____8. Which of the following tissues has cells residing in lacunae?
   a. adipose tissue
   b. fibrous connective tissue
   c. hyaline cartilage
   d. loose connective tissue

_____9. Osteocytes are residents of _____ tissue.
   a. cartilage
   b. bone
   c. muscle
   d. pseudostratified columnar epithelium

For questions 10–13, match the statements to these blood components:
   a. plasma    b. red blood cells
   c. white blood cells    d. platelets

_____10. Contribute to blood clotting process
_____11. Carry oxygen to cells
_____12. Fluid matrix component of the connective tissue called blood
_____13. Participate in immunity

In questions 14–17, match the statements to these components of skin:
   a. epidermis    b. arrector pili    c. melanocytes
   d. oil glands

_____14. When these become blocked, a blackhead forms
_____15. Give rise to skin pigment
_____16. Muscle attached to a hair follicle
_____17. Keratinized layer of skin

_____18. Cells in the nervous tissue that support and protect neurons are called
   a. neurons.
   b. fibers.
   c. axons.
   d. neuroglia.

_____19. Cardiac muscle fibers are characterized by which traits?
   a. striated; intercalated disks
   b. smooth; single nucleus
   c. striated; multiple nuclei; intercalated disks
   d. smooth; tapered; multiple nuclei

_____20. Which of these is located in the thoracic cavity?
   a. small intestine
   b. urinary bladder
   c. lungs
   d. kidneys

_____21. The thoracic cavity is lined with
   a. mucous membrane.
   b. serous membrane.
   c. cutaneous membrane.
   d. synovial membrane.

_____22. Which type of membrane lines movable joint cavities?
   a. mucous membrane
   b. serous membrane
   c. cutaneous membrane
   d. synovial membrane

_____23. Which of these is an example of homeostasis?
   a. Muscle tissue is specialized to contract.
   b. The skin sunburns.
   c. Normal body temperature almost always stays at 37°C.
   d. All of these are examples of homeostasis.

____24. In a negative feedback control system,
- a. homeostasis is impossible.
- b. there is a constancy of the internal environment.
- c. a wide fluctuation occurs continuously.
- d. None of the above is true.

____25. When body temperature rises, sweat glands become
- a. active, and blood vessels constrict.
- b. inactive, and blood vessels dilate.
- c. active, and blood vessels dilate.
- d. inactive, and blood vessels constrict.

## THOUGHT QUESTIONS

Answer in complete sentences.

26. How do the nervous system and endocrine system interact to maintain the homeostasis?

27. How do the blood vessels, the digestive tract, the lungs, and the kidneys work together to maintain homeostasis?

**Test Results:** _____ number correct ÷ 27 = _____ × 100 = _____%

## ANSWER KEY

### STUDY QUESTIONS

**1. a.** See Figure 11.1, in text. **b.** air sacs of the lungs **c.** absorption **d.** stratified **2. a.** cell type **b.** Cuboidal **c.** columnar **3.** This type of tissue appears to be layered but is not, and cells have small, hairlike projections called cilia. **4. a.** Gap **b.** Adhesion **c.** Tight **5. a.** fibroblasts **b.** collagen **c.** elastin **6. a.** dense fibrous connective **b.** muscle **c.** bone **7. a.** Hyaline **b.** lacunae **c.** matrix **8. a.** elastic **b.** elastic **c.** Fibrocartilage **9.** See Figure 11.3*d*, in text. **10. a.** plasma **b.** Red **c.** White **d.** platelets **11. a.** striated; skeleton; voluntary **b.** spindle-shaped; internal organs; involuntary **c.** striated; heart; involuntary **12. a.** neurons **b.** nerves **c.** conduct nerve impulses **d.** neuroglial (neuroglia) **e.** protection **f.** nine **g.** one-half **13. a.** A **b.** P **c.** P **d.** T **e.** T **f.** A **g.** A **h.** A **14. a.** coelom **b.** thoracic **c.** abdominal **15. a.** meninges **b.** Mucous membrane **c.** Synovial **d.** synovial fluid **16. a.** serous **b.** watery serous fluid **17. a.** M **b.** C **c.** S **d.** R **e.** M **18. a.** digestive **b.** muscular **c.** nervous **d.** endocrine **19. a.** organ **b.** epidermis **c.** keratin **20. a.** hair follicle **b.** Arrector pili **c.** Oil **21. a.** epidermis **b.** dermis **c.** subcutaneous layer **d.** hair shaft **e.** arrector pili muscle **f.** sensory receptors **g.** oil gland **h.** hair root **i.** adipose tissue **j.** fibrous connective tissue **k.** blood vessels **l.** sweat gland **m.** nerve **22. a.** Blood vessels constrict, sweat glands are inactive, hair stands on end, and shivering may occur. **b.** Blood vessels dilate, sweat glands are active, and hair lies next to skin. **23. a.** relative constancy of the internal environment **b.** body temperature remains around 37° C **24. a.** sensor **b.** regulatory center **c.** effector **d.** reversal **25.** Sensory receptors in aortic and carotid sinuses communicate with a regulatory center in the brain when there is a fall in blood pressure. This center sends nerve impulses to the arteries, and they constrict. When the blood pressure rises, the sensory receptors no longer communicate with the brain center. **26. a.** F **b.** T **c.** T **d.** F **27. a.** refreshes tissue fluid **b.** provides nutrient molecules **c.** removes carbon dioxide and adds oxygen to the blood **d.** eliminates wastes and salts **28. a.** nervous **b.** endocrine **29. a.** digestive **b.** respiratory **c.** nervous **d.** muscular **e.** integumentary **f.** endocrine **g.** reproductive **h.** cardiovascular **i.** lymphatic **j.** urinary **k.** skeletal

```
E           D E R M I S
P           P
I           L I G A M E N T
T           D
H     B O N E
E           R N O R U E N
L C O E L O M       D
I O         I       N
A M         S T R I A T E D
L P         L
C A R T I L A G E   G
  C
  T E N D O N
```

**a.** epithelial   **b.** dermis   **c.** ligament   **d.** bone   **e.** epidermis   **f.** gland   **g.** striated   **h.** cartilage   **i.** tendon   **j.** compact   **k.** coelom   **l.** neuron

**1.** a   **2.** d   **3.** d   **4.** b   **5.** b   **6.** d   **7.** c   **8.** c   **9.** b   **10.** d   **11.** b   **12.** a   **13.** c   **14.** d   **15.** c   **16.** b   **17.** a   **18.** d   **19.** a   **20.** c   **21.** b   **22.** d   **23.** c   **24.** b   **25.** c   **26.** The nervous system is the ultimate source of control over homeostasis—it registers changes and decides what action to take. Nervous impulses can be sent to trigger an appropriate response to maintain homeostasis. The endocrine system releases hormones into the bloodstream when needed to maintain homeostasis. Hormones last longer, but nervous impulses can travel faster. The nervous system controls the endocrine system.   **27.** The digestive tract adds nutrients, and the lungs add oxygen to the blood. The blood vessels bring nutrients and oxygen to tissue fluid and take away waste molecules, including carbon dioxide. The kidneys excrete metabolic wastes, and the lungs excrete carbon dioxide.

# 12

# DIGESTIVE SYSTEM AND NUTRITION

The **salivary glands** send saliva into the mouth, where the teeth chew the food, and the tongue forms a bolus for swallowing.

The air passage and the food passage cross in the **pharynx.** When a person swallows, the air passage is usually blocked off and food must enter the **esophagus,** where **peristalsis** begins.

The **stomach** expands and stores food. While food is in the stomach, it churns, mixing food with the acidic gastric juices.

The walls of the **small intestine** have fingerlike projections called **villi** where nutrient molecules are absorbed into the cardiovascular and lymphatic systems.

The **large intestine** consists of the **cecum; colons** (ascending, transverse, descending, and sigmoid); and the rectum, which ends at the **anus.** The large intestine does not produce digestive enzymes; it does absorb water, salts, and some vitamins.

The three accessory organs of digestion—the **pancreas, liver,** and **gallbladder**—send secretions to the **duodenum** via ducts. The pancreas produces pancreatic juice, which contains digestive enzymes for carbohydrate, protein, and fat.

The liver produces bile, which is stored in the gallbladder. The liver receives blood from the small intestine by way of the hepatic portal vein. It has numerous important functions, and any malfunction of the liver is a matter of considerable concern.

Digestive enzymes are present in digestive juices, and they break down food into the nutrient molecules: glucose, amino acids, fatty acids, and glycerol (see Table 12.2 in the text). Glucose and amino acids are absorbed into the blood capillaries of the villi. Fatty acids and glycerol rejoin to produce fat, which enters the **lacteals.**

Digestive enzymes have the usual enzymatic properties. They are specific to their substrate and speed up specific reactions at body temperature and optimum pH.

The nutrients released by the digestive process should provide us with an adequate amount of energy, **essential amino acids** and fatty acids, and all necessary **vitamins** and **minerals.**

The bulk of the diet should be carbohydrates (like bread, pasta, and rice) and fruits and vegetables. These are low in saturated fatty acids and cholesterol molecules, whose intake is linked to cardiovascular disease. The vitamins A, E, and C are antioxidants that protect cell contents from damage due to free radicals.

Study the text section by section. Answer the study questions so that you can fulfill the learning objectives for each section.

## 12.1 THE DIGESTIVE TRACT (PP. 214–21)

The learning objectives for this section are:
- Trace the path of food from mouth to anus during digestion.
- Describe the features of the mouth that prepare food for swallowing.
- Explain how the structure of the pharynx ensures the passage of food into the esophagus.
- Discuss the features of the esophagus and how peristalsis moves food along this structure.
- Explain how the structure of the stomach is adapted to high acidity.
- Describe the structure and function of the four layers of the digestive tract.
- Relate the function of the small intestine to its unique structure.
- Explain how villi and microvilli enhance absorption.
- Discuss how digestive secretions are under hormonal control, and list the hormones involved.
- Describe the sections of the large intestine, and know its general function.

In questions 1–7, fill in the blanks using this diagram.

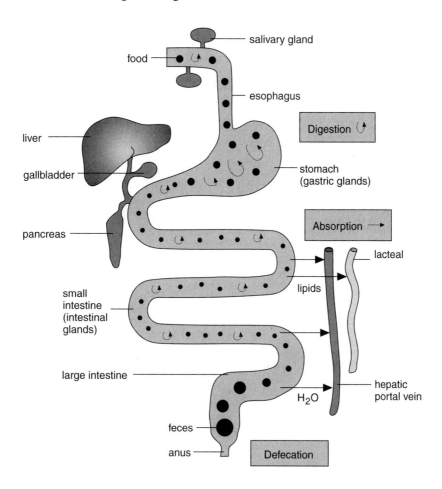

1. Food is received by the ᵃ·_____, which possesses ᵇ·_____ to grind and break up
   particles. ᶜ·_____ glands secrete ᵈ·_____ that moistens food and binds it together for
   swallowing. The taste of food is detected by ᵉ·_____ located on the ᶠ·_____.
2. At the back of the mouth, the muscular ᵃ·_____ is involved in swallowing food. The epiglottis
   folds over the glottis during the swallowing ᵇ·_____.
3. The esophagus propels food toward the ᵃ·_____, using muscular contractions called
   ᵇ·_____. At the upper end of the esophagus, a muscular ring called a(n) ᶜ·_____
   regulates the entrance of food to the esophagus.
4. The stomach is lined with ᵃ·_____ that secretes protective ᵇ·_____. Otherwise, the
   lining would be injured by the strong ᶜ·_____ acid secreted by ᵈ·_____ glands. Gastric
   juice also contains ᵉ·_____ that breaks down proteins.
5. The layer of the digestive tract that houses blood vessels is the ᵃ·_____. The ᵇ·_____
   layer contains two layers of smooth muscle. The outside layer, ᶜ·_____, secretes a fluid that
   enables organs to slide against one another.
6. The first section of the small intestine is the ᵃ·_____. It receives secretions from the ᵇ·_____
   and ᶜ·_____ and also receives food known as ᵈ·_____ from the stomach.
7. The mucosa of the small intestine is folded into ᵃ·_____, which, in turn, have projections from
   individual cells called ᵇ·_____.
   ᶜ·_____ and ᵈ·_____ are absorbed directly into the bloodstream, and fats must be
   reconstructed so they can travel into vessels of the lymphatic system called ᵉ·_____.

**105**

8. State the purpose of microvilli. _____

_____

In questions 9–10, fill in the blanks.

9. A number of hormones control the secretions of digestive juices. a. _____ is secreted in response to protein in foods and enhances gastric gland output, while b. _____ inhibits gastric secretion. c. _____ is secreted in response to acidic chyme. When fats are present in chyme, d. _____ triggers the release of bile from the gallbladder.

10. The large intestine functions to store and compact a. _____. Unusual outgrowths of the lining of the colon, called b. _____, can be either benign or cancerous. Colon cancer incidence may increase for people who do not have enough c. _____ in their diets.

## 12.2 THREE ACCESSORY ORGANS (PP. 222–23)

The learning objectives for this section are:
- Explain how the secretions from the pancreas aid digestion.
- List several functions of the liver, and describe the role of the gallbladder in digestion.
- Discuss the major diseases of the liver and their causes.

In questions 11–13, fill in the blanks.

11. Pancreatic juice contains a mix of a. _____ solution to neutralize stomach acid, and digestive b. _____ to further break down food.

12. The liver produces a greenish substance called a. _____, which is stored and concentrated by the b. _____. The liver has been called the c. _____ to the blood because it detoxifies substances entering the blood from the d. _____. Other functions of the liver include (list three):

e. _____

f. _____

g. _____

13. a. _____ is an inflammatory disease of the liver caused by a viral infection, while b. _____ is damage caused by chronic alcohol abuse.

## 12.3 DIGESTIVE ENZYMES (PP. 224–25)

The learning objective for this section is:
- Name the major digestive enzymes and the types of nutrients they digest, and state in which organs they are produced.

14. Complete this table.

| Major Digestive Enzymes | | | | | |
|---|---|---|---|---|---|
| **Food** | **Digestion** | **Enzyme** | **Optimum pH** | **Produced By** | **Site of Action** |
| Starch | Starch + $H_2O \longrightarrow$ maltose | a. _____ Pancreatic amylase | Neutral c. _____ | Salivary glands d. _____ | b. _____ Small intestine |
| | Maltose + $H_2O \longrightarrow$ glucose + glucose | e. _____ | Basic | Small intestine | f. _____ |
| Protein | Protein + $H_2O \longrightarrow$ peptides | Pepsin | g. _____ | h. _____ | Stomach |
| | Peptide + $H_2O \longrightarrow$ amino acids | i. _____ Peptidases | Basic k. _____ | Pancreas l. _____ | j. _____ Small intestine |
| Nucleic acid | RNA and DNA + $H_2O \longrightarrow$ nucleotides | m. _____ | Basic | Pancreas | n. _____ |
| | Nucleotide + $H_2O \longrightarrow$ base + sugar + phosphate | Nucleosidases | o. _____ | p. _____ | Small intestine |
| Fat | Fat droplet + $H_2O \longrightarrow$ glycerol + fatty acids | q. _____ | Basic | Pancreas | r. _____ |

15. Match the enzyme to the food or breakdown product.

salivary amylase/pancreatic amylase    pepsin/trypsin    lipase    nuclease    maltase
peptidase    nucleosidases

a. _____ protein

b. _____ maltose

c. _____ nucleic acid

d. _____ starch

e. _____ fat

f. _____ nucleotides

g. _____ peptides

## 12.4 NUTRITION (PP. 226–35)

The learning objectives for this section are:
- Describe how the different classes of nutrients enter into general circulation within the body.
- Discuss proper nutrition and how carbohydrates, protein, and fat should be proportioned in the diet.
- Discuss the vitamin and mineral requirements in the diet.

16. Fill in the blanks of the food guide pyramid. Include the following food groups, and indicate the recommended number of daily servings from each group:

> bread, rice, pasta
> dairy
> fruit
> meat, poultry, fish, and beans
> sweets, fats, and oils
> vegetables

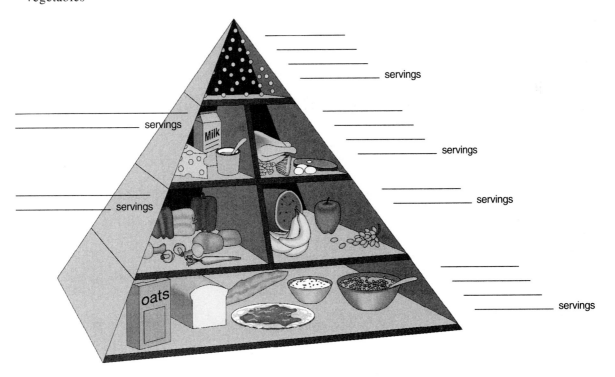

17. Place a check beside the most nutritious food.

    a. _____ fat

    b. _____ beef

    c. _____ vegetables

18. Place a check beside the type of food that should be the bulk of any diet.

    a. _____ vegetables

    b. _____ fruits

    c. _____ dairy

    d. _____ complex carbohydrates

19. Indicate whether these statements are true (T) or false (F).

    a. _____ Complex carbohydrates provide energy and fiber.

    b. _____ Meat provides proteins, but it also contains saturated fats.

    c. _____ The body can make all the essential amino acids it needs.

    d. _____ Saturated fats, whether in butter or margarine, can lead to more LDL, and this can cause plaque buildup in the arteries.

In questions 20–21, fill in the blanks.

20. Many vitamins function as a. _____ in various metabolic pathways. Vitamin b. _____ is important as a visual pigment, and vitamin c. _____ becomes a compound that enhances calcium absorption. Vitamins d. _____ are antioxidants.

21. Too much of the mineral a. _____ can lead to hypertension. The mineral b. _____ is a major component of bones and teeth.

22. Place the appropriate letter next to each statement.

O—obesity    B—bulimia nervosa    A—anorexia nervosa

    a. _____ Body weight is normal, but weight is regulated by purging after eating.

    b. _____ Body weight is too low, and person may binge and purge.

    c. _____ Body weight is 20% or more above appropriate weight for height.

    d. _____ Exercise is usually minimal.

    e. _____ Often accompanied by excessive exercise.

Review key terms by completing this crossword puzzle using the following alphabetized list of terms.

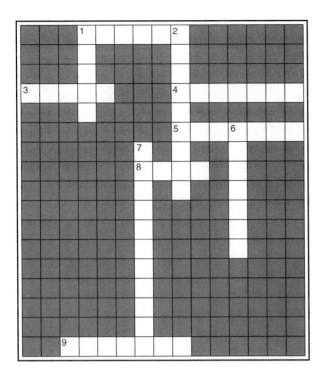

*amylase*
*anus*
*esophagus*
*fiber*
*gallbladder*
*gastric*
*lacteal*
*lipase*
*liver*
*pharynx*

## Across

1. Fat-digesting enzyme secreted by the pancreas.
3. Indigestible plant material that may lower risk of colon cancer.
4. Muscular passageway at the back of the mouth where swallowing occurs.
5. Starch-digesting enzyme.
8. Final canal of the large intestine.
9. Type of gland found in the stomach.

## Down

1. Large, multifunctioned organ in the abdominal cavity; secretes bile.
2. Muscular tube leading from the pharynx to the stomach.
6. Lymphatic vessel inside a villus of the small intestine.
7. Muscular sac that stores bile.

## OBJECTIVE QUESTIONS

Do not refer to the text when taking this test.

In questions 1–6, match the statements to these digestive system structures:

    a. pharynx   b. soft palate   c. esophagus
    d. stomach   e. small intestine   f. large intestine

_____ 1. Blocks food from entering nasal passages

_____ 2. Conducts food from throat to stomach

_____ 3. Common passageway for food and air

_____ 4. Longest portion of the digestive tract

_____ 5. Contains cecum, colon, rectum, and anus

_____ 6. Has sphincter at one end

_____ 7. The layer of the digestive tract that keeps it from sticking to surrounding internal organs is the
    a. serous layer.
    b. submucosa.
    c. mucosa.
    d. muscularis.

In questions 8–12, match these organs to the appropriate function.

    a. large intestine   b. stomach   c. liver
    d. pancreas   e. gallbladder

_d_ 8. Secretes digestive enzymes and bicarbonate

_e_ 9. Releases bile into the duodenum

_____ 10. Absorbs water and alcohol; initiates protein breakdown

_c_ 11. Detoxifies substances removed from blood

_a_ 12. Absorbs water and houses bacteria

_____ 13. Cholecystokinin is a(n) _____ secreted in response to _____ in food.
    a. enzyme, protein
    b. hormone, fats
    c. lipid, carbohydrates
    d. hormone, carbohydrates

_____ 14. Which enzyme digests starch?
    a. lipase
    b. carboxypeptidase
    c. trypsin
    d. amylase

_____ 15. HCl
    a. is an enzyme.
    b. creates the acid environment necessary for pepsin to work.
    c. is found in the intestines.
    d. All of these are correct.

_____ 16. Two enzymes involved in the digestion of protein are
    a. salivary amylase and lipase.
    b. trypsin and hydrochloric acid.
    c. pancreatic amylase and trypsin.
    d. pepsin and trypsin.

_____ 17. What is absorbed into the lacteals?
    a. proteins
    b. fats
    c. carbohydrates
    d. water and amino acids

_____ 18. Villi
    a. are found in the small intestine.
    b. increase the absorptive surface area.
    c. contain capillaries.
    d. All of these are correct.

_____ 19. The enzymes for digestion are referred to as hydrolytic because they require
    a. hydrogen.
    b. HCl.
    c. energy.
    d. water.

_____ 20. What type of food has the highest fat content?
    a. meat
    b. bread
    c. potatoes
    d. fruits

_____ 21. What is the best way to ensure that you are obtaining plenty of vitamins and minerals in your diet?
    a. Take mega-multiple vitamin supplements.
    b. Be sure you eat five fruit and vegetable servings daily.
    c. Eat lots of red meat.
    d. Eat white bread.

_____ 22. Which type of food group supplies quick energy and should comprise the bulk of the diet?
    a. milk and cheese
    b. lipids
    c. carbohydrates
    d. proteins

_____ 23. An eating disorder characterized by binging on food, then inducing vomiting or other purging is
    a. bulimia nervosa.
    b. anorexia nervosa.
    c. obesity.
    d. cirrhosis.

____24. Which type of fat is best to consume in quantity?
   a. saturated fats
   b. monounsaturated fats
   c. polyunsaturated fats
   d. None of these should be consumed in quantity.

____25. Which of the following are considered antioxidants?
   a. calcium and sodium
   b. B vitamins and selenium
   c. vitamin E and iron
   d. vitamins C, E, and A

## THOUGHT QUESTIONS

Answer in complete sentences.

26. Why is digestion a necessary process for humans?

27. What are LDL and HDL cholesterol?

28. How do the digestive and reproductive systems interact?

**Test Results:** _____ number correct ÷ 28 = _____ × 100 = _____%

## ANSWER KEY

### STUDY QUESTIONS

**1. a.** mouth **b.** teeth **c.** Salivary **d.** saliva **e.** taste buds **f.** tongue **2. a.** pharynx **b.** reflex **3. a.** stomach **b.** peristalsis **c.** sphincter (constrictor) **4. a.** mucosa **b.** mucus **c.** hydrochloric **d.** gastric **e.** pepsin **5. a.** submucosa **b.** muscularis **c.** serosa **6. a.** duodenum **b.** pancreas **c.** gallbladder **d.** chyme **7. a.** villi **b.** microvilli **c.** Glucose **d.** amino acids **e.** lacteals **8.** To increase the surface area available for absorption in the small intestine. **9. a.** gastrin **b.** gastric inhibitory peptide **c.** Secretin **d.** cholecystokinin **10. a.** feces **b.** polyps **c.** fiber **11. a.** bicarbonate **b.** enzymes **12. a.** bile **b.** gallbladder **c.** gatekeeper to blood **d.** small intestine **e.** storing iron and fat-soluble vitamins **f.** storing glycogen **g.** making blood plasma proteins **13. a.** Hepatitis **b.** cirrhosis **14. a.** salivary amylase **b.** mouth **c.** basic **d.** pancreas **e.** maltase **f.** small intestine **g.** acidic **h.** gastric glands **i.** trypsin **j.** small intestine **k.** basic **l.** small intestine **m.** nuclease **n.** small intestine **o.** basic **p.** small intestine **q.** lipase **r.** small intestine **15. a.** 2 **b.** 5 **c.** 4 **d.** 1 **e.** 3 **f.** 7 **g.** 6 **16.** See Figure 5.13 in text. **17.** c **18.** d **19. a.** T **b.** T **c.** F **d.** T **20. a.** coenzymes **b.** A **c.** D **d.** C, E, and A **21. a.** sodium **b.** calcium **22. a.** B **b.** A **c.** O **d.** O **e.** A

### DEFINITIONS CROSSWORD

**Across**
**1.** lipase **3.** fiber **4.** pharynx **5.** amylase **8.** anus **9.** gastric

**Down**
**1.** liver **2.** esophagus **6.** lacteal **7.** gallbladder

**1.** b  **2.** c  **3.** a  **4.** e  **5.** f  **6.** d  **7.** a  **8.** d  **9.** e
**10.** b  **11.** c  **12.** a  **13.** b  **14.** d  **15.** b  **16.** d
**17.** b  **18.** d  **19.** d  **20.** a  **21.** b  **22.** c  **23.** a
**24.** d  **25.** d  **26.** Like all other organisms, humans function at the level of the cell. Digestion occurs so that nutrients in foods can be processed to the point that they are available to individual cells.  **27.** LDL refers to low-density lipoproteins—the kind that carry cholesterol away from the liver to the body where it can be deposited. This is bad for the health. HDL refers to high-density lipoproteins. These deposit cholesterol in the liver where it can be processed to bile salts and eventually eliminated from the body.  **28.** The digestive system provides nutrients needed for reproduction and the growth of an unborn fetus while it develops in the uterus. Later, the newborn can be supplied with nutrients from its mother during nursing, all supplied by the digestive system.

# 13

# CARDIOVASCULAR SYSTEM

Blood vessels include **arteries** (and **arterioles**) that take blood away from the heart; **capillaries,** where exchange of substances with the tissues occurs; and **veins** (and **venules**) that take blood to the heart.

The movement of blood in the cardiovascular system is dependent on the beat of the **heart.** During the **cardiac cycle,** the **SA node (pacemaker)** initiates the beat and causes the atria to contract.

The **AV node** conveys the stimulus and initiates contraction of the ventricles. The heart sounds, lub-dup, are due to the closing of the **atrioventricular valves,** followed by the closing of the **semilunar valves.**

The cardiovascular system is divided into the **pulmonary circuit** and the **systemic circuit.** In the pulmonary circuit, two **pulmonary arteries** take blood from the right ventricle to the lungs, and four **pulmonary veins** return it to the left atrium. To trace the path of blood in the systemic circuit, start with the **aorta** from the left ventricle. Follow its path until it branches to an artery going to a specific organ. It can be assumed that the artery divides into arterioles and capillaries and that the capillaries lead to venules. The vein that takes blood to the **vena cava** most likely has the same name as the artery that delivered blood to the organ. In the adult systemic circuit, unlike the pulmonary circuit, the arteries carry oxygen-rich blood and the veins carry oxygen-poor blood.

**Blood pressure** accounts for the flow of blood in the arteries, but because blood pressure drops off in the capillaries, it cannot cause blood flow in the veins. Skeletal muscle contraction pushes blood past a venous valve, which then shuts, preventing backward flow. The velocity of blood flow is slowest in the capillaries, where exchange of nutrients and wastes takes place.

Blood has two main parts: **plasma** and cells. Plasma contains mostly water (90–92%) and proteins (7–8%), but it also contains nutrients and wastes.

The **red blood cells** contain **hemoglobin** and function in oxygen transport. Defense against disease depends on the various types of **white blood cells.** Granular neutrophils and monocytes are phagocytic. Agranular lymphocytes are involved in the development of immunity to disease.

The **platelets** and two plasma proteins, **prothrombin** and **fibrinogen,** function in blood clotting, an enzymatic process that results in fibrin threads.

When blood reaches a capillary, water moves out at the arterial end, due to blood pressure. At the venule end, water moves in, due to osmotic pressure. In between, nutrients diffuse out and wastes diffuse in.

**Hypertension** and **atherosclerosis** are two cardiovascular disorders that lead to heart attack and to stroke. Medical and surgical procedures are available to control cardiovascular disease, but the best policy is prevention by following a heart-healthy diet, getting regular exercise, maintaining a proper weight, and not smoking cigarettes.

**STUDY QUESTIONS**

Study the text section by section. Answer the study questions so that you can fulfill the learning objectives for each section.

## 13.1 THE BLOOD VESSELS (PP. 240–41)

The learning objectives for this section are:
- Name and describe the structure and function of arteries, capillaries, and veins.
- Explain how blood flow may bypass certain capillary beds.

1. Label the blood vessels in this diagram, using the following alphabetized list of terms.
    arterioles
    artery
    capillaries
    heart
    vein
    venules

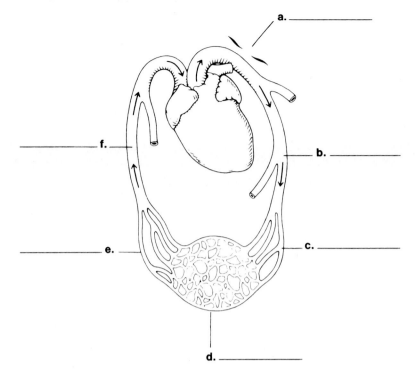

a. _____

f. _____

b. _____

e. _____

c. _____

d. _____

2. Match the statements to these terms.
    artery   vein   capillary

a. _____      Has the thickest walls.
b. _____      Has valves.
c. _____      Takes blood away from the heart.
d. _____      Takes blood to the heart.
e. _____      Exchanges carbon dioxide and oxygen with tissues.
f. _____      Nervous stimulation causes these to constrict during hemorrhaging; they
                               also act as a blood reservoir.

3. Explain how it is possible to bypass capillary beds by shunting blood directly from arteriole to venule.

   _____

   _____

## 13.2 THE HEART (PP. 242–45)

The learning objectives for this section are:
- Name the parts of the heart and their functions.
- Trace the path of blood flow through the heart.
- Describe how impulses from the nervous system and hormones can modify the heart rate.
- Explain how the conduction system of the heart controls the heartbeat.
- Label and explain a normal electrocardiogram.

4. a. Trace the path of blood through the heart from the vena cava to the lungs. _____

   _____

   b. Trace the path of blood from the lungs to the aorta. _____

   _____

5. Label the parts of the heart, using the following alphabetized list of terms.

aorta
aortic semilunar valve
atrioventricular (mitral) valve
atrioventricular (tricuspid) valve
AV node
chordae tendineae  ———————— a. ——————————
inferior vena cava  ———————— b. ——————————
left atrium
left ventricle                                    j. ——————————
pulmonary artery
right atrium  ———————— c.                  k. ——————————
right ventricle                                  l. ——————————
SA node  ———————— d.
septum        ———————— e.
superior vena cava                               m. ——————————
       ———————— f.                    n. ——————————

       ———————— g.

       ———————— h.                    o. ——————————

       ———————— i.

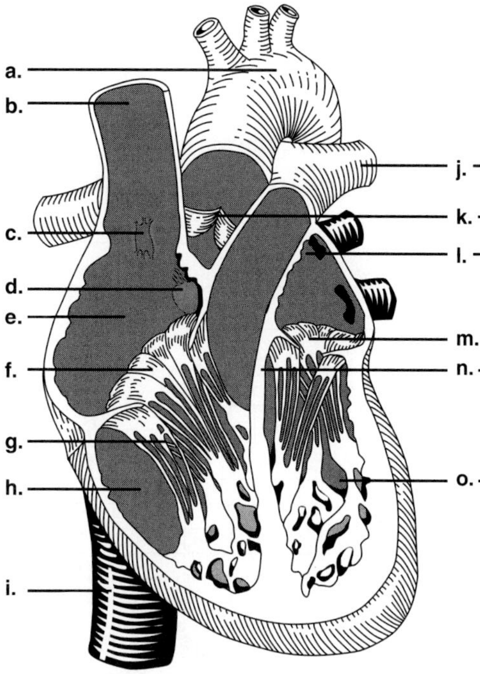

6. How does the thickness of the walls of the ventricles relate to their functions? _____
_____

7. Fill in the following table with the words *systole* (contraction) and *diastole* (relaxation) to show what happens during the 0.85 seconds of one heartbeat.

| Time | Atria | Ventricles |
| --- | --- | --- |
| 0.15 sec | a. | d. |
| 0.30 sec | b. | e. |
| 0.40 sec | c. | f. |

In question 8, fill in the blanks.

8. Heart sounds. When the atria contract, this forces blood through the  a._____ valves into the chambers called the  b._____. The closing of these valves is the lub sound. Next, the ventricles contract and force the blood into the arteries. Now the  c._____ valves close, making the dup sound.

9. Match the phrases to these nodes:

SA node    AV node

a. _____ pacemaker

b. _____ contraction of ventricles

c. _____ base of right atrium near the septum

d. _____ Purkinje fibers

115

10. Match the actions to these divisions of the nervous system:

     parasympathetic system    sympathetic system

    a. _____ normal body functions

    b. _____ active under times of stress

    c. _____ releases norepinephrine to speed up heart

    d. _____ slows heart rate

11. Does the adrenal gland hormone, epinephrine, speed or slow the heart rate? _____

12. What is the significance of each of the following in an electrocardiogram?

    a. *P* wave _____

    b. *QRS* wave _____

    c. *T* wave _____

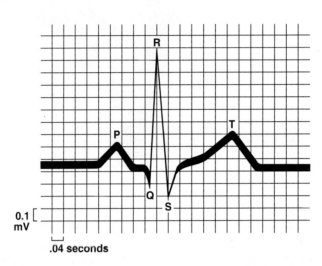

0.1 mV

.04 seconds

## 13.3 THE VASCULAR PATHWAYS (PP. 246–48)

The learning objectives for this section are:
- Describe the pulmonary circuit of circulation.
- Identify the major vessels of the systemic circuit.

13. Trace the path of blood to the left atrium:
    right ventricle

    a. _____

    b. _____

    lungs

    c. _____

    left atrium

From the legs:
legs

    c. _____

    d. _____

right atrium

14. Trace the path of blood from the aorta to the liver:
    aorta

    a. _____

    digestive tract

    b. _____

    liver

From the liver:
liver

    c. _____

    vena cava

Questions 15–17 are based on this diagram. Use the space provided to answer them in complete sentences.

15. What force accounts for blood flow in arteries?

_____

_____

16. Why does this force fluctuate?

_____

_____

17. What causes the blood pressure and velocity to drop off? _____

_____

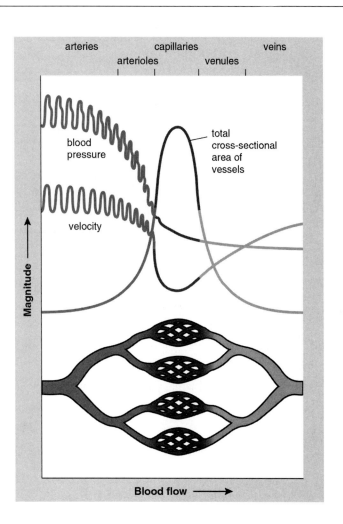

18. What three factors account for blood flow in the veins? _____

_____

19. What keeps blood from flowing backward in veins? _____

_____

## 13.4 BLOOD (PP. 249–54)

The learning objectives for this section are:
- Describe the composition of blood.
- Explain the process of blood clotting.
- Explain how the exchange of materials occurs between blood and tissue fluid.
- Explain how exchange of substances between blood and tissue fluid across capillary walls supplies cells with nutrients and removes wastes.

20. Plasma is mostly  a._____ and  b._____.
21. Place the correct plasma protein in the blank: fibrinogen, albumin, globulin, or all plasma proteins.

    a._____ transports cholesterol.

    b._____ helps blood clot.

    c._____ transports bilirubin.

    d._____ helps maintain the pH and osmotic pressure of the blood.

22. The red blood cells, scientifically called  a._____, are made in the  b._____. Upon maturation, they are biconcave disks that lack a(n)  c._____ and contain  d._____. After about 120 days, red blood cells are destroyed in the  e._____ and _____. The condition of  f._____ is characterized by an insufficient number of red blood cells or not enough hemoglobin.

23. Circle the items that describe hemoglobin correctly:
    a. heme contains iron
    b. globin contains iron
    c. becomes oxyhemoglobin in the tissues
    d. becomes deoxyhemoglobin in the tissues
    e. makes red blood cells red
    f. makes eosinophils red

24. White blood cells, scientifically called  a._____, are made in the  b._____.

25. Name three differences between red blood cells and white blood cells. White blood cells are

    a._____ in size than red blood cells; they do have a  b._____; and they do not contain

    c._____.

26. Match the description to the correct white blood cell.

    neutrophil    eosinophil    basophil    lymphocyte    monocyte

    a. _____ An agranular cell with a large, round nucleus that occurs in two versions. The B lymphocytes produce antibodies, and the T lymphocytes destroy cells that contain viruses.

    b. _____ An abundant granular cell with a multilobed nucleus that phagocytizes pathogens.

    c. _____ A large agranular cell that takes up residence in the tissues and differentiates into a voracious macrophage.

    d. _____ A cell with blue-staining granules that takes up residence in the tissues; these become mast cells and release histamine.

    e. _____ A cell with a bilobed nucleus and red-staining granules that becomes abundant during allergies and parasitic infections.

27. A type of cancer called  a._____ occurs when abnormally large numbers of immature  b._____ fill red bone marrow. The patient becomes both anemic and incapable of fighting disease organisms.

28. The following shows the reactions that occur as blood clots:

platelets ─────────────→ prothrombin activator
prothrombin ─────────────→ thrombin
fibrinogen ─────────────→ fibrin threads

Does the left-hand side or the right-hand side list substances that are always present in the blood? a._____

Which substances function as enzymes? b._____ Which substance is the actual clot? c._____

29. Several nutrients are necessary for clotting to occur. Vitamin a._____ is needed for the production of

prothrombin. The element b._____ is needed for the conversion of prothrombin to thrombin.

30. Label this diagram using these terms:
      amino acid
      arterial end
      blood pressure (two times)
      carbon dioxide
      glucose
      net pressure in
      net pressure out
      osmotic pressure (two times)
      oxygen
      tissue fluid
      venous end
      wastes
      water (two times)

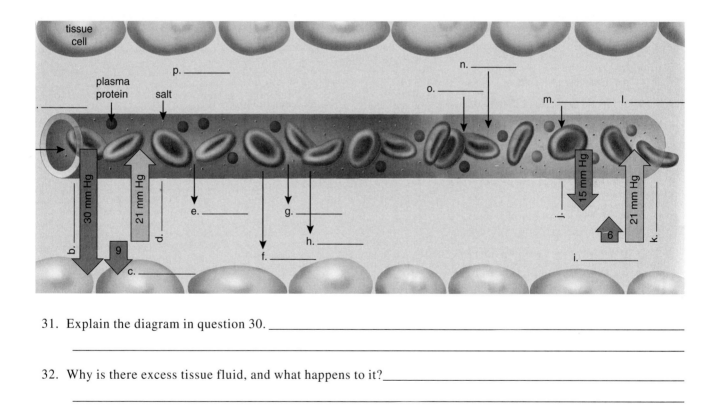

31. Explain the diagram in question 30. _____

_____

32. Why is there excess tissue fluid, and what happens to it?_____

_____

## 13.5 Cardiovascular Disorders (pp. 255–56)

The learning objective for this section is:

- Discuss the factors that can lead to cardiovascular disease and how they can be prevented.

33. Match the statements to these items:

    artificial pacemaker needed   dietary restriction of salt and/or cholesterol   donor heart transplant   coronary bypass

    a. ______________________________ Clearing clogged arteries was unsuccessful.
    b. ______________________________ Blood pressure is 200/140.
    c. ______________________________ Heartbeat is irregular.
    d. ______________________________ Congestive heart failure is present.

34. Match the phrases to these items:

    thrombus and embolus   atherosclerosis and hypertension   varicose veins   hemorrhoids

    a. ______________________________ stroke and heart attack
    b. ______________________________ varicose veins in rectum
    c. ______________________________ weakened valves of veins
    d. ______________________________ development of aneurysm

## Definitions Wordsearch

Review key terms by using the following alphabetized list of terms to fill in the blanks below. Then complete the wordsearch.

| | | | | | | | | | | | | |
|---|---|---|---|---|---|---|---|---|---|---|---|---|
| M | Y | O | C | A | R | D | I | U | M | L | I | M |
| O | U | I | L | T | R | G | H | D | C | V | U | N |
| Q | R | G | T | R | G | J | K | I | N | D | E | I |
| I | C | A | P | I | L | L | A | R | Y | R | E | E |
| R | A | T | R | O | A | L | O | P | E | R | V | V |
| D | E | C | V | V | S | G | E | T | H | E | L | Y |
| E | Y | S | D | E | I | Y | X | Z | E | R | Y | R |
| E | R | N | J | N | H | U | S | J | L | O | P | A |
| L | E | G | U | T | A | S | D | T | A | N | D | N |
| O | T | G | B | R | M | H | K | I | O | G | O | O |
| T | R | D | E | I | N | U | N | S | U | L | L | M |
| S | A | G | Y | C | G | A | Y | D | O | T | E | L |
| A | Y | S | E | U | S | A | I | D | F | U | G | U |
| I | R | R | T | L | D | V | S | K | O | P | L | P |
| D | A | D | A | A | R | T | E | R | I | O | L | E |
| M | N | A | M | R | A | W | P | I | L | L | L | A |
| U | O | R | A | V | J | I | T | M | J | E | N | N |
| I | R | F | E | A | R | M | U | A | E | D | E | A |
| N | O | E | I | L | C | H | M | A | R | D | M | A |
| E | C | M | I | V | M | I | C | O | T | T | O | N |
| A | L | C | H | E | C | H | R | W | E | N | D | Y |

*aorta*
*arteriole*
*atrioventricular valve*
*capillary*
*coronary artery*
*diastole*
*myocardium*
*pulmonary vein*
*septum*
*systole*

a. _____ Valve located between the atrium and the ventricle.

b. _____ Vessel that takes blood from an artery to capillaries.

c. _____ Cardiac muscle in the wall of the heart.

d. _____ Partition in heart that divides it into left and right halves.

e. _____ Contraction of a heart chamber.

f. _____ Relaxation of a heart chamber.

g. _____ Blood vessel returning from lungs to heart.

h. _____ Large artery leaving heart with blood from the left ventricle.

i. _____ Microscopic vessel connecting arterioles to venules.

j. _____ Artery that supplies blood to the wall of the heart.

# CHAPTER TEST

## OBJECTIVE QUESTIONS

Do not refer to the text when taking this test.

_____ 1. Which type of blood vessel allows exchange of material between the blood and the tissues?
   a. arteries
   b. arterioles
   c. capillaries
   d. veins
   e. venules

_____ 2. What is the function of the heart valves?
   a. to push blood
   b. to prevent backflow
   c. to stimulate the heart
   d. to give support to the heart

_____ 3. Arteries
   a. carry blood away from the heart.
   b. carry blood toward the heart.
   c. have valves.
   d. Both *a* and *b* are correct.

_____ 4. Shunting of blood is possible because of a thoroughfare channel that joins _____ to _____ directly.
   a. arterioles; venules
   b. venules; capillaries
   c. arteries; veins
   d. arterioles; veins

_____ 5. Which of these vessels have thinner walls?
   a. arteries
   b. veins
   c. Both are the same.

_____ 6. The venae cavae
   a. carry blood to the right atrium.
   b. carry blood away from the right atrium.
   c. join with the aorta.
   d. have a high blood pressure.

_____ 7. The chamber of the heart that receives blood from the pulmonary veins
   a. is the right atrium.
   b. is the left atrium.
   c. contains $O_2$-rich blood.
   d. contains $O_2$-poor blood.
   e. Both *a* and *c* are correct.
   f. Both *b* and *c* are correct.
   g. Both *b* and *d* are correct.

_____ 8. The coronary arteries carry blood
   a. from the aorta to the heart tissues.
   b. from the heart to the brain.
   c. directly to the heart from the pulmonary circuit.
   d. from the lungs directly to the left atrium.

_____ 9. Which of these chambers has the thickest walls?
   a. right atrium
   b. right ventricle
   c. left atrium
   d. left ventricle

_____10. When the atria are contracting, the ventricles are
   a. contracting.
   b. relaxing.
   c. in diastole.
   d. in systole.
   e. Both *a* and *c* are correct.
   f. Both *b* and *c* are correct.
   g. Both *b* and *d* are correct.

_____11. The SA node
   a. works only when it receives a nerve impulse.
   b. is located in the left atrium.
   c. initiates the heartbeat.
   d. All of these are correct.

____12. The first wave (the P wave) of an ECG occurs prior to
  a. atrial contraction.
  b. ventricular contraction.
  c. ventricular relaxation.
  d. atrial relaxation.
____13. The heart sounds are due to
  a. blood flowing.
  b. the closing of the valves.
  c. the heart muscle contracting.
  d. blood pressure in the aorta.
____14. Blood flows in veins because of
  a. contraction of valves.
  b. arterial blood pressure.
  c. capillary blood pressure.
  d. skeletal muscle contraction.
____15. Systole refers to the contraction of the
  a. major arteries.
  b. SA node.
  c. atria and ventricles.
  d. major veins.
____16. Blood pressure falls off drastically in the capillaries because the capillaries
  a. contain valves.
  b. become veins.
  c. have a large cross-sectional area.
  d. All of these are correct.
____17. Which of these correctly traces the path of blood from the left ventricle to the head?
  a. left ventricle, subclavian artery, head
  b. left ventricle, pulmonary artery, head
  c. left ventricle, aorta, carotid artery, head
  d. left ventricle, vena cava, jugular vein, head
____18. Blood pressure
  a. is the same in all blood vessels.
  b. is highest in the aorta.
  c. is measured by taking an ECG.
  d. never rises above normal.

____19. The major portion of the cardiovascular system is called the
  a. systemic circuit.
  b. pulmonary circuit.
  c. hepatic portal circuit.
  d. coronary circuit.
____20. Blood flowing to the lungs leaves the heart via the _____ and returns to the heart via the _____.
  a. aorta; superior vena cava
  b. superior vena cava; aorta
  c. pulmonary arteries; pulmonary veins
  d. aorta; pulmonary veins
____21. The jugular vein carries blood from the
  a. head.
  b. vena cava.
  c. arm.
  d. aorta.
____22. Blood moves slowly in capillaries,
  a. which facilitates tissue exchange.
  b. because they have valves.
  c. because they have thick walls.
  d. because venules are smaller than capillaries.
____23. People with atherosclerosis often experience
  a. high blood pressure.
  b. a heart attack.
  c. a thrombus.
  d. a stroke.
  e. Any of these is correct.
____24. A heart attack is due to a blocked
  a. pulmonary artery.
  b. coronary artery.
  c. aorta.
  d. vena cava.
____25. Phlebitis and hemorrhoids are conditions involving
  a. arteries.
  b. capillaries.
  c. veins.
  d. arterioles.

## THOUGHT QUESTIONS

Answer in complete sentences.
26. Why can arteries expand without rupturing?


27. Explain how the digestive system and cardiovascular system benefit each other.


**Test Results:** _____ number correct ÷ 27 = _____ × 100 = _____%

## STUDY QUESTIONS

**1. a.** heart **b.** artery **c.** arterioles **d.** capillaries **e.** venules **f.** vein **2. a.** artery **b.** vein **c.** artery **d.** vein **e.** capillary **f.** vein **3.** The shunting of blood around capillary beds is possible because each bed has a thoroughfare channel that allows blood to flow directly from arteriole to venule. Sphincter muscles prevent blood from flowing into the capillaries. **4. a.** vena cava, right atrium, atrioventricular valve, right ventricle, pulmonary semilunar valve, pulmonary artery, lungs **b.** lungs, pulmonary veins, left atrium, atrioventricular valve, left ventricle, aortic semilunar valve, aorta **5. a.** aorta **b.** superior vena cava **c.** SA node **d.** AV node **e.** right atrium **f.** atrioventricular (tricuspid) valve **g.** chordae tendineae **h.** right ventricle **i.** inferior vena cava **j.** pulmonary artery **k.** aortic semilunar valve **l.** left atrium **m.** atrioventricular (mitral) valve **n.** septum **o.** left ventricle **6.** The left ventricle is thicker-walled than the right ventricle because the left one must pump blood the greater distance to the entire body. The right ventricle only pumps the shorter distance to the lungs. **7. a.** systole **b.** diastole **c.** diastole **d.** diastole **e.** systole **f.** diastole **8. a.** atrioventricular **b.** ventricles **c.** semilunar **9. a.** SA node **b.** AV node **c.** AV node **d.** AV node **10. a.** parasympathetic system **b.** sympathetic system **c.** sympathetic system **d.** parasympathetic system **11.** speed **12. a.** associated with atrial systole **b.** associated with ventricular systole **c.** associated with ventricular recovery **13. a.** pulmonary trunk **b.** pulmonary arteries **c.** pulmonary veins **d.** iliac vein **e.** inferior vena cava **14. a.** mesenteric arteries **b.** hepatic portal vein **c.** hepatic vein **15.** blood pressure **16.** systole and diastole of the left ventricle of the heart **17.** distance from heart and increase in cross-sectional area of blood vessels **18.** skeletal muscle contraction, the presence of valves in veins, and respiratory movements. **19.** valves **20. a.** water **b.** plasma proteins **21. a.** globulin **b.** fibrinogen **c.** albumin **d.** all plasma proteins **22. a.** erythrocytes **b.** red bone marrow **c.** nucleus **d.** hemoglobin **e.** liver and spleen **f.** anemia **23.** a, d, e, **24. a.** leukocytes **b.** red bone marrow **25. a.** larger **b.** nucleus **c.** hemoglobin **26. a.** lymphocyte **b.** neutrophil **c.** monocyte **d.** basophil **e.** eosinophil **27 a.** leukemia **b.** white blood cells (leukocytes) **28. a.** left-hand side **b.** prothrombin activator and thrombin **c.** fibrin threads **29. a.** K **b.** calcium **30. a.** arterial end **b.** blood pressure **c.** net pressure out **d.** osmotic pressure **e.** water **f.** oxygen **g.** amino acid **h.** glucose **i.** net pressure in **j.** blood pressure **k.** osmotic pressure **l.** venous end **m.** water **n.** wastes **o.** carbon dioxide **p.** tissue fluid **31.** At the arterial end of a capillary, blood pressure is higher than osmotic pressure. Therefore, water, nutrients, and oxygen leave a capillary. At the venous end of a capillary, osmotic pressure is higher than blood pressure; therefore, water and wastes enter a capillary. In this way, tissue fluid is refreshed. **32.** This system never retrieves all the water that leaves capillaries, and excess tissue fluid is picked up by lymphatic vessels and returned to the bloodstream. **33. a.** coronary bypass **b.** dietary restriction of salt and/or cholesterol **c.** artificial pacemaker needed **d.** donor heart transplant **34. a.** thrombus and embolus **b.** hemorrhoids **c.** varicose veins **d.** atherosclerosis and hypertension

## DEFINITIONS WORDSEARCH

```
M Y O C A R D I U M
      T               N
      R               I
    C A P I L L A R Y E
    A T R O A         V
      V S             Y
    Y E Y             R
  E R N     S         A
  L E T     T         N
  O T R         O     O
  T R I         L     M
  S A C         E L   U
  A Y U             U
  I R L     S       P
  D A   A R T E R I O L E
  A N   R   P
  N O   V   T
  O R   A   U
  R A   L   M
  O L   M
  C V
      E
```

**a.** atrioventricular valve **b.** arteriole **c.** myocardium **d.** septum **e.** systole **f.** diastole **g.** pulmonary vein **h.** aorta **i.** capillary **j.** coronary artery

## CHAPTER TEST

**1.** c **2.** b **3.** a **4.** a **5.** b **6.** a **7.** f **8.** a **9.** d **10.** f **11.** c **12.** a **13.** b **14.** d **15.** c **16.** c **17.** c **18.** b **19.** a **20.** c **21.** a **22.** a **23.** e **24.** b **25.** c **26.** The wall of an artery has a middle layer of elastic tissue and smooth muscle and an outer layer of fibrous connective tissue. This arrangement allows for both strength and flexibility. **27.** The cardiovascular system transports nutrients from the digestive tract to the rest of the body and also services the organs of the digestive tract. The digestive tract provides nutrients for blood cell formation and the formation of plasma proteins. The liver detoxifies the blood, makes plasma proteins, and destroys old red blood cells.

# 14

# LYMPHATIC AND IMMUNE SYSTEMS

The **lymphatic system** consists of lymphatic vessels and lymphoid organs. The lymphatic vessels absorb fat molecules at intestinal villi and excess tissue fluid at the cardiovascular capillaries. Eventually, two main ducts empty into the subclavian veins.

Lymphocytes are produced and accumulate in the lymphoid organs (**lymph nodes, spleen, thymus gland, tonsils,** and **red bone marrow**).

**Immunity** involves nonspecific and specific defenses. Nonspecific defenses include barriers to entry, the **inflammatory reaction,** natural killer cells, and protective proteins.

Specific defenses require lymphocytes, which are produced in the bone marrow. **B lymphocytes** mature in the bone marrow and undergo clonal selection in the lymph nodes and the spleen. **T lymphocytes** mature in the thymus.

B cells are responsible for **antibody-mediated immunity.** An **antibody** is a Y-shaped molecule that has two binding sites. Each antibody is specific for a particular **antigen.** Activated B cells become antibody-secreting **plasma cells** and memory B cells. Memory B cells respond if the same antigen enters the body at a later date.

There are four types of T cells. **Cytotoxic T cells** kill cells on contact; helper T cells stimulate other immune cells and produce lymphokines; suppressor T cells suppress the immune response; and memory T cells remain in the body to provide long-lasting immunity.

Immunity can be induced in various ways. **Vaccines** are available to promote long-lived, active immunity, and antibodies sometimes are available to provide an individual with short-lived, passive immunity.

**Cytokines,** notably **interferon** and **interleukins,** are used in an attempt to promote the body's ability to recover from cancer and to treat AIDS.

**Allergies** result when an overactive immune system forms antibodies to substances not normally recognized as foreign. Cytotoxic T cells attack transplanted organs as nonself; therefore, immunosuppressive drugs must be administered. **Autoimmune** illnesses occur when antibodies and T cells attack the body's own tissues.

Blood transfusions require compatible blood types. The antigens (A and B) are on the red blood cells and the antibodies (anti-A and anti-B) are in the plasma. The Rh antigen is also particularly important because an Rh-negative mother may produce anti-Rh antibodies that will attack the red blood cells of an Rh-positive fetus.

## STUDY QUESTIONS

Study the text section by section. Answer the study questions so that you can fulfill the learning objectives for each section.

### 14.1 THE LYMPHATIC SYSTEM (PP. 262–64)

The learning objectives for this section are:
* Describe the structure of lymphatic vessels.
* List three functions of the lymphatic system.
* Explain the structure and purpose of lymph nodes.
* Discuss the functions of the spleen and of the white and red pulp it contains.
* Describe how the thymus gland and red bone marrow participate in immunity.

1. Give three functions of the lymphatic system.

   a. _____

   b. _____

   c. _____

2. Indicate whether these statements about the structure/function of lymphatic vessels and lymphoid organs are true (T) or false (F).
   a. _____ Bone marrow lacks lymphoid tissue.
   b. _____ Lymph lobules are subdivided into sinus-containing nodes.
   c. _____ The contraction of skeletal muscles blocks the return of lymph to the bloodstream.
   d. _____ The sinuses of the spleen are filled with lymph.
   e. _____ Lymphatic vessels are similar to cardiovascular veins.
   f. _____ Lymphatic vessels contain valves.

3. Indicate whether the following statements are true (T) or false (F). Rewrite any false statements to make them true.
   a. _____ Vessels of the lymphatic system begin with cardiovascular capillaries. Rewrite: _____
   _____
   b. _____ Lymph most closely resembles arterial blood. Rewrite: _____
   c. _____ The right thoracic duct serves the lower extremities, abdomen, one arm, and one side of the head and neck. Rewrite: _____
   d. _____ Lymphatic capillaries merge directly to form a particular lymphatic duct. Rewrite: _____
   _____

4. Place the appropriate letter(s) next to each statement.

   T— thymus     S—spleen     RBM—red bone marrow

   a. _____ contains red pulp and white pulp
   b. _____ contains stem cells
   c. _____ is located along the trachea
   d. _____ is located in the upper left abdominal cavity
   e. _____ produces hormones believed to stimulate the immune system
   f. _____ cleanses blood
   g. _____ site of origin for all types of blood cells

## 14.2 NONSPECIFIC DEFENSES (PP. 264–66)

The learning objective for this section is:
- Describe barriers to entry, the inflammatory reaction, natural killer cells, and protective proteins as agents of nonspecific immunity.

5. Match the descriptions to these defense mechanisms:

   1 barrier to entry     2 inflammatory response     3 complement system     4 natural killer cells

   a. _____ accompanied by swelling and redness
   b. _____ cilia action in the respiratory tract
   c. _____ produces holes in bacterial cell walls
   d. _____ stomach secretions
   e. _____ histamine increases capillary permeability
   f. _____ injured cells release bradykinin
   g. _____ vagina is inhabited by nonpathogenic bacteria
   h. _____ stimulates inflammatory response, binds to some bacteria and punches holes in others
   i. _____ neutrophils and macrophages carry out phagocytosis
   j. _____ secretions of the oil, or sebaceous, glands
   k. _____ kills cells infected with a virus and tumor cells

## 14.3 SPECIFIC DEFENSES (PP. 266–71)

The learning objectives for this section are:
- Compare the origin, maturation, and function of B cells and T cells.
- Describe the general structure of an antibody, and state a function for the variable and constant regions.
- Explain how and where B cells undergo clonal selection and expansion.
- Tell how a T cell recognizes an antigen.
- List the different types of T cells, and give the action of each.

6. Fill in this table with *yes* or *no:*

|  | Cytotoxic T Cell | B Cell |
| --- | --- | --- |
| Ultimately derived from stem cells in bone marrow | a. | b. |
| Pass through thymus | c. | d. |
| Carry antigen receptors on membrane | e. | f. |
| Cell-mediated immunity | g. | h. |
| Antibody-mediated immunity | i. | j. |

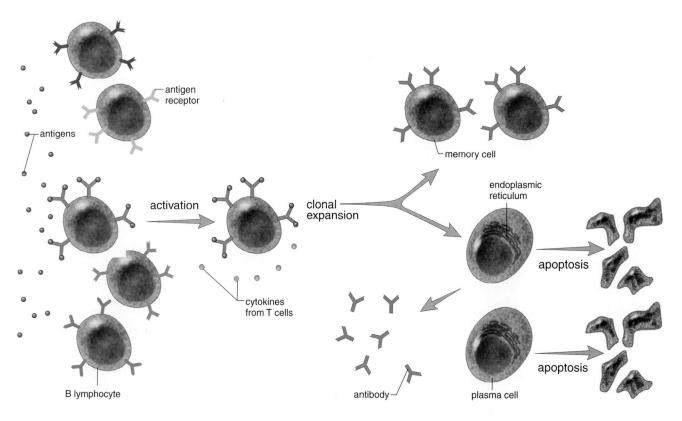

Study the diagram and answer questions 7–10.

7. Which of these makes a B cell undergo clonal expansion?
   a. when an antigen binds to its antigen receptor
   b. when fever is present

8. Explain the expression *clonal selection theory*. _____

   _____

9. Which of these are cells that result from the clonal expansion of B cells?
   a. plasma cells
   b. memory cells
   c. both types of cells

10. What happens to these cells once the infection is under control?
    a. memory cells _____
    b. plasma cells _____

11. Label this diagram of an antibody molecule, using these terms:

  antigen-binding site
  constant region
  heavy chain
  light chain
  variable region

a. _____

b. _____

e. _____

c. _____

d. _____

  f. What is the function of antibodies? _____

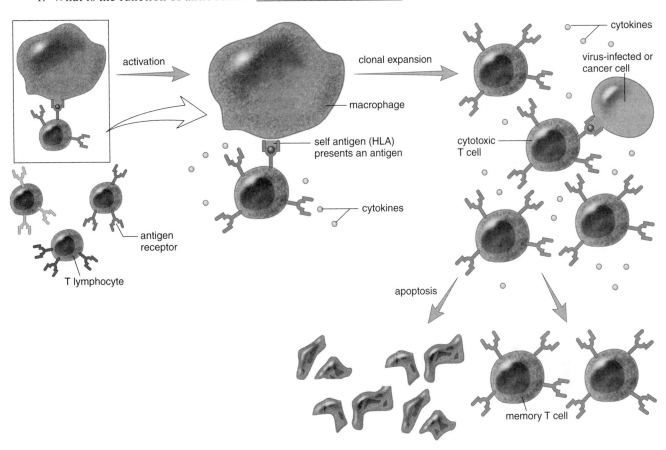

Study the diagram and answer questions 12–14.

  12. Which of these makes a particular T cell undergo clonal expansion?
      a. when the antigen is presented to the T cells by an APC
      b. when a T cell encounters an antigen

  13. a. What is the significance of HLA antigens? _____
      _____

      b. Why are they called antigens? _____
      _____

  14. a. What happens after a helper T cell recognizes an antigen? _____
      _____

      b. What happens after a cytotoxic T cell recognizes an antigen? _____
      _____

      c. What happens to T cells (except for memory T cells) after the infection is past? _____
      _____

## 14.4 INDUCED IMMUNITY (PP. 272–74)

The learning objective for this section is:
- Differentiate between active and passive immunity.

15. Label this diagram, using the following alphabetized list of terms.

first exposure to antigen     plasma antibody concentration     primary response
secondary response        second exposure to antigen

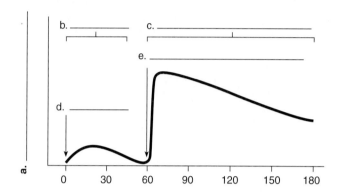

In questions 15e–g, fill in the blanks.

A good secondary response can be related to the e. _____, dependent on the number of

f. _____ and g. _____ cells capable of responding to a particular antigen.

In questions 16–17, fill in the blanks.

16. When an individual receives antibodies from another, as when a baby breast-feeds, it is called _____.

17. List five diseases for which children should be immunized. _____

_____

18. Why is it better to immunize children in advance than to wait until the disease has been contracted?

_____

19. Name two types of cytokines, and tell what role they play in immunotherapy.

a. _____

b. _____

20. What are monoclonal antibodies? _____

_____

## 14.5 IMMUNITY SIDE EFFECTS (PP. 274–77)

The learning objective for this section is:
- List three types of immunological side effects, and relate them to the function of the immune system.

21. Relate the immune response to each of these:

   a. allergy _____

   b. tissue rejection _____

   c. autoimmune disease _____

22. The following table indicates the blood types. Fill in the fourth and fifth columns by using this formula: The donor's antigen(s) must not be of the same type as the recipient's antibody (antibodies).

| Blood Type | Antigen | Antibody | Can Receive From | Can Donate To |
|------------|---------|----------|------------------|---------------|
| A | A | Anti-B | a. | b. |
| B | B | Anti-A | c. | d. |
| AB | A, B | _____ | e. | f. |
| O | None | Anti-A and B | g. | h. |

23. This diagram shows the results of typing someone's blood. What is the blood type? _____

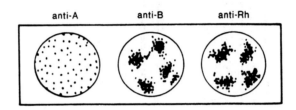

24. Draw a similar diagram showing the results if someone has AB-negative blood.

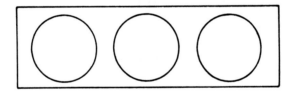

25. Consider these possible combinations of mates:
   Rh⁺ mother and Rh⁻ father    Rh⁻ mother and Rh⁻ father
   Rh⁺ mother and Rh⁺ father    Rh⁻ mother and Rh⁺ father
   Which of these combinations can cause pregnancy difficulties and why?

   _____

   _____

# DEFINITIONS WORDSEARCH

Review key terms by using the following alphabetized list of terms to fill in the blanks below. Then complete the wordsearch.

```
L E N I M A T S I H E R T M
I G S D C O M P L E M E N T
E A N T I B O D Y D L P A O
N H U I P R K J N E G H U D
I P H L L A R N O M J U T V
K O C E A D H E R A L O O B
O R M C H Y K D E H E L I Y
H C E E H K U F F P M E M M
P A P P L I O R R S A V M B
M M M U D N G H E E M Y U D
Y Y D O G I E T T S H I N C
L E A S E N T Y N P P L E C
M U D S L I D E I Y H N T D
```

*antibody*
*autoimmune*
*bradykinin*
*complement*
*edema*
*histamine*
*interferon*
*lymph*
*lymphokine*
*macrophage*

a. _____  Swelling due to the accumulation of tissue fluid.

b. _____  A system of plasma proteins that are a nonspecific defense.

c. _____  A disease caused by the immune system attacking the person's own body.

d. _____  A substance found in damaged tissues that initiates nerve impulses, triggering pain.

e. _____  A molecule released by T cells that enhances the abilities of other immune system cells.

f. _____  Large phagocytic cell.

g. _____  Protein released by cells infected with viruses.

h. _____  A protein produced by B cells in response to foreign antigens.

i. _____  Substance produced by basophils or mast cells in an allergic reaction.

j. _____  Tissue fluid inside lymphatic vessels.

# CHAPTER TEST

## OBJECTIVE QUESTIONS

Do not refer to the text when taking this test.

____ 1. The two collecting ducts of the lymphatic system empty into
   a. cardiovascular arteries.
   b. cardiovascular veins.
   c. pulmonary arteries.
   d. pulmonary veins.

____ 2. The structure of a lymphatic vessel is most similar to that of a
   a. cardiovascular artery.
   b. cardiovascular arteriole.
   c. cardiovascular vein.
   d. skeletal muscle fiber.

____ 3. Lymph nodes house
   a. neutrophils and monocytes.
   b. lymphocytes and macrophages.
   c. granular leukocytes.
   d. red blood cells.

____ 4. Lymph is _____ in lymphatic vessels.
   a. blood
   b. serum
   c. tissue fluid
   d. plasma

_____ 5. All of the following are functions of macrophages EXCEPT
   a. liberate a colony-stimulating hormone to increase leukocytes.
   b. phagocytize bacteria.
   c. scavenge dead and decaying tissue.
   d. transport oxygen in the blood.

In questions 6–9, match the descriptions to these terms:
   a. thymus gland    b. spleen
   c. lymph node    d. red bone marrow

_____ 6. Causes maturation of T cells
_____ 7. Purifies lymph
_____ 8. Purifies blood
_____ 9. Formation of agranular and granular leukocytes
_____10. The spleen
   a. contains stem cells from the bone marrow.
   b. is located along the trachea.
   c. produces a hormone believed to stimulate the immune system.
   d. contains red pulp and white pulp.
_____11. The thymus
   a. contains all types of stem cells from the bone marrow.
   b. is located along the trachea.
   c. produces a hormone believed to stimulate the immune system.
   d. Both _b_ and _c_ are correct.
_____12. Activity of the complement system is an example of nonspecific defense by
   a. barriers to entry.
   b. phagocytic cells.
   c. protective proteins.
   d. Both _a_ and _c_ are correct.
_____13. Secretions of the oil glands are an example of nonspecific defense by a(n)
   a. barrier to entry.
   b. protective protein.
   c. phagocytic cell.
   d. acidic pH.
_____14. Interferon is produced by cells in response to the presence of
   a. chemical irritants.
   b. viruses.
   c. bacterial infection.
   d. malarial parasite in blood.
_____15. The most active white blood cell phagocytes are
   a. neutrophils and macrophages.
   b. neutrophils and eosinophils.
   c. lymphocytes and macrophages.
   d. lymphocytes and neutrophils.

_____16. The white blood cells that are primarily responsible for specific immunity are
   a. neutrophils.
   b. eosinophils.
   c. macrophages.
   d. lymphocytes.
_____17. Which of these is NOT a valid contrast between T cells and B cells?

|  | **T cells** | **B cells** |
|---|---|---|

   a. matures in the thymus–matures in bone marrow
   b. antibody-mediated immunity–cell-mediated immunity
   c. antigen must be presented by APC–direct recognition
   d. cytokines–do not produce cytokines
_____18. A particular antibody can
   a. attack any type of antigen.
   b. attack only a specific type of antigen.
   c. be produced by any B lymphocyte.
   d. be produced by any T lymphocyte.
_____19. The clonal selection theory refers to the
   a. presence of four different types of T lymphocytes in the blood.
   b. response of only one type of B lymphocyte to a specific antigen.
   c. occurrence of many types of plasma cells, each producing many types of antigens.
_____20. The portions of an antibody molecule that pair up with the foreign antigens are the
   a. heavy chains.
   b. light chains.
   c. variable regions.
   d. constant regions.
_____21. Which of these pairs is incorrect?
   a. helper T cells–orchestrate the immune response
   b. cytotoxic T cells–stimulate B cells to produce antibodies
   c. memory T cells–long-lasting active immunity
   d. suppressor T cells–shut down the immune response
_____22. A person receiving an injection of gamma globulin as a protection against hepatitis is an example of
   a. naturally acquired active immunity.
   b. naturally acquired passive immunity.
   c. artificially acquired passive immunity.
   d. artificially acquired active immunity.
_____23. A person vaccinated to produce immunity to the flu is an example of
   a. naturally acquired active immunity.
   b. naturally acquired passive immunity.
   c. artificially acquired passive immunity.
   d. artificially acquired active immunity.

___24. Allergies are caused by
  a. strong toxins in the environment.
  b. autoimmune diseases.
  c. the overproduction of IgE.
  d. the receipt of IgA in breast milk.

___25. Which is NOT true of an autoimmune response?
  a. responsible for such diseases as multiple sclerosis and myasthenia gravis and perhaps type I diabetes
  b. occurs when self-antibodies attack self-tissues
  c. interferes with the transplantation of organs between one person and another
  d. All of these are correct.

## THOUGHT QUESTIONS

Answer in complete sentences.

26. How do we know that one aspect of specific immunity is the ability of the body to recognize self as opposed to nonself?

27. Describe how the lymphatic system aids the activities of the integumentary system, and vice versa.

**Test Results:** _____ number correct ÷ 27 = _____ × 100 = _____%

## STUDY QUESTIONS

**1. a.** return of excess tissue fluid to bloodstream **b.** receive lipoproteins at intestinal villi **c.** defense against disease **2. a.** F **b.** F **c.** F **d.** F **e.** T **f.** T **3. a.** F, . . . begin with lymph capillaries **b.** F, . . . resembles tissue fluid that has entered the lymph vessels **c.** F, The right lymphatic duct serves the right arm, the right side of the head and neck, and the right thoracic area. **d.** F, . . . capillaries form lymphatic vessels first, and these merge before entering a particular lymphatic duct **4. a.** S **b.** RBM **c.** T **d.** S **e.** T **f.** S **g.** RBM **5. a.** 2 **b.** 1 **c.** 3 **d.** 1 **e.** 2 **f.** 2 **g.** 1 **h.** 3 **i.** 2 **j.** 1 **k.** 4 **6.**

| Cytotoxic T Cell | | B Cell | |
|---|---|---|---|
| a. | yes | b. | yes |
| c. | yes | d. | no |
| e. | yes | f. | yes |
| g. | yes | h. | no |
| i. | no | j. | yes |

**7.** a **8.** antigen selects the B cell that will clone **9.** c **10. a.** remain in body ready to produce more antibodies when needed **b.** undergo apoptosis and die off **11. a.** antigen-binding site **b.** variable region **c.** constant region **d.** heavy chain **e.** light chain **f.** combine with antigens and mark them for destruction **12.** a **13. a.** identify cell as belonging to an individual **b.** They are antigenic in someone else's body. **14. a.** secretes cytokines that stimulate other immune cells **b.** destroys cells that are infected with a virus **c.** undergo apoptosis and die **15. a.** See Figure 14.9 in text. **b.** immunological memory **c.** memory B **d.** memory T **16.** passive immunity **17.** tetanus, whooping cough, diphtheria, hepatitis B, measles, mumps, rubella, polio, *Haemophilis influenzae* (type b) **18.** It is better to prevent a disease than to try to treat it with antibiotics. Resistant strains of bacteria and allergies to antibiotics are two side effects, and viruses do not respond to antibiotic therapy. **19. a.** Interleukins activate and maintain killer activity of T cells. **b.** Interferon causes other cells to resist a viral infection. **20.** Same-type antibodies produced by the same lymphocyte. **21. a.** Antigen attaches to IgE antibodies on mast cells, and histamine release causes allergic response. **b.** Antibodies and cytotoxic T cells attack foreign antigens. **c.** Viral infection tricks immune cells into attacking tissues of self. **22. a.** A, O **b.** A, AB **c.** B, O **d.** B, AB **e.** A, B, AB, O **f.** AB **g.** O **h.** A, B, AB, O **23.** B⁺ **24.** Clumping should occur for anti-A and anti-B. **25.** Rh⁻ mother and Rh⁺ father because the mother might form antibodies to destroy red blood cells of this or a future baby who is Rh⁺.

## DEFINITIONS WORDSEARCH

```
    E N I M A T S I H
    G   C O M P L E M E N T
E A N T I B O D Y D     A
N H       R   N E     U
I P       A   O M     T
K O       D   R A     O
O R       Y   E       I
H C   H K     F       M
P A   P I     R       M
M M M     N   E       U
Y Y       I   T       N
L         N   N       E
                I
```

**a.** edema **b.** complement **c.** autoimmune **d.** bradykinin **e.** lymphokine **f.** macrophage **g.** interferon **h.** antibody **i.** histamine **j.** lymph

## CHAPTER TEST

**1.** b **2.** c **3.** b **4.** c **5.** d **6.** a **7.** c **8.** b **9.** d **10.** d **11.** d **12.** c **13.** a **14.** b **15.** a **16.** d **17.** b **18.** b **19.** b **20.** c **21.** b **22.** c **23.** d **24.** c **25.** c **26.** Ordinarily, antibodies and T cells attack only foreign antigens. If and when they attack the body's own cells, illness results. **27.** The lymphatic system drains excess tissue fluid from the skin and protects against infections. The skin serves as a barrier to entry of pathogens.

# 15

# RESPIRATORY SYSTEM

The respiratory tract consists of the nose (**nasal cavities**), the nasopharynx, the **pharynx,** the **larynx** (which contains the **vocal cords**), the **trachea,** the bronchi, and the **bronchioles.** The **bronchi,** along with the pulmonary arteries and veins, enter the lungs, which consist of the **alveoli,** air sacs surrounded by a capillary network.

**Inspiration** begins when the **respiratory center** in the medulla oblongata sends excitatory nerve impulses to the **diaphragm** and the muscles of the **rib cage.** As they contract, the diaphragm lowers and the rib cage moves upward and outward; the lungs expand, creating a partial vacuum, which causes air to rush in. The respiratory center now stops sending impulses to the diaphragm and muscles of the rib cage. As the diaphragm relaxes, it resumes its dome shape, and as the rib cage retracts, air is pushed out of the lungs during **expiration.**

**External respiration** occurs when $CO_2$ leaves blood and $O_2$ enters blood at the alveoli. Oxygen is transported to the tissues in combination with **hemoglobin** as **oxy-hemoglobin** ($HbO_2$). **Internal respiration** occurs when $O_2$ leaves blood and $CO_2$ enters blood at the tissues. Carbon dioxide is mainly carried to the lungs within the plasma as the bicarbonate ion ($HCO_3^-$). Hemoglobin combines with hydrogen ions and becomes **reduced (HHb).**

A number of illnesses are associated with the respiratory tract. In addition to colds and flu, the lungs may be infected by the more serious **pneumonia** and tuberculosis. Two illnesses that have been attributed to breathing polluted air are **emphysema** and **lung cancer.**

Study the text section by section. Answer the study questions so that you can fulfill the learning objectives for each section.

## 15.1 THE RESPIRATORY SYSTEM (PP. 282—85)

The learning objective for this section is:
• Describe the pathway air takes in and out of the lungs and the structures involved that are designed to filter, warm, and moisten air.

1. Complete this table. Refer to Table 15.1 in the textbook as needed.

| Structure | Function |
| --- | --- |
| a. _____ | Filter, warm, and moisten air |
| Glottis | b. _____ |
| c. _____ | Sound production |
| Trachea | d. _____ |
| e. _____ | Passage of air to each lung |
| Bronchioles | f. _____ |
| g. _____ | Gas exchange |

2. Label this diagram, using the
   following alphabetized list of terms.
   epiglottis
   glottis
   hard palate
   larynx
   nasal cavity
   soft palate
   trachea

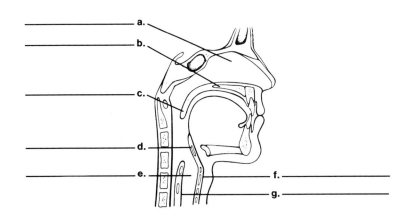

In question 3, fill in the blanks.

3. The nasal cavities contain a._____, and each, as well as the trachea, is lined with

   b._____ to screen the incoming air. Mucus, dust, and other material are moved into the

   c._____ for swallowing or expectoration. During swallowing, the d._____ folds

   down over the glottis to keep food from entering the trachea. The lungs of premature infants often lack a film

   called e._____ that keeps their lung tissues from sticking together.

## 15.2 MECHANISM OF BREATHING (PP. 286–89)

The learning objectives for this section are:
- Relate the respiratory volumes to a diagram showing the amount of air that is moved in and out of the lungs when breathing occurs.
- Describe the mechanism by which breathing occurs, including inspiration and expiration.

4. Label this diagram, using the following alphabetized list of terms.
   expiratory reserve volume
   inspiratory reserve volume
   residual volume (used twice)
   tidal volume

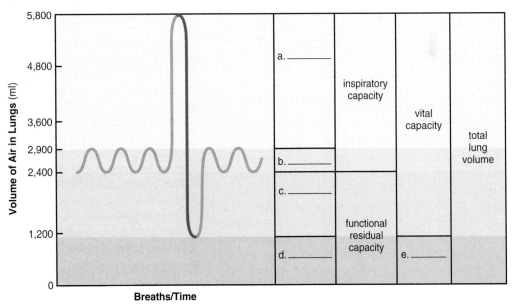

5. In the preceding diagram, the sum of the volumes labeled *a, b,* and *c* is termed the _____.
6. Place the appropriate letter next to each phrase.

   I—inspiration     E—expiration

   a. _____ lungs expanded
   b. _____ muscles (diaphragm and ribs) relaxed
   c. _____ diaphragm dome-shaped
   d. _____ chest enlarged
   e. _____ less air pressure in lungs than in the environment

7. What is the proper sequence for these statements? (Indicate by letters.) _____
   a. Respiratory center stops sending nerve impulse to diaphragm and rib cage.
   b. Respiratory center sends nerve impulse to diaphragm and rib cage.
   c. Diaphragm relaxes and becomes dome-shaped, and rib cage moves down and inward.
   d. Lungs expand as diaphragm lowers and rib cage moves upward and outward.
   e. Air goes rushing out as lungs recoil.
   f. Air comes rushing in as lungs expand.

## 15.3 GAS EXCHANGES IN THE BODY (PP. 290–92)

The learning objectives for this section are:
- Describe the events that occur during external and internal respiration.
- Show that hemoglobin is well suited to its role as a respiratory pigment.

8. Match the statements to these terms:

   internal respiration     cellular respiration     inspiration and expiration     external respiration

   a. _____ entrance and exit of air into and out of lungs
   b. _____ exchange of gases between blood and tissue fluid
   c. _____ production of ATP in cells
   d. _____ exchange of gases between lungs and blood
   e. Next, place the terms in the proper sequence.

   First _____

   Second _____

   Third _____

   Last _____

9. Give the equation that describes how oxygen is transported in the blood. Label one arrow *lungs* and the reverse arrow *tissues*.

10. a. Give the equation that describes how most of the carbon dioxide is transported in the blood. Label one arrow *lungs* and the reverse arrow *tissues*.

In questions 10*b–d,* fill in the blanks.

   b. What is the name of the enzyme that speeds up this reaction? _____

   c. Carbon dioxide transport produces hydrogen ions. Why does the blood not become acidic? _____

   _____

   d. By what process does carbon dioxide move from the blood to the alveoli? _____

11. After studying Figure 15.8 in the text, fill in the blanks.

   a. Where does oxygen enter the blood? _____

   b. Where does oxygen exit the blood? _____

   c. Where does carbon dioxide enter the blood? _____

   d. Where does carbon dioxide exit the blood? _____

   e. In Figure 15.8, what two types of vessels are high in oxygen?

   _____

   f. In the figure, what two types of vessels are high in carbon dioxide? _____

   _____

12. a. Hemoglobin is remarkably suited to the transport of oxygen. Why? _____

   _____

   b. Why does a person die from carbon monoxide poisoning? _____

   _____

   c. How does hemoglobin help with the transport of carbon dioxide? _____

   _____

## 15.4 RESPIRATION AND HEALTH (PP. 293–96)

The learning objectives for this section are:
* List the names, symptoms, and causes of various diseases of the respiratory tract.
* Explain why women now have an incidence of lung cancer equal to that of men.

13. Match the descriptions to these terms:

   tonsillitis   pneumonia   tuberculosis   emphysema   pulmonary fibrosis   lung cancer

   a. _____   Cells build a protective capsule around the bacteria. X rays can detect the presence of these capsules.

   b. _____   Fibrous connective tissue builds up in the lungs of a person who has inhaled particles.

   c. _____   A first line of defense against an invasion of the body.

   d. _____   This most often begins in a bronchus and is caused by smoking cigarettes.

   e. _____   Lungs balloon because air is trapped in the alveoli.

   f. _____   Lobules of the lungs fill with fluid, caused by a pathogen.

14. Why do women now suffer from lung cancer rates equivalent to those of men when in the past they did not?

   _____

Review key terms by using the following alphabetized list of terms to fill in the blanks below. Then complete the wordsearch.

```
A G L O T T I S A G A I
E X I A L V E O L U S N
P P T O R S T O P T O D
Y G I T Q Y B E G U N I
T L D G R E N L I V I A
I D A S L C I X I S N P
C A L V W O J H R O O H
A D V I I L T C I L I R
P G O D L L A T O X T A
A A L A V A A K I E A G
C V V G R L E I V S R M
L D M O I S V E M Y I J
A G E T A Y E R L B P V
T B N J J N C M P O X M
I E I M E N H A I R E I
V O C A L C O R D O O N
```

*alveolus*
*diaphragm*
*epiglottis*
*expiration*
*glottis*
*larynx*
*tidal volume*
*ventilation*
*vital capacity*
*vocal cords*

a. _____ Layer of muscle separating thoracic and abdominal cavities.

b. _____ Contains the vocal cords.

c. _____ Opening for airflow into the larynx.

d. _____ Act of expelling air.

e. _____ Amount of air involved in normal inhale/exhale cycle.

f. _____ The process of breathing.

g. _____ Fold of tissue in larynx that creates sounds.

h. _____ Structure that covers glottis during swallowing.

i. _____ Air sac in the lung.

j. _____ Maximum amount of air moved in or out of lungs during breathing.

# CHAPTER TEST

## OBJECTIVE QUESTIONS

Do not refer to the text when taking this test.

____ 1. Why is oxygen needed by the body?
   a. to aerate the lungs
   b. to cleanse the blood
   c. to produce ATP
   d. Both *a* and *b* are correct.

____ 2. The structure(s) that receive(s) air after the trachea is(are) the
   a. pharynx.
   b. bronchi.
   c. bronchiolus.
   d. alveoli.

____ 3. Which structure carries both air and food?
   a. larynx
   b. pharynx
   c. trachea
   d. esophagus

____ 4. How are foreign particles trapped to prevent them from entering the lungs?
   a. coarse hair just inside nasal cavities
   b. mucus in the nasal cavity
   c. cilia in the trachea
   d. All of these are correct.

_____ 5. Which of these constricts during an asthma attack?
   a. trachea
   b. bronchus
   c. bronchiole
   d. pharynx

_____ 6. The alveoli
   a. are sacs in the lungs.
   b. contain capillaries.
   c. are where gas exchange occurs.
   d. All of these are correct.

_____ 7. Which of these contains the vocal cords?
   a. glottis
   b. epiglottis
   c. pharynx
   d. larynx

_____ 8. Before oxygen is picked up in the lungs by hemoglobin, it first diffuses through (a) alveolar cells, (b) blood plasma, (c) red blood cell plasma membranes, and (d) capillary walls, though not necessarily in this order. What is the correct order?
   a. a, b, d, c
   b. a, d, b, c
   c. d, a, c, b
   d. d, b, a, c
   e. a, b, c, d

_____ 9. When the lungs recoil,
   a. inspiration occurs.
   b. external respiration occurs.
   c. internal respiration occurs.
   d. expiration occurs.
   e. All of these are correct.

_____ 10. The respiratory center
   a. is stimulated by carbon dioxide.
   b. is located in the chest.
   c. sends nerve impulses to lung tissue.
   d. is stimulated by oxygen levels.

_____ 11. The amount of air that enters or leaves the lungs during a normal respiratory cycle is the
   a. tidal volume.
   b. respiratory volume.
   c. residual volume.
   d. vital capacity.

_____ 12. The maximum amount of air a person can exhale after taking the deepest breath possible is a measure of the
   a. residual volume.
   b. tidal volume.
   c. vital capacity.
   d. inspiratory reserve volume.

_____ 13. External respiration is defined as
   a. an exchange of gases in the lungs.
   b. breathing.
   c. an exchange of gases in the tissues.
   d. cellular respiration.

_____ 14. Which gas is carried partially by the plasma?
   a. $O_2$
   b. $CO_2$
   c. both $O_2$ and $CO_2$
   d. neither $O_2$ nor $CO_2$

_____ 15. $CO_2$ enters the blood as a result of
   a. active transport.
   b. diffusion.
   c. blood pressure.
   d. air pressure.

_____ 16. The enzyme carbonic anhydrase causes
   a. carbon dioxide to react with water.
   b. carbon dioxide to react with bicarbonate ions.
   c. water to react with hydrogen ions.
   d. Both _b_ and _c_ are correct.

_____ 17. Hemoglobin combines with
   a. oxygen more readily in the lungs.
   b. carbon dioxide more readily in the tissues.
   c. oxygen more readily in the tissues.
   d. carbon dioxide more readily in the lungs.
   e. Both _a_ and _b_ are correct.

_____ 18. Hemoglobin carries
   a. $O_2$.
   b. $CO_2$.
   c. hydrogen ions.
   d. All of these are correct.

_____ 19. Which lung disorder is NOT caused by a pathogen?
   a. pneumonia
   b. tuberculosis
   c. emphysema
   d. laryngitis

_____ 20. Smoking cigarettes
   a. causes tuberculosis.
   b. leads to emphysema and cancer.
   c. increases the vital capacity of the lungs.
   d. leads to good health and longer life.

In questions 21–23, match the statements to these items:
   a. pneumonia   b. lung cancer   c. infant respiratory distress

_____ 21. Nonfunctional tissues interfere with gas exchange
_____ 22. Fluid-filled aveoli
_____ 23. Alveolar collapse due to high surface tension

_____ 24. Which body system does the respiratory system aid by providing oxygen so neurons can function properly?
   a. nervous system
   b. lymphatic system
   c. cardiovascular system
   d. integumentary system

_____ 25. Which body system helps the respiratory system by protecting the lungs and providing points for breathing muscle attachment?
   a. cardiovascular system
   b. skeletal system
   c. muscular system
   d. urinary system

Answer in complete sentences.

26. Explain how expiration occurs once the lungs have filled with air.

27. Relate the large surface area provided by the alveoli to the process by which external respiration occurs.

**Test Results:** _____ number correct ÷ 27 = _____ × 100 = _____%

## ANSWER KEY

### STUDY QUESTIONS

**1. a.** nasal cavities **b.** passage of air into larynx **c.** larynx **d.** passage of air to bronchi **e.** bronchi **f.** passage of air to lungs **g.** lungs **2. a.** nasal cavity **b.** hard palate **c.** soft palate **d.** epiglottis **e.** glottis **f.** larynx **g.** trachea **3. a.** coarse hairs **b.** cilia **c.** pharynx **d.** epiglottis **e.** surfactant **4. a.** inspiratory reserve volume **b.** tidal volume **c.** expiratory reserve volume **d.** residual volume **e.** residual volume **5.** vital capacity **6. a.** I **b.** E **c.** E **d.** I **e.** I **7.** b, d, f, a, c, e **8. a.** inspiration and expiration **b.** internal respiration **c.** cellular respiration **d.** external respiration **e.** inspiration and expiration, external, internal, cellular

**9.** $Hb + O_2 \underset{tissues}{\overset{lungs}{\rightleftharpoons}} HbO_2$

**10. a.** $CO_2 + H_2O \underset{lungs}{\overset{tissues}{\rightleftharpoons}} H_2CO_3 \underset{lungs}{\overset{tissues}{\rightleftharpoons}} H^+ + HCO_3^-$
**b.** carbonic anhydrase **c.** Hemoglobin combines with excess hydrogen ions. **d.** diffusion **11. a.** lungs **b.** tissues **c.** tissues **d.** lungs **e.** pulmonary vein and aorta (systemic arteries) **f.** venae cavae (systemic veins) and pulmonary artery **12. a.** It easily combines with oxygen in the lungs and easily gives it up in the tissues. **b.** Hemoglobin combines with carbon monoxide preferentially to oxygen. **c.** It combines with carbon dioxide to a degree, called carbaminohemoglobin, and picks up hydrogen ions from the equation of 10a. **13. a.** tuberculosis **b.** pulmonary fibrosis **c.** tonsillitis **d.** lung cancer **e.** emphysema **f.** pneumonia **14.** Women now smoke cigarettes as frequently as men. In the past, it was less acceptable for women to smoke.

### DEFINITIONS WORDSEARCH

```
    G L O T T I S
E     A L V E O L U S
  P T   R               D
Y   I     Y             I
T   D G     N           A
I   A   L     X   N     P
C   L     O       O O H
A   V       T   I     I R
P   O         T     T A
A   L         A   I   A G
C   V     L       S R M
L   M   I             I
A   E T               P
T   N                 X
I E                   E
V O C A L C O R D
```

**a.** diaphragm **b.** larynx **c.** glottis **d.** expiration **e.** tidal volume **f.** ventilation **g.** vocal cords **h.** epiglottis **i.** alveolus **j.** vital capacity

### CHAPTER TEST

**1.** c **2.** b **3.** b **4.** d **5.** c **6.** d **7.** d **8.** b **9.** d **10.** a **11.** a **12.** c **13.** a **14.** b **15.** b **16.** b **17.** e **18.** d **19.** c **20.** b **21.** b **22.** a **23.** c **24.** a **25.** b **26.** The diaphragm relaxes when the respiratory center stops sending messages to contract. Once relaxation occurs, expiration is passive. Air leaves the lungs with the elastic recoil of the lungs. Muscle contraction can force additional air from the lungs. **27.** Since oxygen enters the capillaries of the alveoli by the process of diffusion, a passive process, a large surface area is required.

# 16

# URINARY SYSTEM AND EXCRETION

The **kidneys** excrete ammonia, **urea, uric acid,** and **creatinine,** all nitrogenous wastes. Urine is composed primarily of nitrogenous waste products and salts in water.

The path of urine is through the kidneys, **ureters, urinary bladder,** and finally, the **urethra.**

Macroscopically, the kidneys are divided into the cortex, medulla, and pelvis. Microscopically, they contain the **nephrons.**

Each nephron has its own blood supply; the afferent arteriole approaches the **glomerular capsule** and divides to become the **glomerulus.** The spaces between the **podocytes** of the glomerular capsule allow small molecules to enter the capsule from the glomerulus, a capillary tuft. The efferent arteriole leaves the capsule and immediately branches into the **peritubular capillaries.**

Each region of the nephron is anatomically suited to its task in urine formation. The spaces between the podocytes of the glomerular capsule allow small molecules to enter the capsule from the glomerulus, a capillary knot. The cuboidal epithelial cells of the **proximal convoluted tubule** have many mitochondria and microvilli to carry out active transport (following passive transport) from the tubule to blood. In contrast, the cuboidal epithelial cells of the **distal convoluted tubule** have numerous mitochondria but lack microvilli. They carry out active transport from the blood to the tubule.

The steps in urine formation are **glomerular filtration, tubular reabsorption,** and **tubular secretion,** as explained in Figure 16.6 in your textbook. Water is reabsorbed from all parts of the tubule, and the **loop of the nephron** establishes an osmotic gradient that draws water from the descending loop of the nephron and also the **collecting duct.**

The kidneys contribute to homeostasis not only by ridding the body of nitrogenous wastes but also by helping control the pH and salt/water balance of the blood. The latter also determines blood volume, which is controlled by several hormones, including **ADH, aldosterone,** and **ANH.**

Various types of problems, including repeated urinary infections, can lead to kidney failure, which necessitates receiving a kidney from a donor or undergoing dialysis by utilizing a kidney machine or CAPD.

Study the text section by section. Answer the study questions so that you can fulfill the learning objectives for each section.

## 16.1 URINARY SYSTEM (PP. 302–4)

The learning objectives for this section are:
- List the organs involved in the urinary system.
- State, in general, the contents of urine.
- Trace the path of urine, and describe the general structure and function of each organ mentioned.
- Describe how urination is controlled by the nervous system.
- Explain how the kidneys function to maintain homeostasis of the body's internal environment.

1. Match the functions to these urinary organs:

   kidney    ureter    urinary bladder    urethra

   a. _____ muscular tube leading from kidneys to urinary bladder

   b. _____ tube leading from bladder to the outside

   c. _____ hollow, muscular organ that stores urine

   d. _____ bean-shaped organ that filters blood

2. What triggers urination? _____
_____
_____

3. List five functions of the kidneys.

a. _____

b. _____

c. _____

d. _____

e. _____

## 16.2 KIDNEY STRUCTURE (PP. 305–8)

The learning objectives for this section are:
- Describe the macroscopic structure of the kidney.
- State the parts of a kidney nephron, and relate these to the macroscopic anatomy.

4. Macroscopically, the kidney is composed of these three parts:

a. _____

b. _____

c. _____

5. Label this diagram of the parts of the nephron, using the following alphabetized list of terms.

| | | |
|---|---|---|
| afferent arteriole | efferent arteriole | loop of the nephron |
| collecting duct | glomerular capsule | peritubular capillaries |
| distal convoluted tubule | glomerulus | proximal convoluted tubule |

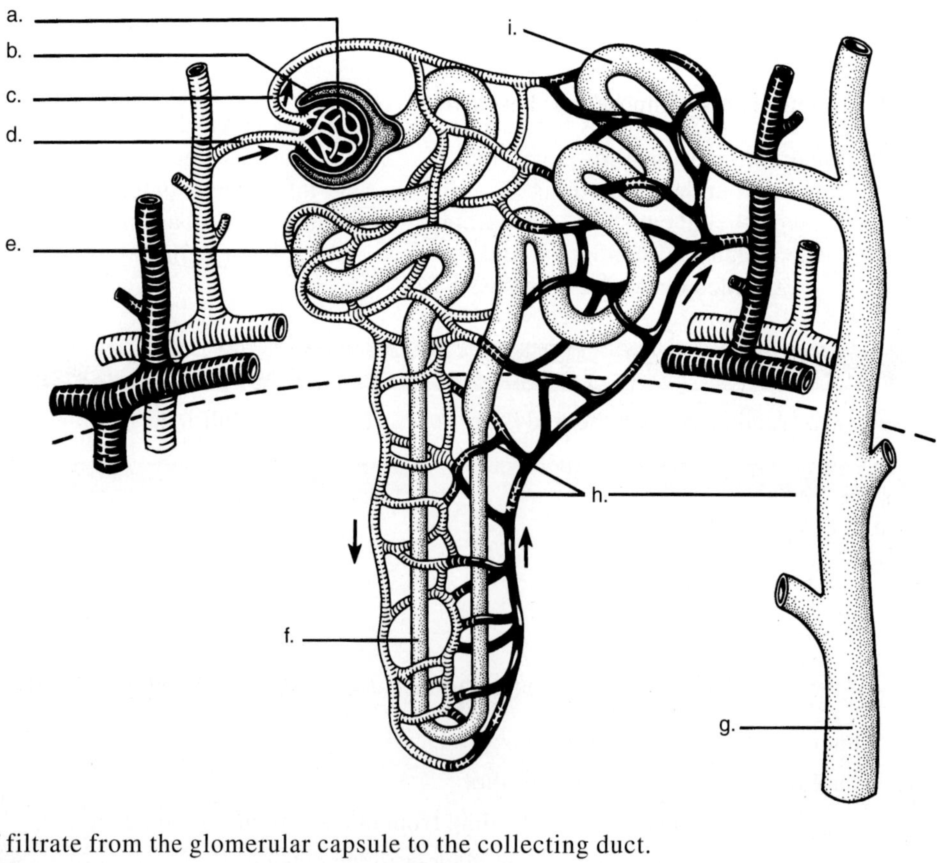

6. Trace the path of filtrate from the glomerular capsule to the collecting duct.
   Glomerular capsule

a. _____

b. _____

c. _____

   Collecting duct

The learning objective for this section is:
* Describe the three steps in urine formation, and relate these to parts of a nephron.

In questions 7–9, fill in the blanks.

7. In this diagram, add the three steps in urine formation.

b. _____   c. _____

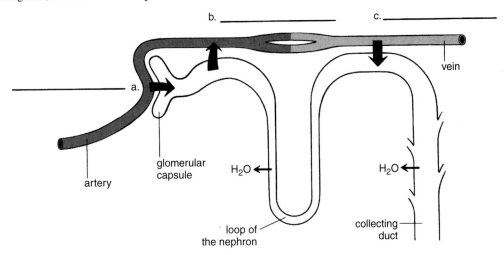

_____  a.

artery

glomerular capsule

$H_2O$ ←

loop of the nephron

$H_2O$ ←

collecting duct

vein

d. Where does the first step take place? _____

e. Where does the second step take place? _____

f. Where does the third step take place? _____

8. Blood in the glomerulus is composed of these two portions:

**Small Molecules**
Nutrients (glucose, amino acids)
Wastes (urea, uric acid)
Salts
Water

**Large Molecules, etc.**
Proteins
Formed elements

a. Which portion will undergo filtration?

_____

b. Which of the small molecules will maximally undergo reabsorption?

_____

c. Which of the small molecules will minimally undergo reabsorption?

_____

9. Give an example of a molecule that undergoes tubular secretion. _____

## 16.4 MAINTAINING WATER-SALT BALANCE (PP. 310–12)

The learning objectives for this section are:
* Describe how the loop of the nephron contributes to water reabsorption.
* Name three hormones involved in maintaining blood volume, and explain how they function.

In question 10, fill in the blanks.

10. The presence of which parts of the nephron accounts for maximal reabsorption of water in humans?

a. _____ What causes water to leave these parts of the nephron?

b. _____

Because water is maximally reabsorbed, humans excrete a c. _____ urine.

11. Complete this table.

| ADH | Urine Quantity |
|---|---|
| Increased amount | a. |
| Reduced amount | b. |

12. In question 12*a–f,* match the descriptions to these items:

blood pressure rises   adrenal cortex   converting enzyme   renin   aldosterone   atrial natriuretic hormone

a. _____ changes angiotensinogen to angiotensin I. b. _____

changes angiotensin I to angiotensin II. Angiotensin II acts on c. _____ to secrete

d. _____. Kidneys absorb $Na^+$ and e. _____. When

blood pressure rises, the heart secretes f. _____, which causes the kidneys to excrete $Na^+$.

## 16.5 MAINTAINING ACID-BASE BALANCE (P. 313)

The learning objective for this section is:
- Describe how the kidneys excrete hydrogen ions and reabsorb bicarbonate ions to regulate the pH of blood.

In question 13, fill in the blanks.

13. If the blood is acidic, a. _____ ions are excreted in combination with

b. _____, while c. _____ are reabsorbed. If the blood is basic,

fewer d. _____ ions are excreted, and fewer e. _____ are reabsorbed.

## Excretion Elimination

In the table, place an X beside the component of blood if the following descriptions pertain to it:
a. in the afferent arteriole
b. in the filtrate
c. in the efferent arteriole
d. reabsorbed into the peritubular capillary
e. secreted from the peritubular capillary
f. present in urine
g. absent from urine
h. in venous blood

| | a | b | c | d | e | f | g | h |
|---|---|---|---|---|---|---|---|---|
| 1. plasma proteins | | | | | | | | |
| 2. red blood cells | | | | | | | | |
| 3. white blood cells | | | | | | | | |
| 4. glucose | | | | | | | | |
| 5. amino acids | | | | | | | | |
| 6. sodium chloride | | | | | | | | |
| 7. water | | | | | | | | |
| 8. urea | | | | | | | | |
| 9. uric acid | | | | | | | | |
| 10. penicillin | | | | | | | | |

There are 45 correct answers. There are 80 possible errors of omission or commission. Any 10 errors and you're ELIMINATED!

Review key terms by completing this crossword puzzle, using the following alphabetized list of terms.

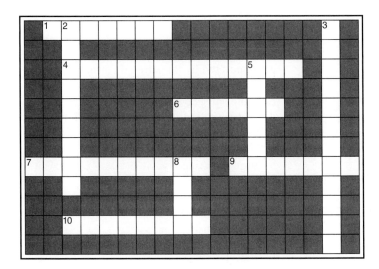

*antidiuretic*
*collecting duct*
*excretion*
*glomerulus*
*kidney*
*nephron*
*proximal*
*urea*
*ureter*
*urethra*

*Across*

1　Anatomical and functional unit of the kidney.
4　Tube that receives urine from the distal convoluted tubules of several nephrons.
6　Organ of the urinary system that produces and excretes urine.
7　The ball of capillaries that is surrounded by a glomerular capsule of a nephron.
9　Tube conveying urine from the bladder to outside the body.
10　Convoluted tubule nearest the glomerular capsule.

*Down*

2　Process of removing metabolic wastes from the body.
3　Type of hormone from the posterior pituitary that promotes the reabsorption of water from the collecting duct.
5　One of two tubes leading from the kidneys to the urinary bladder.
8　Main nitrogenous waste derived from the breakdown of amino acids.

# CHAPTER TEST

## OBJECTIVE QUESTIONS

Do not refer to the text when taking this test.

____ 1. Kidneys are organs of homeostasis because they
a. regulate the blood volume.
b. regulate the pH of the blood.
c. help maintain the correct concentration of ions in the blood.
d. excrete nitrogenous wastes.
e. All of these are correct.

____ 2. Which of these contains urine?
a. urethra
b. uterus
c. intestine
d. gallbladder
e. All of these are correct.

____ 3. Which portion of the urinary tract varies significantly in length or size between males and females?
a. kidneys
b. ureters
c. urinary bladder
d. urethra

____ 4. Urination is triggered by
a. contraction of the bladder and relaxation of sphincter muscles.
b. relaxation of the bladder and contraction of sphincter muscles.
c. contraction of kidney and ureter muscles.
d. contraction of kidneys and relaxation of the urinary bladder.

_____ 5. Which portion is NOT part of the kidney?
  a. renal cortex
  b. renal urethra
  c. renal medulla
  d. renal pelvis
_____ 6. The collecting ducts are primarily in the
  a. renal cortex.
  b. renal medulla.
  c. renal pelvis.
  d. afferent arteriole.

In questions 7–11, match the function with these structures:
  a. glomerulus    b. glomerular capsule
  c. renal cortex    d. loop of the nephron
  e. collecting duct
_____ 7. Often extends into the medulla
_____ 8. A knot of capillaries
_____ 9. Variably permeable to water
_____ 10. Site of afferent/efferent arterioles
_____ 11. Blind end of the proximal convoluted tubule

_____ 12. Urine collects in the _____
  before entering the ureter.
  a. renal medulla
  b. renal cortex
  c. renal pelvis
  d. capsule
_____ 13. Glomerular filtration should be associated with
  a. the glomerular capsule.
  b. the distal convoluted tubule.
  c. the collecting duct.
  d. All of these are correct.
_____ 14. Sodium is removed from the nephron by
  a. passive reabsorption.
  b. active reabsorption.
  c. an attraction to $Cl^-$.
  d. secretion.
_____ 15. Tubular secretion occurs at
  a. the glomerular capsule.
  b. the proximal convoluted tubule.
  c. the loop of the nephron.
  d. the distal convoluted tubule.
_____ 16. In humans, water is
  a. found in the glomerular filtrate.
  b. reabsorbed from the nephron.
  c. in the urine.
  d. All of these are correct.
_____ 17. Glucose
  a. is in the filtrate and urine.
  b. is in the filtrate and not in urine.
  c. undergoes tubular secretion and is in urine.
  d. undergoes tubular secretion and is not in urine.

_____ 18. The loop of the nephron is characteristic of animals that excrete
  a. a diluted urine.
  b. a concentrated urine.
  c. no urine.
  d. too much urine.
_____ 19. Aldosterone
  a. is secreted by the adrenal cortex.
  b. causes the blood volume to lower.
  c. is the same as renin.
  d. causes the kidneys to excrete sodium.
_____ 20. Which organ excretes bile pigments?
  a. kidney
  b. large intestine
  c. liver
  d. All of these are correct.
_____ 21. ADH is necessary for
  a. water reabsorption.
  b. glucose reabsorption.
  c. protein reabsorption.
  d. All of these are correct.
_____ 22. If the nephrons do not function,
  a. urea accumulates in the blood.
  b. edema occurs.
  c. water and salt retention occur.
  d. All of these are correct.
_____ 23. Renin is an enzyme that converts
  a. angiotensinogen to angiotensin I.
  b. angiotensin I to angiotensin II.
  c. angiotensin II to converting enzyme.
  d. All of these are correct.
_____ 24. Which body system removes excess tissue fluid, thus maintaining blood pressure and ensuring proper kidney function?
  a. respiratory system
  b. endocrine system
  c. lymphatic system
  d. reproductive system
_____ 25. The kidneys provide active vitamin D for calcium absorption that directly benefits which body system?
  a. skeletal system
  b. cardiovascular system
  c. digestive system
  d. integumentary system

Answer in complete sentences.

26. Explain the difference between defecation and excretion.

27. What role is played by the high concentration of salt and urea in the renal medulla?

**Test Results:** _____ number correct ÷ 27 = _____ × 100 = _____ %

## ANSWER KEY

### STUDY QUESTIONS

**1. a.** ureter **b.** urethra **c.** urinary bladder **d.** kidney
**2.** Stretch receptors send impulses to the spinal cord, which sends nerve impulses back to muscles controlling the urinary bladder. Contraction of the bladder and relaxation of sphincter muscles occur, expelling urine to the outside.
**3. a.** regulation of salt and water balance **b.** regulation of blood volume and blood pressure **c.** stimulation of red blood cell production **d.** regulation of the pH of the blood **e.** removal of metabolic wastes, including nitrogenous wastes **4. a.** renal cortex **b.** renal medulla **c.** renal pelvis **5. a.** glomerulus **b.** glomerular capsule **c.** efferent arteriole **d.** afferent arteriole **e.** proximal convoluted tubule **f.** loop of the nephron **g.** collecting duct **h.** peritubular capillaries **i.** distal convoluted tubule **6. a.** proximal convoluted tubule **b.** loop of the nephron **c.** distal convoluted tubule **7. a.** glomerular filtration **b.** tubular reabsorption **c.** tubular secretion **d.** between the glomerulus and the glomerular capsule **e.** between the proximal convoluted tubule and blood **f.** between blood and the distal convoluted tubule **8. a.** small molecules **b.** nutrients, salts, water **c.** wastes **9.** uric acid, hydrogen ions, ammonia, penicillin, creatinine **10. a.** loop of the nephron and the collecting duct **b.** hypertonic medulla due to presence of salt (from ascending limb of loop of the nephron) and urea (from collecting duct) **c.** hypertonic **11. a.** little urine **b.** much urine **12. a.** renin **b.** converting enzyme **c.** adrenal cortex **d.** aldosterone **e.** blood pressure rises **f.** atrial natriuretic hormone **13. a.** hydrogen **b.** ammonia **c.** bicarbonate ions **d.** hydrogen **e.** bicarbonate ions

### GAME: EXCRETION ELIMINATION

|  | a. | b. | c. | d. | e. | f. | g. | h. |
|---|---|---|---|---|---|---|---|---|
| plasma proteins | X |  | X |  |  |  | X | X |
| red blood cells | X |  | X |  |  |  | X | X |
| white blood cells | X |  | X |  |  |  | X | X |
| glucose | X | X |  | X |  |  | X | X |
| amino acids | X | X |  | X |  |  | X | X |
| sodium chloride | X | X |  | X |  | X |  | X |
| water | X | X | X | X |  | X |  | X |
| urea | X | X |  |  |  | X |  | X |
| uric acid | X | X |  |  | X | X |  |  |
| penicillin | X |  | X |  | X | X |  |  |

### DEFINITIONS CROSSWORD

**Across**
1. nephron   4. collecting duct   6. kidney   7. glomerulus   9. urethra   10. proximal

**Down**
2. excretion   3. antidiuretic   5. ureter   8. urea

### CHAPTER TEST

**1.** e **2.** a **3.** d **4.** a **5.** b **6.** b **7.** d **8.** a **9.** e **10.** c **11.** b **12.** c **13.** a **14.** b **15.** d **16.** d **17.** b **18.** b **19.** a **20.** c **21.** a **22.** d **23.** a **24.** c **25.** a **26.** Defecation is the elimination of nondigested material from the gut, and excretion is the elimination of end products of metabolism by the kidneys. **27.** The high concentration of salt and urea in the renal medulla draws water out of the loop of the nephron and the collecting duct.

# 17

# NERVOUS SYSTEM

## CHAPTER REVIEW

The anatomical unit of the nervous system is the **neuron,** of which there are three types: sensory, motor, and interneuron. Each of these is made up of a **cell body,** an **axon,** and a **dendrite(s).**

When an axon is not conducting a **nerve impulse,** the **resting potential** indicates that the inside of the axon is negative compared to the outside. Because of the **sodium-potassium pump,** there is a concentration of $Na^+$ ions outside an axon and $K^+$ ions inside an axon.

When an axon is conducting a nerve impulse, an **action potential** (i.e., electrochemical change) travels along a neuron. Depolarization occurs (the inside becomes positive) due to the movement of $Na^+$ to the inside of the axon. Then repolarization occurs (the inside becomes negative again) due to the movement of $K^+$ to the outside of the axon.

Transmission of the nerve impulse from one neuron to another takes place across a **synapse.** In humans, synaptic vesicles release a chemical, known as a **neurotransmitter,** into the **synaptic cleft.** The binding of the neurotransmitter to receptors in the postsynaptic membrane can either increase the chance of a nerve impulse (stimulation) in the next neuron or decrease the chance of a nerve impulse (inhibition) in the next neuron, depending on the type of neurotransmitter and/or the type of receptor.

The **CNS** consists of the spinal cord and brain. The gray matter of the cord contains cell bodies; the white matter contains tracts that consist of the long axons of interneurons. These run from all parts of the cord, even up to the cerebrum.

The **brain** integrates all nervous system activity and commands all voluntary activities. The **cerebrum,** which is responsible for consciousness, can be mapped, and each lobe seems to have particular functions. In the **diencephalon,** the **hypothalamus,** in particular, controls homeostasis, and the **thalamus** specializes in sense reception. The **cerebellum** coordinates muscle contractions. In the **brain stem,** the **medulla oblongata** and **pons** have centers for visceral functions.

Neurological drugs, although quite varied, have been found to affect the **limbic system** by either promoting or preventing the action of neurotransmitters.

The **peripheral nervous system** contains the **somatic system** and the **autonomic system. Reflexes** are automatic, and some do not require the involvement of the brain. A simple reflex requires the use of neurons that make up a reflex arc. In the somatic system, a **sensory neuron** conducts the nerve impulse from a receptor to an **interneuron,** which in turn transmits the impulse to a **motor neuron,** which conducts it to an effector.

While the motor portion of the somatic division of the PNS controls skeletal muscle, the motor portion of the autonomic division controls smooth muscle of the internal organs and glands. The **sympathetic division,** which is often associated with those reactions that occur during times of stress, and the **parasympathetic division,** which is often associated with those activities that occur during times of relaxation, are both parts of the autonomic system.

Study the text section by section. Answer the study questions so that you can fulfill the learning objectives for each section.

## 17.1 NERVOUS TISSUE (PP. 318–23)

The learning objective for this section is:
• Describe the structure and function of the three major types of neurons.

1. Every neuron has the three parts listed here. What is the function of each?

   a. dendrite _____

   b. cell body _____

   c. axon _____

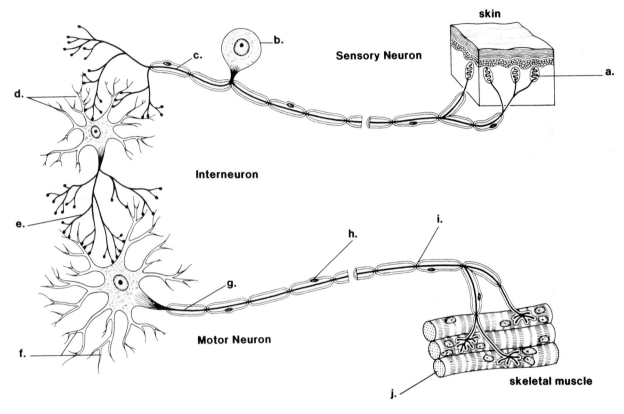

2. Label the parts of the sensory neuron, the interneuron, and the motor neuron, using the following alphabetized list of terms. *Note:* Some terms may be used more than once.

   axon    axon bulb    cell body    dendrite    effector    node of Ranvier
   nucleus of Schwann cell    sensory receptor

3. State the function of the complete sensory neuron. _____

   _____

   _____

4. State the function of the complete interneuron. _____

   _____

   _____

5. State the function of the complete motor neuron. _____

   _____

   _____

6. The following drawings represent axons. On the left, place (+) or (−) to indicate the polarity asked for. On the right, indicate the distribution of ions that produces this polarity.

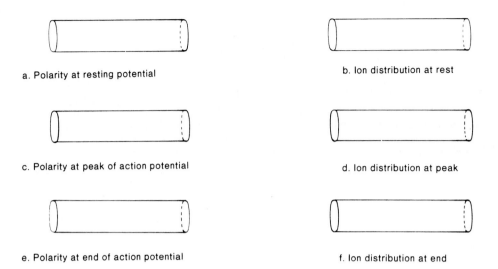

a. Polarity at resting potential

b. Ion distribution at rest

c. Polarity at peak of action potential

d. Ion distribution at peak

e. Polarity at end of action potential

f. Ion distribution at end

7. On this drawing of the trace that appears on the oscilloscope screen during the time of the action potential, label on one side $Na^+$ (sodium ions) *gates open* and on the other side $K^+$ (potassium ions) *gates open*. Write in the appropriate values to indicate the resting potential and the peak potential.

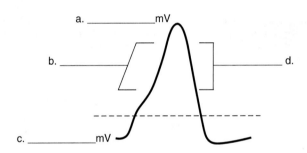

a. _____ mV

b. _____

d. _____

c. _____ mV

In question 8, fill in the blanks.

8. During the time of rest, the _____-_____ pump restores the original distribution of ions across the membrane of a nerve fiber.

9. Label the following diagrams, using these terms:

axon    dendrite    neurotransmitter    postsynaptic membrane    synapse
synaptic cleft    synaptic vesicle

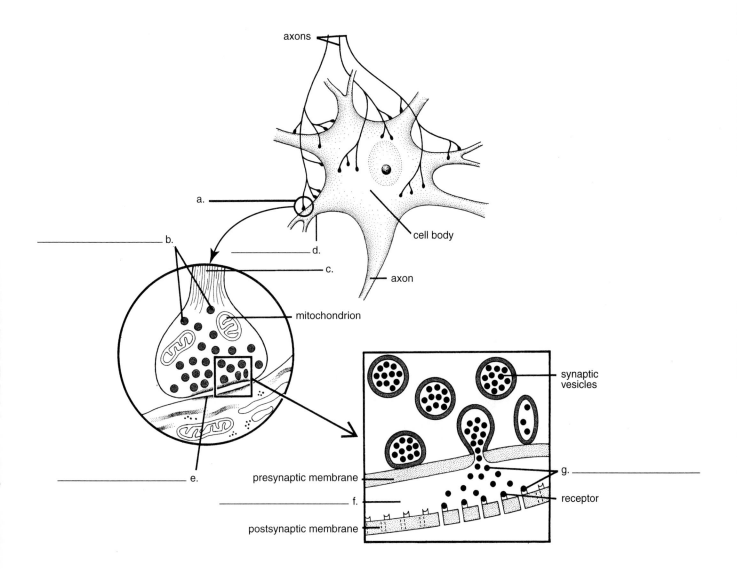

10. What causes the transmission of the nerve impulse across a synapse? _____
_____
_____

11. Indicate whether these statements are true (T) or false (F).
   a. _____ A single neuron synapses with one other neuron.
   b. _____ A single neuron synapses with many other neurons.
   c. _____ Excitatory signals have a hyperpolarizing effect, and inhibitory signals have a depolarizing effect.
   d. _____ Integration is the summing up of excitatory and inhibitory signals.
   e. _____ The more inhibitory signals received, the more likely the axon will conduct a nerve impulse.

The learning objectives for this section are:
- Label a cross section of the spinal cord, and give two functions of the spinal cord.
- Describe, in general, the anatomy of the brain, name five major parts, and give a function of each.
- Name the lobes of the cerebrum, and give a function of each.

12. Label the parts of the spinal cord, using the following alphabetized list of terms.

central canal
dorsal root
gray matter
ventral root
white matter

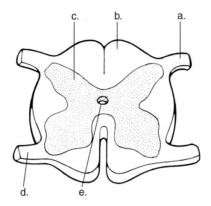

13. Indicate whether the following statements are true (T) or false (F). Rewrite all false statements as true statements.

a. _____ White matter is white because it contains cell bodies of interneurons that run together in bundles called tracts. Rewrite: _____

b. _____ The spinal cord carries out the integration of incoming information before sending signals to other parts of the nervous system. Rewrite: _____

c. _____ When the spinal cord is severed, we suffer a loss of sensation but not a loss of voluntary control. Rewrite: _____

_____

14. Label the parts of the brain, using the following alphabetized list of terms.

cerebellum    cerebrum    corpus callosum    medulla oblongata    pituitary gland    pons    thalamus

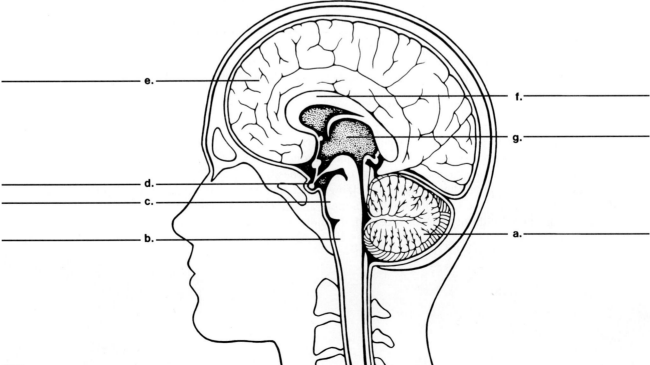

15. Fill in the following table to indicate the functions of the parts of the brain.

| Brain Part | Function |
| --- | --- |
| Cerebrum | a. |
| Thalamus | b. |
| Hypothalamus | c. |
| Cerebellum | d. |
| Medulla oblongata | e. |

16. Match each description to the correct name of the lobe, using these terms:

    all lobes    frontal lobe    occipital lobe    parietal lobe    temporal lobe

    *Note:* Some terms may be used more than once.

    a. _____ Contains primary motor area, which controls voluntary motions.
    b. _____ Contains primary somatosensory area, which receives sensory information from the skin and
        skeletal muscles.
    c. _____ Contains a primary visual area.
    d. _____ Contains a primary auditory area.
    e. _____ Contains a primary association area.
    f. _____ Carries on higher mental functions such as reasoning and critical thinking.

## 17.3  THE LIMBIC SYSTEM AND HIGHER MENTAL FUNCTIONS (PP. 329–32)

The learning objectives for this section are:
*   Describe in general the structure and the function of the limbic system.
*   List and describe the different types of memory.
*   Tell how the limbic system assists memory storage and retrieval.
*   Describe our present understanding of the human ability to use language and speak.

17. Place a check beside those structures that are a part of the limbic system.
    a. _____ tracts that join portions of the cerebral lobes, subcortical nuclei, and the diencephalon
    b. _____ hippocampus, which functions in retrieving memories
    c. _____ amygdala, which adds emotional overtones to memories

18. Which of these best describes the limbic system? _____
    a. a system that involves reasoning
    b. a system that involves emotions
    c. a system that involves memories
    d. All of these are correct.

In questions 19*a–e*, fill in the blanks.

19. Study the following diagram and answer the questions.

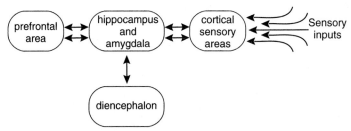

The hippocampus and amygdala are in contact with what areas of the brain? a. _____,

b. _____, c. _____

In which of these are memories stored for later retrieval? d. _____

In which of these areas are memories used to plan future actions? e. _____

For questions 19*f–h*, answer using a complete sentence.

What is the role of the hippocampus? f. _____

_____

What is the role of the amygdala? g. _____

_____

Why are there two sets of double-headed arrows in the diagram? h. _____

_____

20. Indicate whether these statements are true (T) or false (F).
    _____ a. Broca's area is a motor speech area.
    _____ b. Damage to Wernicke's area results in the inability to comprehend speech.
    _____ c. Only the left side of the brain contains a Broca's area and a Wernicke's area.

## 17.4 THE PERIPHERAL NERVOUS SYSTEM (PP. 334–37)

The learning objectives for this section are:
• Define a nerve, and distinguish between spinal and cranial nerves.
• Describe the path of a reflex arc in the somatic system.
• Contrast the somatic system with the autonomic system.
• Describe the autonomic system, and cite similarities as well as differences in the structure and function of the two divisions.

In question 21, fill in the blanks.

21. Name two types of nerves in the peripheral nervous system: a. _____ and b. _____ nerves. What

is a nerve? c. _____. Why is a spinal nerve called

a mixed nerve? d. _____

22. Label this diagram of the reflex arc, using the following alphabetized list of terms.
    dorsal root ganglion
    effector
    interneuron
    motor neuron
    sensory neuron
    sensory receptor

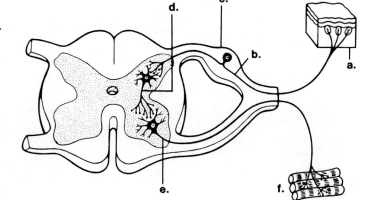

In question 23, fill in the blanks.

23. A stimulus is received by a. _____ in the skin and they generate nerve impulses in the fibers of b. _____. Next the impulses are picked up by c. _____ in the spinal cord. Then they are received by the fibers of d. _____ which stimulate a muscle to contract causing a reaction to the stimulus.

24. a. Explain how the brain becomes aware of automatic reflex actions. _____
    _____
    _____

    b. Explain why the left side of the brain controls the right side of the body. _____
    _____
    _____

25. Indicate three ways in which the sympathetic and parasympathetic systems are similar.
    a. _____
    b. _____
    c. _____

26. Indicate ways in which the sympathetic and parasympathetic systems differ by filling in the following table.

|  | Sympathetic | Parasympathetic |
| --- | --- | --- |
| Type of situation | a. | b. |
| Neurotransmitter | c. | d. |
| Ganglia near cord, or ganglia near organ? | e. | f. |
| Spinal nerves only, or spinal nerves plus vagus? | g. | h. |

## 17.5 DRUG ABUSE (PP. 338–39)

The learning objective for this section is:
• Describe drug action in general, and discuss the effects of alcohol, marijuana, cocaine, and heroin.

In questions 27–29, fill in the blanks.

27. Drugs are believed to affect, in particular, what part of the brain? a. _____ There are both inhibitory and excitatory neurotransmitters in the brain. If a drug blocks the action of an inhibitory neurotransmitter, what psychological effect will it have? b. _____ If a drug blocks the action of an excitatory neurotransmitter, what psychological effect will it have? c. _____

28. The drug a. _____ in tobacco products causes neurons to release dopamine, which reinforces dependence on this drug. Occasional b. _____ users experience euphoria, alterations in vision and judgment, and distortions of space and time. Heavy use of c. _____ often leads to liver damage. d. _____ is a ready-to-smoke form of cocaine.

29. According to endorphin research, what causes heroin withdrawal symptoms? _____
    _____

For each correct answer, Simon says, "You may move one step forward." Total possible number of steps forward is 10 steps.

1. Which of these would NOT be used when studying nerve conduction?
   a. voltmeter
   b. oscilloscope
   c. electron microscope
   d. electrodes
   e. electric current

2. Which one is NOT directly needed for nerve conduction?
   a. dendrites
   b. axons
   c. plasma membrane
   d. nucleus
   e. cytoplasm of the axon
   f. ions

3. Which one does NOT move during nerve conduction?
   a. sodium
   b. potassium
   c. plus charges
   d. minus charges

4. Which one does NOT accurately describe a resting neuron?
   a. positive on the outside of the membrane and negative on the inside
   b. $Na^+$ on the outside of the membrane and $K^+$ on the inside
   c. –65 mV inside
   d. negative on both sides of the membrane

5. Which one is NOT involved with an action potential?
   a. resting potential
   b. permeability
   c. sodium-potassium pump
   d. plasma membrane
   e. acetylcholine
   f. ions
   g. glycogen

6. Which one does NOT conduct a nerve impulse?
   a. sensory neurons
   b. osteocytes
   c. motor neurons
   d. sensory nerves
   e. motor nerves

7. Which pair is improperly matched?

8. Which number could NOT be associated with an action potential?
   a. –65 mV
   b. 0 mV
   c. +40 mV
   d. –40 watts

9. Which pair is improperly matched?
   a. ($e^-$) (nerve impulse)
   b. (sodium-potassium pump) (resting potential)
   c. (+) ($Na^+$)
   d. (–) ($K^+$)
   e. (plasma membrane) (semipermeable)

10. Which one is NOT true?

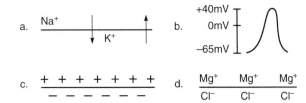

a. Na⁺ / K⁺

b. +40mV / 0mV / −65mV

c. + + + + + + / − − − − −

d. Mg⁺ Mg⁺ Mg⁺ / Cl⁻ Cl⁻ Cl⁻

How many steps were you allowed by Simon? _____

# DEFINITIONS WORDSEARCH

Review key terms by using the following alphabetized list of terms to fill in the blanks below. Then complete the wordsearch.

```
G A N G L I O N P O L I C
C D E T H A L A M U S E C
N G N L L I M B I C Y Y I
S E O K I M D F C H N H T
C E R E B E L L U M A T A
H T U X E L F E R L P N M
U D E N D R I T E O S K O
N K N G O D L O V E E I S
A C E T Y L C H O L I N E
```

*acetylcholine*
*cerebellum*
*dendrite*
*ganglion*
*limbic*
*neuron*
*reflex*
*somatic*
*synapse*
*thalamus*

a. _____ Portion of brain in third ventricle; integrates sensory input.
b. _____ Nerve cell.
c. _____ Posterior section of brain that coordinates graceful skeletal muscle movement.
d. _____ Fiber of a neuron that conducts signals toward the cell body.
e. _____ Automatic, involuntary response to a stimulus.
f. _____ Region between two neurons where impulses pass.
g. _____ System in brain concerned with emotions and memory.
h. _____ A neurotransmitter.
i. _____ Portion of peripheral nervous system leading to skeletal muscle.
j. _____ Collection of cell bodies within the peripheral nervous system.

# CHAPTER TEST

## OBJECTIVE QUESTIONS

Do not refer to the text when taking this test.

_____ 1. Sensory neurons
   a. take impulses to the CNS.
   b. take impulses away from the CNS.
   c. have a cell body in the dorsal root ganglion.
   d. Both *a* and *c* are correct.

_____ 2. Which of the following neurons would be found in the autonomic division of the peripheral nervous system?
   a. motor neurons ending in skeletal muscle
   b. motor neurons surrounding the esophagus
   c. sensory neurons at the surface of the skin
   d. sensory neurons attached to olfactory receptors
   e. interneurons in the spinal cord

_____ 3. The neuron that is found wholly and completely within the CNS is the
   a. motor neuron.
   b. sensory neuron.
   c. interneuron.
   d. All of these are correct.

_____ 4. Which of these contains the nucleus?
   a. axon
   b. dendrite
   c. cell body
   d. Any of these may contain the nucleus.

_____ 5. The downswing of the nervous impulse is caused by the movement of
    a. sodium ions to the inside of a neuron.
    b. sodium ions to the outside of a neuron.
    c. potassium ions to the inside of a neuron.
    d. potassium ions to the outside of a neuron.

_____ 6. The resting potential is maintained by the sodium-potassium pump.
    a. true
    b. false

_____ 7. Rapid conduction of a nerve impulse in vertebrates is due to
    a. the large diameters of the axons.
    b. openings in the myelin sheath.
    c. an abundance of synapses.
    d. the high permeability of neuronal membranes to ions.
    e. All of these are correct.

_____ 8. What is involved in a nerve impulse?
    a. ions
    b. electrons
    c. atoms
    d. molecules

_____ 9. Synaptic vesicles are
    a. at the ends of dendrites and axons.
    b. at the ends of axons only.
    c. along the lengths of long fibers.
    d. All of these are correct.

_____ 10. Acetylcholine
    a. is a neurotransmitter.
    b. crosses the synaptic cleft.
    c. is broken down by acetylcholinesterase.
    d. All of these are correct.

_____ 11. A spinal nerve is a
    a. motor nerve.
    b. sensory nerve.
    c. mixed nerve.
    d. All of these are correct.

_____ 12. Automatic responses to specific external stimuli require
    a. rapid impulse transmission along the spinal cord.
    b. the involvement of the brain.
    c. simplified pathways called reflex arcs.
    d. the involvement of the autonomic nervous system.

_____ 13. Which portion of the nervous system is required for a reflex arc?
    a. mixed spinal nerve
    b. gray matter of spinal cord
    c. cerebrum
    d. Both _a_ and _b_ are correct.
    e. _a, b,_ and _c_ are correct.

_____ 14. The autonomic system has two divisions called the
    a. CNS and peripheral nervous system.
    b. somatic and skeletal systems.
    c. efferent and afferent.
    d. sympathetic and parasympathetic.

_____ 15. Motor axons of the somatic system release
    a. acetylcholine.
    b. noradrenalin.
    c. dopamine.
    d. serotonin.

_____ 16. Which system is active during stress?
    a. parasympathetic
    b. sympathetic
    c. somatic
    d. All of these are correct.

_____ 17. The neurotransmitter of the parasympathetic system is
    a. noradrenalin.
    b. acetylcholine.
    c. cholinesterase.
    d. Both _a_ and _b_ are correct.

_____ 18. Which is the largest part of the human brain?
    a. cerebrum
    b. cerebellum
    c. medulla
    d. thalamus

_____ 19. The function of the cerebellum is
    a. consciousness.
    b. motor coordination.
    c. homeostasis.
    d. sense reception.

_____ 20. Which portion of the brain is involved in judgement?
    a. cerebellum
    b. frontal lobe of cerebrum
    c. medulla
    d. parietal lobe of cerebrum

_____ 21. The drug that is classified as a hallucinogen is
    a. marijuana.
    b. alcohol.
    c. caffeine.
    d. nicotine.

_____ 22. Drugs of abuse primarily affect the
    a. cerebellum.
    b. medulla oblongata.
    c. limbic system.
    d. thalamus.

_____ 23. A ready-to-smoke, highly addictive form of cocaine is
    a. heroin.
    b. crack.
    c. marijuana.
    d. alcohol.

Answer in complete sentences.

24. Contrast the way the nerve impulse travels along an axon with the way it travels across a synapse.

25. In either of what two ways would you expect an inhibitory psychoactive drug to affect transmission across a synapse?

**Test Results:** _____ number correct ÷ 25 = _____ × 100 = _____%

## ANSWER KEY

### STUDY QUESTIONS

**1. a.** sends signal to cell body **b.** control center **c.** takes impulse away from cell body **2. a.** sensory receptor **b.** cell body **c.** axon **d.** dendrites **e.** axon **f.** dendrite **g.** axon **h.** nucleus of Schwann cell **i.** node of Ranvier **j.** effector **3.** to take nerve impulses to CNS **4.** to take nerve impulses from one part of CNS to another **5.** to take nerve impulses away from CNS **6. a.** plus on outside; minus on inside **b.** Na$^+$ on outside and K$^+$ on inside **c.** minus on outside; plus on inside **d.** Some Na$^+$ have moved to inside **e.** plus on outside; minus on inside **f.** Some K$^+$ have moved to outside. **7. a.** +40 **b.** Na$^+$ (sodium ions) gates open **c.** –65 **d.** K$^+$ (potassium ions) gates open **e.** –65 **8.** sodium-potassium **9. a.** synapse **b.** synaptic vesicles **c.** axon **d.** dendrite **e.** postsynaptic membrane **f.** synaptic cleft **g.** neurotransmitter **10.** reception of neurotransmitter at receptor site **11. a.** F **b.** T **c.** F **d.** T **e.** F **12. a.** dorsal root **b.** white matter **c.** gray matter **d.** ventral root **e.** central canal **13. a.** F, . . .contains myelinated axons. . . **b.** T **c.** F, . . .and a loss of voluntary control. . . **14. a.** cerebellum **b.** medulla oblongata **c.** pons **d.** pituitary gland **e.** cerebrum **f.** corpus callosum **g.** thalamus **15. a.** motor control, higher levels of thought **b.** integrates and sends sensory information to cerebrum **c.** homeostasis **d.** motor coordination **e.** control of internal organs **16. a.** frontal lobe **b.** parietal lobe **c.** occipital lobe **d.** temporal lobe **e.** all lobes **f.** frontal lobe **17.** a, b, c **18.** d **19. a.** cortical sensory areas **b.** prefrontal area **c.** diencephalon **d.** cortical sensory areas **e.** prefrontal area **f.** serves as bridge between prefrontal area and cortical sensory areas **g.** adds emotional overtones to memories **h.** one set of arrows for semantic memory and one set for episodic memory **20. a.** T **b.** T **c.** T **21. a.** cranial **b.** spinal **c.** bundle of fibers (axons) **d.** it contains both sensory and motor fibers **22. a.** sensory receptor **b.** sensory neuron **c.** dorsal root ganglion **d.** interneuron **e.** motor neuron **f.** effector **23. a.** sensory receptors **b.** sensory neurons **c.** interneurons **d.** motor neurons **24. a.** Tracts in CNS take impulses up and down the cord. **b.** Tracts cross over. **25. a.** control internal organs **b.** have motor neurons **c.** have ganglia **26. a.** fight or flight **b.** normal activity **c.** norepinephrine (NE)

**d.** acetylcholine (ACh) **e.** near cord **f.** near organ **g.** spinal nerves only **h.** spinal nerves plus vagus **27. a.** limbic system **b.** increased likelihood of excitation **c.** decreased likelihood of excitation **28. a.** nicotine **b.** marijuana **c.** alcohol **d.** Crack **29.** Body's production of endorphins has decreased.

### GAME: SIMON SAYS ABOUT NERVOUS CONDUCTION

**1.** c **2.** d **3.** d **4.** d **5.** g **6.** b **7.** c **8.** d **9.** a **10.** d

### DEFINITIONS WORDSEARCH

```
G A N G L I O N
      T H A L A M U S   C
  N   L I M B I C Y     I
  O               N     T
C E R E B E L L U M A   A
  U X E L F E R     P   M
  D E N D R I T E   S   O
  N               E     S
A C E T Y L C H O L I N E
```

**a.** thalamus **b.** neuron **c.** cerebellum **d.** dendrite **e.** reflex **f.** synapse **g.** limbic **h.** acetylcholine **i.** somatic **j.** ganglion

### CHAPTER TEST

**1.** d **2.** b **3.** c **4.** c **5.** d **6.** a **7.** b **8.** a **9.** b **10.** d **11.** c **12.** c **13.** d **14.** d **15.** a **16.** b **17.** b **18.** a **19.** b **20.** b **21.** a **22.** c **23.** b **24.** There is an exchange of Na$^+$ and K$^+$ as the nerve impulse travels along an axon, but the release of a neurotransmitter causes the nerve impulse to travel across a synapse. **25.** An inhibitory psychoactive drug could either prevent the action of an excitatory neurotransmitter or promote the action of an inhibitory neurotransmitter at a synapse.

# 18

# SENSES

Each type of **sensory receptor** is sensitive to one of five kinds of stimuli. When stimulation occurs, receptors initiate nerve impulses that are transmitted to the spinal cord and/or brain. Only when nerve impulses reach the brain are we conscious of **sensation.**

**Proprioceptors** in the muscles, joints, and tendons, and in other internal organs maintain the body's equilibrium and posture. The skin contains **cutaneous** sensory receptors for touch, pressure, pain, and temperature (hot and cold).

Taste and smell are due to chemoreceptors that are stimulated by chemicals in the environment. The **taste buds** contain cells in contact with nerve fibers, while the receptors for smell are neurons.

When molecules bind to receptor proteins on the microvilli of taste cells and the cilia of **olfactory cells,** nerve impulses eventually reach the brain, which determines the taste and odor according to the pattern of receptors stimulated.

Vision is dependent on the eye, the **optic nerve,** and the visual cortex of the cerebrum. The eye has three layers. The outer layer, the **sclera,** can be seen as the white of the eye; it also becomes the transparent bulge in the front of the eye called the **cornea.** The **rods,** receptors for vision in dim light, and the **cones,** receptors that depend on bright light and provide color and detailed vision, are located in the **retina,** the inner layer of the eyeball. The cornea, the humors, and especially the **lens** bring the light rays to **focus** on the retina. To see a close object, accommodation occurs as the lens rounds up. Due to the optic chiasma, both sides of the brain must function together to give three-dimensional vision.

When light strikes **rhodopsin** (composed of opsin and retinal), **retinal** changes shape and rhodopsin is activated. Chemical reactions that produce electrochemical changes eventually result in nerve impulses that are carried in the **optic nerve** to the brain.

Hearing is a specialized sense dependent on the ear, the **cochlear nerve,** and the auditory cortex of the cerebrum. The ear is divided into three parts: outer, middle, and inner. The outer ear consists of the pinna and the auditory canal, which direct sound waves to the middle ear. The **middle ear** begins with the **tympanic membrane** and contains the **ossicles (malleus, incus,** and **stapes).** The malleus is attached to the tympanic membrane, and the stapes is attached to the **oval window,** which is covered by membrane. The **inner ear** contains the **cochlea** and the semicircular canals, plus the **utricle** and **saccule.** The outer and middle portions of the ear simply convey and magnify the sound waves that strike the oval window. Its vibrations set up pressure waves within the cochlea, which contains the spiral organ (organ of Corti), consisting of hair cells with the **tectorial membrane** above. When the cilia of the hair cells strike this membrane, electrochemical changes result in nerve impulses that are carried in the cochlear nerve to the brain.

The ear also contains sensory receptors for our sense of balance. **Rotational equilibrium** is dependent on the stimulation of hair cells within the ampullae of the semicircular canals. **Gravitational equilibrium** relies on the stimulation of hair cells by **otoliths** within the utricle and the saccule.

Study the text section by section. Answer the study questions so that you can fulfill the learning objectives for each section.

## 18.1 SENSORY RECEPTORS AND SENSATIONS (PP. 344—45)

The learning objectives for this section are:
- Describe the various receptors, and state the type of stimuli they receive.
- Explain how sensation occurs.

1. Match the descriptions to these terms:

   chemoreceptors  mechanoreceptors  proprioceptors  thermoreceptors  pain receptors  photoreceptors

   a. _____ located only in the eye
   b. _____ monitor the pH of the blood
   c. _____ detect tissue damage
   d. _____ detect stretch in tendons and ligaments
   e. _____ sensitive to changes in heat and cold
   f. _____ hearing
   g. _____ taste and smell

Indicate whether these statements are true (T) or false (F). Rewrite the false statements to make true statements.

2. _____ When stimulated, a sensory receptor generates a nerve impulse that travels in a sensory neuron to the CNS. Rewrite: _____

3. _____ Each type of sensory receptor is sensitive to only one stimulus. Rewrite: _____
   _____

4. _____ Sensation occurs in the brain and not at the sensory receptor. Rewrite: _____
   _____

5. _____ Sensory receptors generate nerve impulses. Rewrite: _____
   _____

6. _____ Sensory receptors are part of a reflex arc. Rewrite: _____
   _____

## 18.2 PROPRIOCEPTORS AND CUTANEOUS RECEPTORS (PP. 346–47)

The learning objectives for this section are:
- Name two types of proprioceptors, and describe their functions.
- Associate the names of cutaneous receptors with their functions.

In question 7, fill in the blanks.

7. Name two types of proprioceptors: a. _____ and b. _____

   When a muscle spindle relaxes, c. _____

   Muscle spindles have two functions, d. _____ and e. _____

8. Match the types of cutaneous receptors with the sense they detect.
   a. pressure   b. heat   c. cold   d. pain   e. touch

   _____ free nerve endings
   _____ Merkel disks
   _____ Krause end bulbs
   _____ Meissner corpuscles
   _____ Pacinian corpuscles
   _____ Ruffini endings

## 18.3 CHEMICAL SENSES (PP. 348–49)

The learning objectives for this section are:
- Describe the senses that rely on chemoreceptors.
- Explain how chemoreceptors operate.

9. Match the descriptions to these terms:

taste receptors    smell receptors    both taste and smell receptors

a. _____ receptor proteins combine with chemical

b. _____ brain senses impulses as a weighted average

c. _____ taste buds with microvilli house receptor proteins

d. _____ salty receptor proteins on tip of tongue

e. _____ olfactory cells

f. _____ are not effective when you have a cold

g. _____ easily adapt to outside stimuli

h. _____ involved in enjoyment of food

i. _____ a characteristic combination of receptor proteins are activated

10. The senses of taste and smell work because specific a._____ in the organs of taste and smell combine with b._____ in the air or food. Both senses thus employ c._____ to detect changes in the environment.

## 18.4 SENSE OF VISION (PP. 350–55)

The learning objectives for this section are:
- Describe the anatomy of the eye and the function of each part.
- Describe the receptors for sight, their mechanism of action, and the mechanism for stereoscopic vision.
- Identify common disorders of sight discussed in the text.

11. Label the parts of the eye using the following alphabetized list of terms. On the answer blanks provided, state the name and function of each part of the eye indicated in the illustration.

choroid
ciliary body
cornea
fovea centralis
iris
lens
optic nerve
retina
sclera

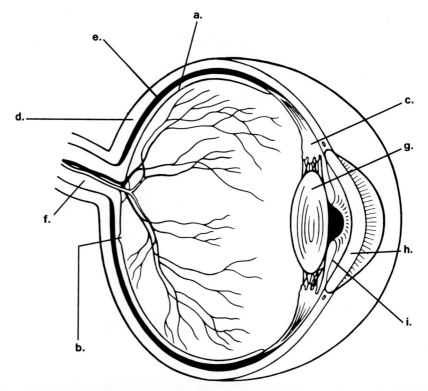

**Structure**                    **Function**

a. _____        _____

b. _____        _____

c. _____        _____

d. _____        _____

e. _____        _____

f. _____        _____

g. _____        _____

h. _____        _____

i. _____        _____

In questions 12–16, fill in the blanks.

12. The lens is a._____ for distant objects and b._____ for close objects.

    This is called c._____.

13. The receptors of sight are classified as a._____. Two kinds exist: b._____

    perceive motion and are responsible for night vision, and c._____ are responsible for color vision.

14. Rod cells have a pigment called a._____, which is made up of the protein b._____

    and a pigment molecule called c._____, a derivative of vitamin A. When light strikes the pigment

    molecule, the rhodopsin is activated. Color vision depends on d._____ kinds of cones, each of which

    has a slightly different structure of e._____ molecule. Each is able to detect a different wavelength,

    or color, of light. Rod cells are located throughout the retina, but cone cells are concentrated in the

    f._____.

15. Considering the layers of the retina—rod cells and cone cells/bipolar cells/ganglion cells—which of these are at the

    back of the retina (closest to the choroid)? a._____ Which of these are receptors for sight?

    b._____ Which of these are fewest in number? c._____

    This supports the belief that d._____ occurs in the retina.

16. With reference to the following figure, the region where the optic nerves cross is the a._____. Each

    primary visual area of the cerebral cortex receives information about b._____ (*the complete* or *one-

    half the*) visual field. Also, it is now known that the visual areas c._____ (*parcel out* or *retain as a

    unified whole*) information regarding color, form, motion, etc. This means that the cerebral cortex has to

    d._____ (*rebuild* or *imagine*) the visual field before we can "see" it.

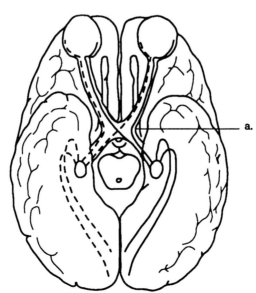

17. Fill in the blanks in this table.

| Name | Description | Image Focused | Correction |
|---|---|---|---|
| Nearsightedness | See nearby objects | a. | Concave lens |
| Farsightedness | b. | c. | d. |
| Astigmatism | Cannot focus | Image not focused | e. |

## 18.5 SENSE OF HEARING (PP. 358–60)

The learning objectives for this section are:
- Describe the anatomy of the ear and the function of each part.
- Discuss the receptors for balance and hearing and their mechanism of action.
- Identify the two types of deafness.

18. Label the parts of the ear, using the following alphabetized list of terms. On the answer blanks provided, state the name and function of each part of the ear indicated in the illustration.

auditory canal
auditory tube
cochlea
cochlear nerve
malleus (hammer)
pinna
semicircular canal
stapes (stirrup)
tympanic membrane
vestibule

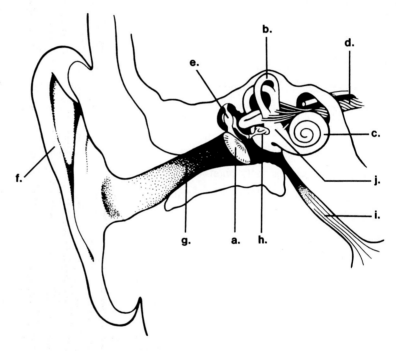

**Structure/Function**

a. _____

c. _____

e. _____

g. _____

i. _____

**Structure/Function**

b. _____

d. _____

f. _____

h. _____

j. _____

19. What is the proper sequence of the events that lead to the formation of an auditory nerve impulse? Indicate by letter.
    a. Vibration is transferred from the malleus to the incus to the stapes.
    b. Basilar membrane moves up and down.
    c. Nerve impulse is transmitted in cochlear nerve to brain.
    d. Sound waves pass through the auditory canal.
    e. Cilia of hair cells rub against tectorial membrane.
    f. Sound waves cause tympanic membrane to vibrate.
    g. Nerve impulse is generated.
    h. Vibrations move from vestibular canal to tympanic canal.
    i. Membrane at oval window vibrates.

## 18.6 SENSE OF EQUILIBRIUM (P. 361)

The learning objective for this section is:
• Explain rotational and gravitational equilibrium.

In question 20, fill in the blanks.

20. a. _____ equilibrium is called for when the entire body is moving. Fluid in the semicircular canals displaces the gelatinous material within the b. _____. For c. _____ equilibrium, movement occurs in one plane (vertical or horizontal), and the otoliths in the d. _____ and saccule are displaced in the gelatinous material, bending the cilia of the hair cells. Hair cells in the ampullae and utricle and saccule synapse with the e. _____ nerve.

## DEFINITIONS WORDSEARCH

Review key terms by using the following alphabetized list of terms to fill in the blanks below. Then complete the wordsearch.

```
A C C O M M O D A T I O N X
M R E H T O P S D N I L B V
P E T U O R E B Y U J D F E
U J I K S R R E D S I C L I
L H U T S C O C H L E A L O
L G Y C I S X I W A S N I U
A B H I C L O P D S I D E D
K O T O L I T H B C Z T Y U
M U J K E R H O D O P S I N
```

accommodation
ampulla
blind spot
choroid
cochlea
ossicle
otolith
rhodopsin

a. _____ Lens adjustment to see close objects.

b. _____ Vascular, pigmented layer of eyeball.

c. _____ Snail-shell portion of inner ear.

d. _____ Area of eye where optic nerve passes through retina.

e. _____ Base of semicircular canal in inner ear.

f. _____ One of the small bones of the middle ear.

g. _____ Calcium carbonate granule within inner ear.

h. _____ Visual pigment found in rod cells.

Do not refer to the text when taking this test.

_____ 1. A receptor
   a. is the first portion of a reflex arc.
   b. initiates a nerve impulse.
   c. responds to only one type of stimulus.
   d. is attached to a dendrite.
   e. All of the above are correct.

_____ 2. There are five sense receptors in the skin. They are
   a. choroid, cochlea, ossicles, and otoliths.
   b. hot, cold, pressure, touch, and pain.
   c. sclera, choroid, cornea, cones, and cochlea.
   d. mechanoreceptors, chemoreceptors, and photoreceptors.

_____ 3. Taste cells and olfactory cells are both
   a. somatic senses.
   b. mechanoreceptors.
   c. pseudociliated epithelium.
   d. chemoreceptors.

_____ 4. The blind spot is
   a. a nontransparent area on the lens.
   b. a nontransparent area on the cornea.
   c. on the retina, where there are no rods or cones.
   d. called the fovea centralis.

In questions 5–8, match the questions to these terms:
   a. retina   b. optic nerve
   c. lens   d. all   e. none

_____ 5. Which of these is (are) necessary to proper vision?
_____ 6. Which of these contain(s) receptors for sight?
_____ 7. Which of these focus(es) light?
_____ 8. In which is the sensation of sight realized?

_____ 9. The current theory of color vision proposes that
   a. there are three primary colors associated with color vision.
   b. cone cells respond selectively to different wavelengths of light.
   c. the rod cells are responsible for nighttime color vision.
   d. Both _a_ and _b_ are correct.
   e. _a, b,_ and _c_ are correct.

_____ 10. Each side of the brain receives information from both eyes due to the
   a. optic chiasma.
   b. fovea centralis.
   c. ciliary muscle.
   d. ciliary body.

_____ 11. If you are nearsighted, the image is focused
   a. in front of the retina.
   b. behind the retina.
   c. on the retina.
   d. at the blind spot.

_____ 12. The disorders of nearsightedness and farsightedness are due to
   a. an eyeball of incorrect length.
   b. a cloudy lens.
   c. pressure increase.
   d. a torn retina.

_____ 13. Vitamin A is needed for the
   a. lens.
   b. rod cells.
   c. cone cells.
   d. cornea.

_____ 14. The cochlear nerve is associated with the
   a. spiral organ.
   b. ossicles.
   c. tympanic membrane.
   d. auditory tube.

In questions 15–17, match the questions to the terms.
   a. ossicles   b. otoliths
   c. cochlea   d. auditory tube

_____ 15. Which of these have nothing to do with hearing?
_____ 16. In which would you find receptors for hearing?
_____ 17. Which of these is concerned with balance?

_____ 18. Equilibrium receptors differ from hearing receptors in that equilibrium receptors _____, while hearing receptors _____.
   a. are located in the outer ear; are located in the inner ear
   b. respond to pressure waves; do not respond to pressure waves
   c. consist of hair cells; do not consist of hair cells
   d. are in the semicircular canals; are in the spiral organ

_____ 19. Which part of the ear is for equilibrium?
   a. outer
   b. middle
   c. inner
   d. All of these are correct.

_____ 20. Nerve deafness may be due to
   a. fused ossicles.
   b. worn stereocilia.
   c. German measles.
   d. Both _a_ and _c_ are correct.

In questions 21–25, match the questions to these terms.
   a. spiral organ   b. rods and cones
   c. pressure receptor   d. olfactory receptor
   e. taste buds   f. utricle and saccule

_____ 21. Which of these are NOT mechanoreceptors?
_____ 22. Which of these is located in the skin?
_____ 23. Which of these help maintain equilibrium?
_____ 24. Which of these are located in the ear?
_____ 25. Which of these are stimulated by chemicals?

Answer in complete sentences.

26. The length of the spiral organ (organ of Corti) is related to what ability in humans?

27. Relate the function of the rod cells to their placement in the retina.

**Test Results:** _____ number correct ÷ 27 = _____ × 100 = _____%

# ANSWER KEY

## STUDY QUESTIONS

**1. a.** photoreceptors **b.** chemoreceptors **c.** pain receptors **d.** proprioceptors, mechanoreceptors **e.** thermoreceptors **f.** mechanoreceptors **g.** chemoreceptors **2.** T **3.** T **4.** T **5.** T **6.** T **7. a.** muscle spindle **b.** Golgi tendon organ **c.** nerve impulses are generated **d.** maintain posture **e.** tell position of limbs **8.** d, e, c, e, a, b **9. a.** both **b.** taste receptors **c.** taste receptors **d.** taste receptors **e.** smell receptors **f.** smell receptors **g.** smell receptors **h.** both **i.** smell receptors **10. a.** protein receptors **b.** chemicals (molecules) **c.** chemoreceptors **11. a.** retina; photoreceptors for sight **b.** fovea centralis; makes acute vision possible **c.** ciliary body; holds lens in place; accommodation **d.** sclera; protects eyeball **e.** choroid; absorbs stray light rays **f.** optic nerve; transmission of nerve impulse **g.** lens; focusing **h.** cornea; refracts light rays **i.** iris; regulates entrance of light **12. a.** flat **b.** rounded **c.** accommodation **13. a.** photoreceptors **b.** rod cells **c.** cone cells **14. a.** rhodopsin **b.** opsin **c.** retinal **d.** three **e.** opsin **f.** fovea centralis **15. a.** rod cells and cone cells **b.** rod cells and cone cells **c.** ganglion cells **d.** integration **16. a.** optic chiasma **b.** one-half the **c.** parcel out **d.** rebuild **17. a.** in front of retina **b.** see distant objects **c.** behind retina **d.** convex lens **e.** irregular lens **18. a.** tympanic membrane; starts vibration of ossicles **b.** semicircular canal; rotational equilibrium **c.** cochlea; contains mechanoreceptors for hearing **d.** cochlear nerve; transmission of nerve impulse **e.** malleus (hammer); transmits vibrations **f.** pinna; reception of sound waves **g.** auditory canal; collection of sound waves **h.** stapes (stirrup); transmits vibrations to oval window **i.** auditory tube; connects middle ear to pharynx **j.** vestibule; gravitational equilibrium **19.** d, f, a, i, h, b, e, g, c **20. a.** rotational **b.** cupula **c.** gravitational **d.** utricle **e.** vestibular

```
A C C O M M O D A T I O N
M       H T O P S D N I L B
P       O
U       S R
L       S C O C H L E A
L       I     I
A       C     D
  O T O L I T H
        E R H O D O P S I N
```

**a.** accommodation  **b.** choroid  **c.** cochlea  **d.** blind spot  **e.** ampulla  **f.** ossicle  **g.** otolith  **h.** rhodopsin

**1.** e  **2.** b  **3.** d  **4.** c  **5.** d  **6.** a  **7.** c  **8.** e  **9.** d  **10.** a  **11.** a  **12.** a  **13.** b  **14.** a  **15.** b and d  **16.** c  **17.** b  **18.** d  **19.** c  **20.** b  **21.** b, d, e  **22.** c  **23.** f  **24.** a, f  **25.** d, e  **26.** The length of the spiral organ (organ of Corti) is related to the ability of humans to hear a range of pitches.  **27.** The rod cells are located throughout the retina, and this is consistent with their ability to detect motion.

# 19

# MUSCULOSKELETAL SYSTEM

A long bone has a shaft and two ends covered by cartilage. The shaft contains a medullary cavity and **compact bone.** The ends contain **spongy bone** in which blood cells are produced.

Bone is constantly being renewed; **osteoclasts** break down bone, and osteoblasts build new bone. **Osteocytes** are in the lacunae of **osteons.**

A rigid skeleton gives support to the body, helps protect internal organs, and assists movement. In humans, the skeleton is also a storage area for calcium and phosphorous salts, as well as the site of blood cell production.

The human skeleton is divided into two parts: (1) the **axial skeleton,** which is made up of the skull, the ribs, the sternum, and the vertebrae; and (2) the **appendicular skeleton,** which is composed of the girdles and their appendages.

**Joints** are classified as immovable, like those of the cranium; slightly movable, like those between the vertebrae; and freely movable or **synovial,** like those in the knee and hip. In synovial joints, **ligaments** bind the two bones together, forming a capsule in which there is synovial fluid.

Whole skeletal muscles can only shorten when they contract; therefore, for a bone to be returned to its original position or the muscle to its original length, muscles must work in antagonistic pairs.

The **sliding filament theory** of muscle contraction says that **myosin** filaments have cross-bridges, which attach to and detach from **actin** filaments, causing them to slide and the **sarcomere** to shorten. The H zone disappears as actin filaments approach one another.

Innervation of a muscle fiber begins at a **neuromuscular junction.** Here, synaptic vesicles release ACh into the synaptic cleft. When the **sarcolemma** receives ACh, a muscle action potential moves down the T system to calcium storage sacs. When calcium ions are released, contraction occurs. When nerve input to the muscle ceases, calcium ions are actively transported back into the storage sacs, and muscle relaxation occurs.

The occurrence of **muscle twitch,** summation, and **tetanus** is dependent on the frequency with which a muscle is stimulated.

Study the text section by section. Answer the study questions so that you can fulfill the learning objectives for each section.

## 19.1 ANATOMY AND PHYSIOLOGY OF BONES (PP. 366—67)

The learning objectives for this section are:
- Describe the structure of bone and tissues associated with bones.
- Describe bone development and growth.
- Explain how adult bones undergo remodeling.

In question 1, fill in the blanks.

1. Compact bone is composed of bone cells arranged in concentric circles called ᵃ·_____. Inside

   compact bone, a lighter type of bone, called ᵇ·_____, has spaces that are often filled with

   ᶜ·_____, the blood-forming tissue. Ligaments, which attach bone to bone, are made up of

   ᵈ·_____ tissue.

2. Name the three types of cartilage and where they can be found.

   a. _____

   b. _____

   c. _____

3. Label this diagram of a long bone, using the following alphabetized list of terms.
     cartilage
     compact bone
     medullary cavity
     spongy bone

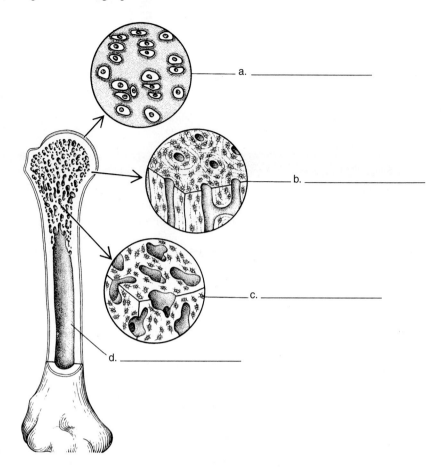

a. _____

b. _____

c. _____

d. _____

In questions 4–6, fill in the blanks.

4. Which part of the long bone shown in the previous illustration is associated with red bone marrow?
   a._____ Which part of bone is the hardest? b._____
   Which part of bone is the most flexible? c._____ Which part of a long bone is associated with
   yellow bone marrow? d._____

5. What is the relationship between osteoblasts and osteocytes? a._____
   What role is played by the cartilaginous disks of long bones? b._____
   What is the function of an osteoclast? c._____
   What is the process of bone formation called? d._____

6. During the continual process of bone remodeling, osteoclasts remove worn-out bone cells. At the same time,
   a._____ is released into the bloodstream. The two factors that affect bone thickness are
   b._____ and c._____.

## 19.2 BONES OF THE SKELETON (PP. 369–76)

The learning objectives for this section are:
  • List the functions of the skeleton.
  • Identify and state a function for the bones of the axial skeleton, including the cranium and face.
  • Identify and state a function for the bones of the appendicular skeleton.
  • Classify joints according to their types, and list the different types of joint movements.

7. Name five functions of the skeleton.

a._____

b._____

c._____

d._____

e._____

8. Axial versus appendicular skeleton. Write *ax* in front of all bones belonging to the axial skeleton; write *ap* in front of all bones belonging to the appendicular skeleton. Write *pec* in front of all bones belonging to the pectoral girdle; write *pel* in front of all bones belonging to the pelvic girdle. Some items have more than one answer.

a._____ coxal bone      g._____ ribs

b._____ sternum         h._____ radius

c._____ humerus         i._____ clavicle

d._____ scapula         j._____ tibia

e._____ skull           k._____ fibula

f._____ femur           l._____ ulna

In question 9, fill in the blanks.

9. The [a]_____ bone forms the forehead, and the [b]_____ bone has the foramen magnum through which the spinal cord passes. The [c]_____ bones have an opening for the ears. The [d]_____ is the only movable portion of the skull and permits us to chew our food. The [e]_____ forms the upper jaw and anterior hard palate. The [f]_____ bones form the cheekbone, and the nasal bones form the bridge of the nose.

10. a. Give a function of vertebrae. _____

b. Give the name of the vertebrae that have ribs. _____

c. Give the name of the vertebrae in the lower back. _____

d. Give the name of the vertebrae in the neck. _____

In question 11, fill in the blanks.

11. Comparisons. The radius and ulna are to the forearm as the [a]_____ and [b]_____ are to the leg. The femur is to the thigh as the [c]_____ is to the upper arm. The metacarpals are to the palm as the [d]_____ are to the foot.

12. Give the scientific terms for the common names.

a. Shinbone _____

b. Collarbone _____

c. Hipbone _____

d. Thighbone _____

13. Match the descriptions to these types of joints:

fibrous joint   cartilaginous joint   synovial joint

a. _____ lined with synovial membrane

b. _____ immovable, as in a suture

c. _____ slightly movable

d. _____ freely movable

e. _____ connected by hyaline cartilage

14. Fill in the table to describe the shoulder and elbow joints.

| Joint | Anatomical Type | Degree of Movement |
|---|---|---|
| Shoulder joint | a. | b. |
| Elbow joint | c. | d. |

## 19.3 SKELETAL MUSCLES (PP. 377–80)

The learning objectives for this section are:
• Explain how muscles work in antagonistic pairs.
• Describe the functions of muscles.

15. List three functions of skeletal muscle.

a. _____

b. _____

c. _____

16. Label this diagram of muscles and bones in the upper limb, using the following alphabetized list of terms.

biceps brachii
humerus
insertion
origin
radius
scapula
triceps brachii
ulna

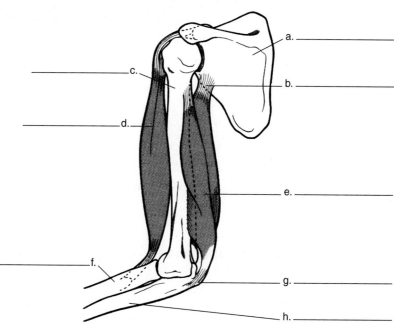

a. _____

b. _____

c.

d.

e. _____

f.

g. _____

h. _____

17. With reference to the muscles in the drawing for question 16, why are the biceps brachii and triceps brachii antagonistic pairs? _____

_____

18. Name the thigh muscles that act as antagonists.

a. front of thigh _____

b. back of thigh _____

19. Name the antagonistic muscles of the leg.

a. front of leg _____

b. back of leg _____

## 19.4 MECHANISM OF MUSCLE FIBER CONTRACTION (PP. 381–83)

The learning objectives for this section are:
- Describe the anatomy of a muscle fiber.
- Explain how the sarcomere shortens during muscle contraction.
- Describe the neuromuscular junction, and tell how impulses are transferred to the muscle sarcolemma.

20. Label this diagram of a portion of a muscle fiber, using the following alphabetized list of terms.

muscle fiber
myofibril
sarcolemma
sarcomere
sarcoplasmic
    reticulum
T tubule
Z line
    (used twice)

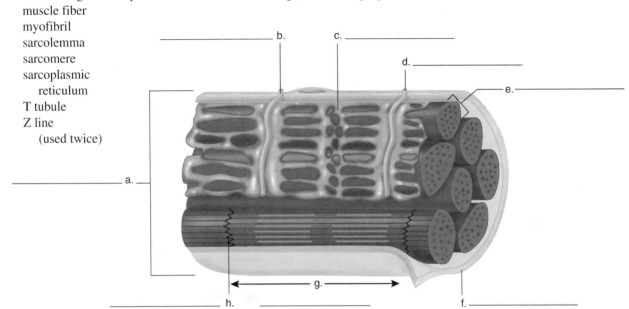

21. Label this diagram of a sarcomere, using the following alphabetized list of terms.

actin filament
H zone
myosin filament
Z line

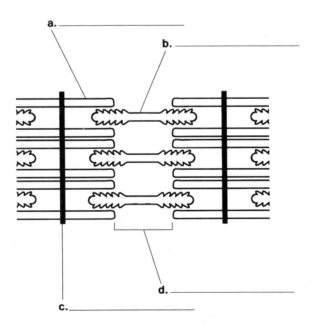

In question 22, fill in the blanks.

22. Which of your labels in question 21 is a thin filament? a._____ Which of your labels is a thick

filament? b._____ Which of your labels is reduced in size when a sarcomere contracts?

c._____ Which component has cross-bridges? d._____ Which of your labels

is the filament that moves when the sarcomere contracts? e._____ What molecule immediately

supplies energy for muscle contraction? f._____

23. What is the proper sequence for these phrases to describe what occurs at the neuromuscular junction to trigger muscle contraction? Indicate by letter. _____
   a. receptor sites on sarcolemma
   b. nerve impulse
   c. release of calcium from sarcoplasmic reticulum
   d. the neurotransmitter acetylcholine is released
   e. sarcomeres shorten
   f. synaptic cleft
   g. spread of impulses over sarcolemma to T tubules

24. Study this diagram and fill in the blanks that follow.

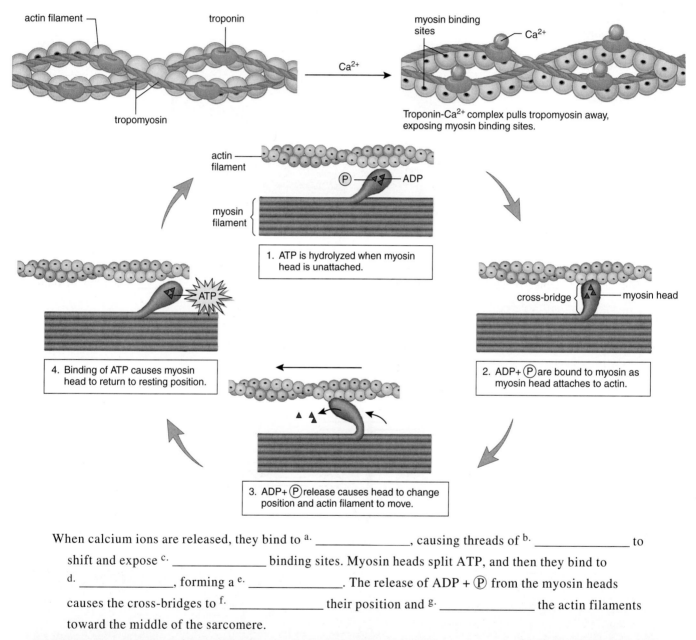

When calcium ions are released, they bind to <sup>a.</sup> _____, causing threads of <sup>b.</sup> _____ to

shift and expose <sup>c.</sup> _____ binding sites. Myosin heads split ATP, and then they bind to

<sup>d.</sup> _____, forming a <sup>e.</sup> _____. The release of ADP + $P$ from the myosin heads

causes the cross-bridges to <sup>f.</sup> _____ their position and <sup>g.</sup> _____ the actin filaments

toward the middle of the sarcomere.

## 19.5 WHOLE MUSCLE CONTRACTION (PP. 384–85)

The learning objectives for this section are:
 • Describe the basic laboratory experiments on whole muscle contraction.
 • Explain how muscle tone is maintained.
 • Name three sources of ATP for muscle contraction.

25. Label this diagram of a single muscle twitch, using the following alphabetized list of terms.

contraction
latent period
relaxation

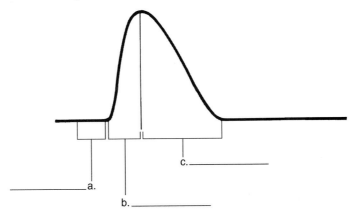

a. _____

b. _____

c. _____

26. Indicate the locations on this myogram where the stimulus was applied, where fatigue begins, where tetanus occurs, and where the time interval is shown.

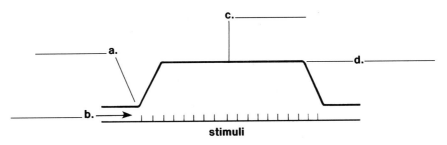

a. _____

b. _____

c. _____

d. _____

**stimuli**

27. How do you recognize fatigue of a muscle? _____

28. Muscle tone is achieved because of special nerve receptors called _____ that send the brain information about when the muscle needs to contract.

29. Stimulation of a muscle fiber results in a contraction that can be described as _____.

30. Compare sources of energy and oxygen for muscle contraction. Match the statements to these terms:

ATP    creatine phosphate    myoglobin    fermentation

a. _____ anaerobic source of energy

b. _____ produced during cellular respiration

c. _____ used to regenerate ATP from ADP

d. _____ stores oxygen in muscle fibers

e. _____ results in the buildup of lactic acid

31. Oxygen debt occurs when a._____ is used up in muscles, and the blood does not supply
b._____ rapidly enough. Why are marathon runners less apt to become exhausted due to
oxygen debt? c._____

## 19.6 EXERCISE AND MUSCLE CONTRACTION (P. 387)

The learning objectives for this section are:
• Describe some of the health benefits of exercise.
• Define slow-twitch and fast-twitch muscle fibers.

32. Muscle wasting is referred to as a._____, while muscle building is called
b._____. To build muscles, a person must lift weights or exercise to achieve what percent of
maximum muscle contraction? c._____ Muscles increase in size because there is an increase in the number of
d._____.

33. Match the statements to these terms:

slow-twitch fibers    fast-twitch fibers

a. _____ more common in marathon runners

b. _____ more common in weight lifters

c. _____ fewer mitochondria

d. _____ white in color

e. _____ abundant reserve of fat and glycogen

f. _____ can contract for longer periods

g. _____ contract with more force

## DEFINITIONS WORDSEARCH

Review key terms by using the following alphabetized list of terms to fill in the blanks below. Then complete the wordsearch.

```
S E E T H M Y O G L O B I N H I M O J K M O
E M O S O M O R H C B R P F A M I P A Y R P
T H O R I M O T R Y C Y L T T O C S T C A K
T Z X S I O T B G T D O A G D H R I D H T E
G I N N T O C Y T O S K M L E T O T E T R N
R E N T O E R D C P T U I O S L T I S H O Y
A Y L I L L O A Q L E Z E U P I U D P P P H
F U N I D A R N T T Y V L K C G B I C I H P
N E T N U G G U S S L O O H G A U O G F Y O
I I T C O E N O Y M E U I U O M L T O D L R
S I L Y T L I U T E R U R O M E E S M M E T
O L A G G R T M C N Y O T F O N T A N E L R
Y O L G E U B U S I K J N O P T Y M P L Y E
M I L P U M N R T S L L E N A E R O A G R P
O U C L A I U B F C T O C J U N P G U M P Y
P L A G G L T M C U Y O T R E T S I E E S H
O O L G A U B S I S O I L O C S Y G P L Y G
R I L I U M N R T E L L E N A G R O A G R O
T U S A R C O M E R E O C J U M P G U M P G
```

atrophy
fontanel
hypertrophy
ligament
mastoiditis
meniscus
myoglobin
osteons
periosteum
sarcomere
scoliosis
tropomyosin

a. _____ Cylindrical units containing bone cells that have surrounded a central canal; Haversian system.

b. _____ Tough cord or band of dense fibrous connective tissue that joins bone to bone at a joint.

c. _____ Fibrous connective tissue covering the surface of bone.

d. _____ Membranous region located between certain cranial bones in the skull of a fetus or infant.

e. _____ Inflammation of the cells of the mastoid.

f. _____ Abnormal lateral curvature of the vertebral column.

g. _____ Cartilaginous wedges that separate the surfaces of bones in synovial joints.

h. _____ One of many units, arranged linearly within a myofibril, whose contraction produces muscle contraction.

i. _____ Protein that blocks muscle contraction until calcium ions are present.

j. _____ Pigmented compound in muscle tissue that stores oxygen.

k. _____ Wasting away or decrease in size of an organ or tissue.

l. _____ Increase in muscle size following long-term exercise.

## OBJECTIVE QUESTIONS

Do not refer to the text when taking this test.

_____ 1. Bone-building cells are _____, while mature bone cells are called _____.
   a. osteoblasts; osteocytes
   b. osteoclasts; osteoblasts
   c. osteoclasts; osteocytes
   d. osteoblasts; lacunae

_____ 2. Where is the blood-forming tissue housed within a bone?
   a. throughout the bone
   b. along the periosteum
   c. within the compact bone
   d. within the spongy bone

_____ 3. Ligaments join
   a. bone to bone.
   b. muscle to muscle.
   c. muscle to bone.
   d. All of these are correct.

_____ 4. Which of these is NOT a bone in the lower limb?
   a. femur
   b. tibia
   c. ulna
   d. fibula

_____ 5. Vertebrae have
   a. immovable joints.
   b. freely movable joints.
   c. slightly movable joints.
   d. joints that vary from one person to the next.

_____ 6. The shoulder joint is an example of a
   a. synovial joint.
   b. freely movable joint.
   c. ball-and-socket joint.
   d. All of these are correct.

_____ 7. The bones making up your palm are the _____.
   a. phalanges
   b. carpals
   c. metacarpals
   d. metatarsals

_____ 8. Which of these is NOT in the appendicular skeleton?
   a. clavicle
   b. coxal bone
   c. metatarsals
   d. vertebrae

_____ 9. Which of these is NOT a facial bone?
   a. frontal bone
   b. occipital bone
   c. mandible
   d. zygomatic bone

_____ 10. Which vertebrae support the lower back?
   a. cervical
   b. lumbar
   c. thoracic
   d. cranial

_____ 11. Fluid-filled sacs within joints are called
   a. bursae.
   b. menisci.
   c. articulations.
   d. synovial fluid.

_____ 12. Which muscle is located on the front thigh and raises (extends) the leg?
   a. trapezius
   b. rectus abdominis
   c. pectoralis major
   d. quadriceps femoris

_____ 13. Which portion of the muscle is on the stationary bone?
   a. the insertion
   b. the origin
   c. the bursa
   d. the belly

_____ 14. Which of these contains the cross-bridges?
   a. the actin
   b. the myosin
   c. the Z line
   d. All of these are correct.

_____ 15. According to the sliding filament theory,
   a. actin moves past myosin.
   b. myosin moves past actin.
   c. both myosin and actin move past each other.
   d. None of these are correct.

_____ 16. The release of calcium from the sarcoplasmic reticulum
   a. causes the sarcomeres to relax.
   b. causes the sarcomeres to contract.
   c. is the end result of a nervous impulse to contract.
   d. Both _b_ and _c_ are correct.
   e. All of these are correct.

_____ 17. When a nervous impulse reaches the synaptic end bulb,
   a. neurotransmitter is released into the synaptic cleft.
   b. calcium is released from the sarcoplasmic reticulum.
   c. calcium is stored in the T tubules.
   d. neurotransmitter is released by the sarcolemma.

_____18. Which of these is a function of skeletal muscle?
   a. produce heat
   b. posture
   c. movement
   d. All of these are correct.
_____19. Choose the muscle that is the antagonist to the triceps brachii.
   a. gastrocnemius
   b. quadriceps femoris
   c. biceps brachii
   d. latissimus dorsi
_____20. Choose the muscle that is NOT in the upper limb.
   a. biceps brachii
   b. triceps brachii
   c. flexor carpi group
   d. iliopsoas
_____21. The Z line of a sarcomere is where the
   a. cross-bridges attach.
   b. actin filaments attach.
   c. myosin filaments attach.
   d. nerve innervates the muscle cell.

_____22. Muscle fatigue
   a. follows summation and tetanus.
   b. involves the buildup of lactic acid.
   c. occurs only in the laboratory.
   d. Both a
   e. All of these are correct.
_____23. Oxygen debt may be associated with
   a. anaerobic cellular respiration.
   b. fermentation.
   c. muscle contraction.
   d. lactic acid metabolism.
   e. All of these are correct.
_____24. Creatine phosphate is
   a. used by sarcomeres.
   b. used to change ADP to ATP.
   c. a molecule found in DNA.
   d. All of these are correct.
_____25. Fast-twitch fibers are
   a. white in color.
   b. associated with strength training.
   c. generally anaerobic.
   d. All of these are correct.

## THOUGHT QUESTIONS

Answer in complete sentences.

26. Why are knee injuries usually quite serious?

27. A nervous action potential eventually causes a neurotransmitter to be secreted at a synapse. What event occurs as a result of a muscle action potential?

**Test Results:** _____ number correct ÷ 27 = _____ × 100 = _____%

## STUDY QUESTIONS

**1. a.** osteons **b.** spongy bone **c.** red bone marrow **d.** fibrous connective **2. a.** hyaline cartilage; ends of bones **b.** fibrocartilage; intervertebral disks **c.** elastic cartilage; ear flaps **3. a.** cartilage **b.** compact bone **c.** spongy bone **d.** medullary cavity **4. a.** spongy bone **b.** compact bone **c.** cartilage **d.** medullary cavity **5. a.** Bone-forming osteoblasts eventually become mature osteocytes. **b.** They increase in length and allow bone to grow longer. **c.** to break down bone **d.** ossification **6. a.** calcium **b.** exercise **c.** hormones **7. a.** supports the body **b.** protects soft body parts **c.** produces blood cells **d.** stores calcium and phosphorus salts **e.** permits body movement **8. a.** ap, pel **b.** ax **c.** ap **d.** ap, pec **e.** ax **f.** ap **g.** ax **h.** ap **i.** ap, pec **j.** ap **k.** ap **l.** ap **9. a.** frontal **b.** occipital **c.** temporal **d.** mandible **e.** maxilla **f.** zygomatic **10. a.** to protect the spinal cord **b.** thoracic **c.** lumbar **d.** cervical **11. a.** tibia **b.** fibula **c.** humerus **d.** metatarsals **12. a.** tibia **b.** clavicle **c.** coxal bone **d.** femur **13. a.** synovial **b.** fibrous **c.** cartilaginous **d.** synovial **e.** cartilaginous **14. a.** ball-and-socket (synovial) **b.** freely movable **c.** hinge (synovial) **d.** freely movable **15. a.** provides posture **b.** movement **c.** provides heat **16. a.** scapula **b.** origin **c.** humerus **d.** biceps brachii **e.** triceps brachii **f.** radius **g.** insertion **h.** ulna **17.** The biceps brachii raises and the triceps brachii lowers the forearm. **18. a.** quadriceps femoris group **b.** hamstring muscles **19. a.** tibialis anterior **b.** gastrocnemius **20. a.** openings to T tubules **b.** sarcolemma **c.** myofibrils **21. a.** actin filament **b.** myosin filament **c.** Z line **d.** H zone **22. a.** actin filament **b.** myosin filament **c.** H zone **d.** myosin **e.** actin filament **f.** ATP **23.** b, d, f, a, g, c, e **24. a.** troponin **b.** tropomyosin **c.** myosin **d.** actin **e.** cross-bridge **f.** change **g.** pull **25. a.** latent period **b.** contraction **c.** relaxation **26. a.** stimulus applied **b.** time interval **c.** tetanus occurs **d.** fatigue begins **27.** The muscle relaxes even though a stimulus has been applied. **28.** muscle spindles **29.** all-or-none **30. a.** fermentation, creatine phosphate **b.** ATP **c.** creatine phosphate **d.** myoglobin **e.** fermentation **31. a.** ATP **b.** oxygen **c.** Marathon runners have more mitochondria. **32. a.** atrophy **b.** hypertrophy **c.** 75% **d.** myofibrils **33. a.** slow-twitch fibers **b.** fast-twitch fibers **c.** fast-twitch fibers **d.** fast-twitch fibers **e.** slow-twitch fibers **f.** slow-twitch fibers **g.** fast-twitch fibers

## DEFINITIONS WORDSEARCH

**a.** osteon **b.** ligament **c.** periosteum **d.** fontanel **e.** mastoiditis **f.** scoliosis **g.** meniscus **h.** sarcomere **i.** tropomyosin **j.** myoglobin **k.** atrophy **l.** hypertrophy

```
            M Y O G L O B I N

     O                              S       A
      S                             I       T
       T              M         T   R       O   Y
        E         U         L   I   P       H
         O    E           I   D       H     P
          N   T         G     I       Y     P
 N         S                 A     O       R
 I          O   M           M     T       T
 S           I   E           E     S       R
 O       R     N       F O N T A N E L   E
 Y     E       I           T   M           P
 M   P         S                           Y
 O             C                           H
 P             U
 O       S I S O I L O C S
 R
 T   S A R C O M E R E
```

## CHAPTER TEST

**1.** a **2.** d **3.** a **4.** c **5.** c **6.** d **7.** c **8.** d **9.** b **10.** b **11.** a **12.** d **13.** b **14.** b **15.** a **16.** d **17.** a **18.** d **19.** c **20.** d **21.** b **22.** d **23.** e **24.** b **25.** d **26.** The knee supports the weight of the body, which can make healing a very slow process. **27.** A muscle action potential causes calcium to be released from calcium storage sacs.

# 20

# ENDOCRINE SYSTEM

**Hormones** are chemical signals that affect the activity of other glands or tissues. Endocrine glands secrete hormones into the bloodstream, and from there they are distributed to target organs or tissues. The endocrine system works with the nervous system to maintain homeostasis. The secretion of hormones involved in maintaining homeostasis is controlled in two ways: by negative feedback and/or by antagonistic hormonal actions.

Neurosecretory cells in the hypothalamus produce **antidiuretic hormone** (ADH) and **oxytocin,** which are stored in axon endings in the **posterior pituitary** until they are released.

The hypothalamus produces **hypothalamic-releasing** and **hypothalamic-inhibiting hormones,** which pass to the anterior pituitary by way of a portal system. The **anterior pituitary** produces at least six types of hormones, and some of these stimulate other hormonal glands to secrete hormones. Therefore, the anterior pituitary is sometimes called the master gland.

The **thyroid gland** produces thyroxine and triiodothyronine, hormones that play a role in the growth and development of immature forms; in mature individuals, they increase the metabolic rate. The thyroid gland also produces **calcitonin,** which helps lower the blood calcium level. The **parathyroid glands** raise the blood calcium and decrease the blood phosphate level.

The adrenal glands respond to stress. Immediately, the adrenal medulla secretes **epinephrine** and **norepinephrine,** which bring about responses we associate with emergency situations. On a long-term basis, the adrenal cortex primarily produces the **glucocorticoids (cortisol)** and the **mineralocorticoids (aldosterone).** Cortisol stimulates the hydrolysis of proteins to amino acids, which are converted to glucose; in this way, it raises the blood glucose level. Aldosterone causes the kidneys to reabsorb sodium ions ($Na^+$) and to excrete potassium ions ($K^+$).

The **islets** of the **pancreas** secrete **insulin,** which lowers the blood glucose level, and **glucagon,** which has the opposite effect. The most common illness due to hormonal imbalance is **diabetes mellitus,** which is caused by the failure of the pancreas to produce insulin or the failure of the cells to take it up.

The gonads produce the sex hormones; the **thymus** secretes thymosins, which stimulate T lymphocyte production and maturation; the **pineal gland** produces **melatonin,** whose function in humans is uncertain—it may be involved in **circadian rhythms** and the development of the reproductive organs. Tissues also produce hormones. Adipose tissue produces leptin, which acts on the hypothalamus, and various tissues produce growth factors. Prostaglandins are produced and act locally.

Chemical signals are active at various levels of communication between body parts, between cells, and even between individuals. Some chemical signals, such as traditional endocrine hormones and secretions of neurosecretory cells, act between body parts. Others, such as prostaglandins, growth factors, and neurotransmitters, act locally. Pheromones, which are well known in other animals, are chemical signals that act at a distance between individuals. Research suggests that humans may also have pheromones.

**Steroid hormones** enter the nucleus and combine with a receptor molecule, and the complex attaches to and activates DNA. Transcription and translation lead to protein synthesis. The peptide hormones are usually received by a receptor located in the plasma membrane. Most often their reception leads to the activation of an enzyme that changes ATP to **cyclic AMP** (cAMP). cAMP then activates another enzyme, which activates another, and so forth.

Study the text section by section. Answer the study questions so that you can fulfill the learning objectives for each section.

## 20.1 ENDOCRINE GLANDS (PP. 392–93)

The learning objectives for this section are:
- State the locations of endocrine glands, and be able to label an illustration that shows the endocrine glands.
- Associate the name of the gland and its hormones.
- Associate hormones with their actions.

1. Label the endocrine glands in this diagram, and name at least one hormone produced by each.

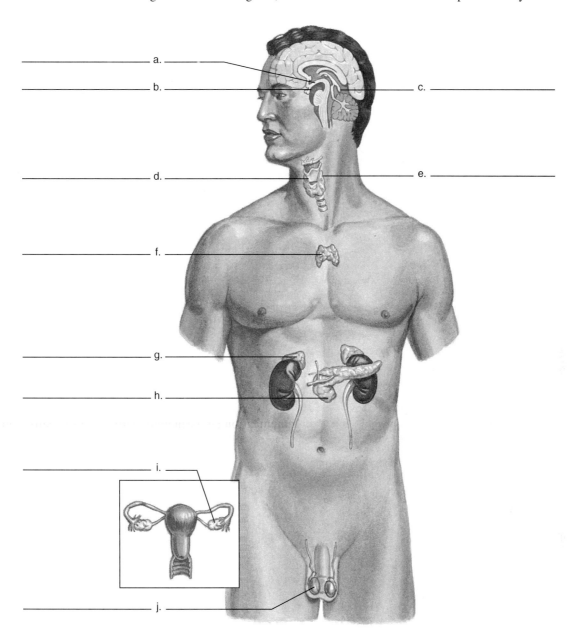

a. _____

b. _____

c. _____

d. _____

e. _____

f. _____

g. _____

h. _____

i. _____

j. _____

In question 2, fill in the blanks.

2. Control of hormone release. Often the release of a hormone is dependent upon the blood level of the substance it is controlling. When the level of a substance increases, generally this causes the hormone secretion to

a. _____. This is an example of  b. _____ feedback. In other instances,

c. _____ hormones oppose each other's actions, thus regulating the target substance in the body.

## 20.2  HYPOTHALAMUS AND PITUITARY GLAND (PP. 394–96)

The learning objectives for this section are:
- Contrast the ways in which the hypothalamus controls the posterior and anterior pituitary.
- List the hormones produced by the anterior and posterior pituitary, and describe their effects.

3. Match the descriptions to the numbers in the diagram.
   a. _____ hypothalamus
   b. _____ anterior pituitary
   c. _____ target gland
   d. _____ feedback that inhibits hypothalamus
   e. _____ feedback that inhibits anterior pituitary
   f. _____ releasing hormone
   g. _____ stimulating hormone
   h. _____ target gland hormone

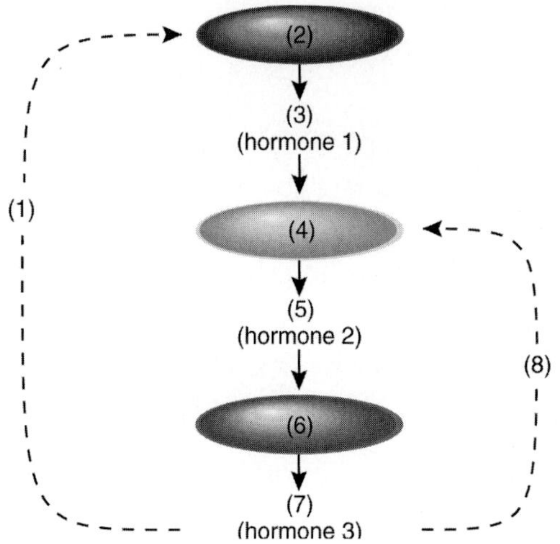

In question 4, fill in the blanks.

4. If the "gland" in the diagram in question 3 is the thyroid, then the first hormone is a._____, the second is b._____, and the third is c._____. If the "gland" in the diagram in question 3 is the adrenal cortex, then the first hormone is d._____, the second is e._____, and the third is f._____.

5. Place the appropriate letters next to each statement.
        AP—anterior pituitary        PP—posterior pituitary
   a. _____ connected to hypothalamus by nerve fibers
   b. _____ connected to hypothalamus by blood vessels
   c. _____ secretes hormones produced by hypothalamus
   d. _____ controlled by releasing hormones produced by hypothalamus

6. To show why the anterior pituitary is sometimes called the master gland, complete this table.

| Anterior Pituitary Produces | Gland Controlled | Hormone Produced by Gland |
|---|---|---|
| TSH | a. | b. |
| ACTH | c. | d. |
| Gonadotropic hormones | | |
| Female | e. | f. |
| Male | g. | h. |

In question 7, fill in the blanks.

7. The anterior pituitary produces three other hormones. The hormone ᵃ· _____ causes the mammary glands to develop and produce milk. ᵇ· _____ hormone causes skin color changes in lower vertebrates. Growth hormone (GH) promotes cell division, protein synthesis, and ᶜ· _____ growth. If too little GH is produced during childhood, the individual becomes a pituitary ᵈ· _____. If too much is produced, the individual is a pituitary ᵉ· _____. If there is an overproduction of GH in the adult, ᶠ· _____ results, and the face, hands, and feet ᵍ· _____.

## 20.3 THYROID AND PARATHYROID GLANDS (PP. 397–98)

The learning objectives for this section are:
- List the hormones of the thyroid gland, and give their functions.
- Explain why iodine is a necessary dietary component for thyroid hormones.
- Give the function of parathyroid hormone.
- Be able to identify thyroid disorders.

8. Match the phrases to these conditions:
   cretinism   exophthalmic goiter   simple goiter   myxedema

   a. _____ hypothyroidism (choose more than one)

   b. _____ hyperthyroidism

   c. _____ hypothyroidism since birth

   d. _____ hypothyroidism in the adult

   e. _____ lack of iodine

9. Match the descriptions to these phrases:
   1. low blood $Ca^{2+}$           2. high blood $Ca^{2+}$
   3. $Ca^{2+}$ is deposited in bones     4. $Ca^{2+}$ is deposited in blood
   a. _____ Calcitonin is present.
   b. _____ Parathyroids are mistakenly removed during an operation.
   c. _____ Calcitonin will be released.
   d. _____ PTH will be released.
   e. _____ PTH is present.
   f. _____ Calcitonin will not be released.
   g. _____ PTH will not be released.

## 20.4 ADRENAL GLANDS (PP. 399–401)

The learning objectives for this section are:
- Describe the hormones of the adrenal medulla, their control, and their effects.
- List the hormones of the adrenal cortex, state their functions, and tell how their release is controlled.

10. Place the appropriate letters next to each description.
    AM—adrenal medulla        AC—adrenal cortex
    a. _____ inner portion of adrenal gland
    b. _____ outer portion of adrenal gland
    c. _____ hypothalamus sends nervous impulses
    d. _____ hypothalamus sends releasing hormone to anterior pituitary, and anterior pituitary sends ACH to target gland
    e. _____ releases glucocorticoids and mineralocorticoids
    f. _____ releases epinephrine and norepinephrine
    g. _____ short-term reaction to stress
    h. _____ long-term reaction to stress

11. Distinguish between cortisol and aldosterone by writing *yes* or *no* on each line.

|  | Cortisol | Aldosterone |
|---|---|---|
| Controlled by ACTH | a._____ | _____ |
| Glucocorticoid | b._____ | _____ |
| Mineralocorticoid | c._____ | _____ |
| Na$^+$/K$^+$ balance | d._____ | _____ |
| Amino acids → glucose | e._____ | _____ |
| Controlled by angiotensin II | f._____ | _____ |

In question 12, fill in the blanks.

12. When there is low blood Na$^+$, the kidneys secrete ^a. _____ , an enzyme that converts angiotensinogen to ^b. _____ , which later becomes ^c. _____ in lung capillaries. The latter causes the adrenal cortex to release ^d. _____ . Blood pressure now ^e. _____ . The heart releases a hormone called ^f. _____ that is antagonistic to aldosterone.

13. Place the appropriate letters next to these symptoms:
    AD—Addison disease       CS—Cushing syndrome
    a. _____ cannot handle bodily stress
    b. _____ cannot maintain blood glucose level
    c. _____ tendency toward diabetes mellitus
    d. _____ low blood pressure because of a low blood sodium level
    e. _____ high blood pressure because of a high blood sodium level
    f. _____ edema because of too much sodium in system
    g. _____ bronzing of skin
    h. _____ thin arms and legs; enlarged trunk

## 20.5 PANCREAS (PP. 402–3)

The learning objectives for this section are:
- Describe the functions of insulin and glucagon.
- Discuss the problems associated with diabetes.

14. Write the word *insulin* or *glucagon* on the appropriate arrow.

    glycogen ^a. ←——————— glucose molecules

    storage in liver ^b. ———————→ in the blood

15. Complete each of the following statements with the term *increases* or *decreases*.

    Glucagon ^a._____ blood sugar concentration. In type I diabetes, insulin production from the pancreas ^b._____. In type II diabetes, the response of body cells to the influence of insulin ^c._____.

## 20.6 OTHER ENDOCRINE GLANDS (PP. 404–5)

The learning objectives for this section are:
- Describe the functions of the testes and ovaries, the thymus, and the pineal gland.
- Describe the function of leptin, growth factors, and prostaglandins, which are produced by various tissues as opposed to glands.

16. Match these numbered items to the glands. There is more than one match for each gland, and answers may be used more than once.

    1. T lymphocytes   2. melatonin   3. testosterone   4. circadian rhythm   5. males
    6. females   7. estrogen and progesterone   8. thymosins   9. secondary sex characteristics

a. _____ testes
b. _____ ovaries
c. _____ thymus
d. _____ pineal gland

17. Match these phrases to the chemical signals listed. Answers can be used more than once.

    1. cause cell division   2. causes a feeling of satiety   3. has various effects   4. act locally

a. _____ leptin
b. _____ growth factors
c. _____ prostaglandins

## 20.7 CHEMICAL SIGNALS (PP. 407–8)

The learning objectives for this section are:
- Tell in general how chemical signals influence the behavior and/or metabolism of cells.
- Distinguish between the effects of steroid and peptide hormones on cells.
- Give examples to show that both endocrine systems and nervous systems utilize chemical signals.
- Tell how the endocrine system works with other systems of the body to maintain homeostasis.

18. Write either *peptide hormone* or *steroid hormone* on the lines above each diagram. Using the following alphabetized list of terms, place an appropriate word or phrase on the lines within each diagram.

    active    cyclic AMP    hormone receptor    hormone-receptor complex    protein synthesis

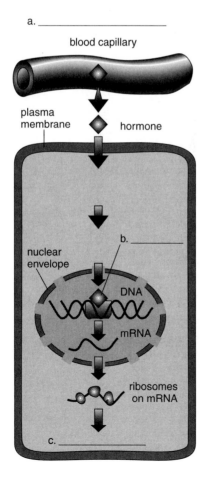
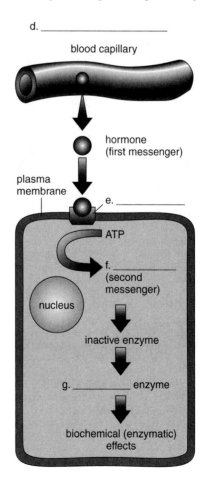

19. Place the appropriate letters next to each description.
    ES—endocrine system      NS—nervous system
    a. _____ uses chemical signals that bind to protein receptors
    b. _____ sometimes acts from a distance and sometimes acts locally
    c. _____ utilizes releasing hormones
    d. _____ utilizes neurotransmitters
    e. _____ utilizes hormones that are distributed in the blood

## Hormone Hockey

For every 5 correct answers in sequence, you have scored one goal.

First goal: Match the hormone to the glands (a–i).

Glands:
    a. anterior pituitary
    b. thyroid
    c. parathyroids
    d. adrenal cortex
    e. adrenal medulla
    f. pancreas
    g. gonads
    h. pineal gland
    i. posterior pituitary

Hormones:
1. ___ insulin
2. ___ oxytocin
3. ___ melatonin
4. ___ cortisol
5. ___ thyroxine

Second goal: Match the condition to the glands (a–i).
Conditions:
6. ___ diabetes mellitus
7. ___ cretinism
8. ___ Addison disease
9. ___ hypertension
10. ___ giantism

Third goal: Match the function to the hormones (a–l).
    a. melatonin
    b. estrogens
    c. androgens
    d. insulin
    e. glucagon
    f. epinephrine
    g. aldosterone
    h. cortisol
    i. parathyroid hormone
    j. thyroxine
    k. calcitonin (lowers)
    l. antidiuretic hormone

Functions:
11. ___ raises blood calcium level
12. ___ reduces stress
13. ___ maintains secondary female sex characteristics
14. ___ involved in circadian rhythms
15. ___ stimulates water reabsorption by kidneys

Fourth goal: Match the glands to the hormones (a–l). Some glands require two answers.

Glands:
16. ___ testes
17. ___ adrenal cortex
18. ___ pancreas
19. ___ thyroid
20. ___ adrenal medulla

Fifth goal: Select five hormones secreted by the anterior pituitary by answering yes or no to each of these.

21. ___ thyroid-stimulating hormone
22. ___ androgens
23. ___ gonadotropic hormones
24. ___ glucagon
25. ___ oxytocin
26. ___ growth hormone
27. ___ prolactin
28. ___ antidiuretic hormone
29. ___ estrogens
30. ___ adrenocorticotropic hormone

How many goals did you make? _____

# Definitions Wordsearch

Review key terms by using the following alphabetized list of terms to fill in the blanks below. Then complete the wordsearch.

| | | | | | | | | | | | | | | |
|---|---|---|---|---|---|---|---|---|---|---|---|---|---|---|
| C | R | E | T | I | N | I | S | M | N | U | J | L | I | P |
| O | A | R | H | G | T | Z | B | E | S | A | I | L | H | R |
| R | A | L | P | O | X | Y | T | O | C | I | N | H | U | O |
| T | M | L | C | F | R | S | H | U | K | N | H | S | R | S |
| I | E | R | G | I | U | M | N | V | Y | S | A | F | T | T |
| S | D | G | T | R | T | D | O | F | W | U | J | H | U | A |
| O | E | P | O | L | I | O | U | N | Y | L | R | E | R | G |
| L | X | H | Y | T | F | G | N | B | E | I | S | E | F | L |
| P | Y | J | T | D | H | S | E | I | D | N | S | G | B | A |
| I | M | P | H | E | R | O | M | O | N | E | B | G | F | N |
| N | O | R | E | P | I | N | E | P | H | R | I | N | E | D |
| S | G | W | Y | T | J | T | U | D | H | R | T | K | F | I |
| P | T | Y | E | R | W | Q | A | S | E | D | F | R | G | N |

*calcitonin*
*cortisol*
*cretinism*
*hormone*
*insulin*
*myxedema*
*norepinephrine*
*oxytocin*
*prostaglandin*

a. ____________________ Thyroid hormone that regulates blood calcium.

b. ____________________ Chemical signal.

c. ____________________ Condition due to improper development of thyroid in infants.

d. ____________________ Posterior pituitary hormone causing uterine contractions and milk letdown.

e. ____________________ Pancreatic hormone that lowers blood glucose.

f. ____________________ Adrenal gland hormone that increases blood glucose.

g. ____________________ Condition caused by lack of thyroid hormone in adult.

h. ____________________ Stress hormone from adrenal medulla.

i. ____________________ Local tissue hormone.

## OBJECTIVE QUESTIONS

Do not refer to the text when taking this test.

_____ 1. All hormones are believed to
   a. have plasma membrane receptors.
   b. affect cellular metabolism.
   c. increase the amount of cAMP.
   d. increase the amount of protein synthesis.

_____ 2. The adrenal glands are
   a. at the base of the brain.
   b. on the trachea.
   c. on the kidney.
   d. beneath the stomach.

_____ 3. Which statement is NOT true about hormones?
   a. Hormones search throughout the bloodstream for their receptors.
   b. They act as chemical signals.
   c. They are released by endocrine glands.
   d. They can affect our appearance, our metabolism, or our behavior.

_____ 4. Hormonal secretions are most often controlled by
   a. negative feedback mechanisms.
   b. positive feedback mechanisms.
   c. the hormone insulin.
   d. the cerebrum of the brain.

_____ 5. Steroid hormones
   a. combine with hormone receptors in the plasma membrane.
   b. pass through the membrane.
   c. activate genes leading to protein synthesis.
   d. Both _b_ and _c_ are correct.

_____ 6. Which gland produces the greatest number of hormones?
   a. posterior pituitary
   b. anterior pituitary
   c. thymus
   d. pineal gland

_____ 7. The hypothalamus controls the anterior pituitary via
   a. nervous stimulation.
   b. the midbrain.
   c. vasopressin.
   d. releasing hormones.

_____ 8. ADH and oxytocin are
   a. secreted by the hypothalamus.
   b. secreted by the posterior pituitary.
   c. secreted by the thyroid gland.
   d. secreted by the parathyroids.

_____ 9. Which hormone is involved with milk production and nursing?
   a. prolactin
   b. androgens
   c. antidiuretic hormone
   d. growth hormone

_____ 10. The anterior pituitary stimulates the
   a. thyroid.
   b. adrenal cortex.
   c. adrenal medulla.
   d. pancreas.
   e. Both _a_ and _b_ are correct.

_____ 11. Too much urine matches too
   a. little ADH.
   b. much ADH.
   c. little ACTH.
   d. much ACTH.

_____ 12. Thyroxine
   a. increases metabolism.
   b. stimulates the thyroid gland.
   c. lowers oxygen uptake.
   d. All of these are correct.

_____ 13. The adrenal cortex produces hormones affecting
   a. glucose metabolism.
   b. amino acid metabolism.
   c. sodium balance.
   d. All of these are correct.

_____ 14. Which hormone regulates blood calcium levels?
   a. calcitonin
   b. parathyroid hormone
   c. cortisol
   d. Both _a_ and _b_ are correct.

_____ 15. Which gland produces sex hormones?
   a. anterior pituitary
   b. posterior pituitary
   c. adrenal cortex
   d. Both _a_ and _b_ are correct.

_____ 16. Tetany occurs when there is too
   a. little calcium in the blood.
   b. much calcium in the blood.
   c. little sodium in the blood.
   d. much sodium in the blood.

_____ 17. Cushing syndrome is due to a malfunctioning
   a. thyroid.
   b. adrenal cortex.
   c. adrenal medulla.
   d. pancreas.

_____ 18. A simple goiter is caused by
   a. too much salt in the diet.
   b. too little iodine in the diet.
   c. too many sweets in the diet.
   d. a bland diet.

_____ 19. Acromegaly might be due to a tumor of the
   a. pancreas.
   b. anterior pituitary.
   c. thyroid.
   d. adrenal cortex.

_____ 20. If a person is suffering from insulin shock, he or she should
   a. be given some sugar.
   b. sit with the head down.
   c. be given insulin.
   d. not eat fatty foods.

_____ 21. Diabetes insipidus is a disease of the
   a. pancreas.
   b. adrenal cortex.
   c. posterior pituitary.
   d. Both *a* and *b* are correct.

_____ 22. In which case is insulin not produced?
   a. type I diabetes
   b. type II diabetes
   c. type III diabetes
   d. diabetes insipidus

_____ 23. Which of these is not a similarity between the nervous and endocrine systems?
   a. Both use chemical signals.
   b. Both utilize hormones.
   c. Both utilize nerve impulses.
   d. Both act from a distance or act locally.

_____ 24. The system most directly affected by the secretion of epinephrine for blood pressure adjustments is the
   a. respiratory system.
   b. cardiovascular system.
   c. urinary system.
   d. reproductive system.

_____ 25. The gonads that produce sex hormones also belong to which system?
   a. lymphatic system
   b. nervous system
   c. reproductive system
   d. urinary system

## THOUGHT QUESTIONS

Answer in complete sentences.

26. Explain the occurrence of a goiter when an individual does not receive enough iodine in the diet.

27. Why does the release of renin by the kidneys cause the blood pressure to rise?

**Test Results:** _____ number correct ÷ 27 = _____ × 100 = _____ %

## ANSWER KEY

### STUDY QUESTIONS

**1. a.** hypothalamus, hypothalamic-releasing hormone **b.** pituitary gland, growth hormone, ACTH **c.** pineal gland, melatonin **d.** thyroid gland, thyroxine, calcitonin **e.** parathyroid, parathyroid hormone **f.** thymus, thymosin **g.** adrenal gland, cortisol, aldosterone, epinephrine, norepinephrine **h.** pancreas, insulin, glucagon **i.** ovary, estrogen, progesterone **j.** testis, testosterone **2. a.** decrease **b.** negative **c.** antagonistic **3. a.** (2) **b.** (4) **c.** (6) **d.** (1) **e.** (8) **f.** (3) **g.** (5) **h.** (7) **4. a.** TRH (thyroid-releasing hormone) **b.** TSH (thyroid-stimulating hormone) **c.** thyroxine **d.** ACRH (adrenocorticoid-releasing hormone) **e.** ACTH (adrenocorticotropic hormone) **f.** cortisol **5. a.** PP **b.** AP **c.** PP **d.** AP

**6. a.** thyroid **b.** thyroxine **c.** adrenal cortex **d.** cortisol **e.** ovaries **f.** estrogen, progesterone **g.** testes **h.** testosterone **7. a.** prolactin **b.** Melanocyte-stimulating **c.** skeletal **d.** dwarf **e.** giant **f.** acromegaly **g.** enlarge **8. a.** cretinism, simple goiter, myxedema **b.** exophthalmic goiter **c.** cretinism **d.** myxedema **e.** simple goiter **9. a.** (1) and (3) **b.** (1) and (3) **c.** (2) **d.** (1) **e.** (2) and (4) **f.** (1) **g.** (2) **10. a.** AM **b.** AC **c.** AM **d.** AC **e.** AC **f.** AM **g.** AM **h.** AC **11. a.** yes, no **b.** yes, no **c.** no, yes **d.** no, yes **e.** yes, no **f.** no, yes **12. a.** renin, **b.** angiotensin I **c.** angiotensin II **d.** aldosterone **e.** rises **f.** atrial natriuretic hormone (ANH) **13. a.** AD **b.** AD **c.** CS **d.** AD **e.** CS **f.** CS **g.** AD **h.** CS **14. a.** insulin **b.** glucagon **15. a.** increases **b.** decreases **c.** decreases **16. a.** 3,

5, 9 **b.** 6, 7, 9 **c.** 1, 8 **d.** 2, 4 **17. a.** 2 **b.** 1, 4 **c.** 3, 4
**18. a.** steroid hormone **b.** hormone-receptor complex
**c.** protein synthesis **d.** peptide hormone **e.** hormone-
receptor complex **f.** cyclic AMP **g.** active **19. a.** ES, NS
**b.** ES, NS **c.** NS **d.** NS **e.** ES

## GAME: HORMONE HOCKEY

First goal: **1.** f **2.** i **3.** h **4.** d **5.** b. Second goal:
**6.** f **7.** b **8.** d **9.** d **10.** a. Third goal: **11.** i **12.** h
**13.** b **14.** a **15.** l. Fourth goal: **16.** c **17.** g and h
**18.** d and e **19.** j **20.** f. Fifth goal: **21.** yes
**22.** no **23.** yes **24.** no **25.** no **26.** yes **27.** yes
**28.** no **29.** no **30.** yes

## DEFINITIONS WORDSEARCH

```
C R E T I N I S M        P
O A   H                  R
R A L   O X Y T O C I N  O
T M   C       R     N    S
I E   I       M   S   U  T
S D       T   O   N   L  A
O E           O N E I    G
L X         N E I        L
  Y             I N      A
  M             N E      N
N O R E P I N E P H R I N E D
                          I
                          N
```

**a.** calcitonin **b.** hormone **c.** cretinism **d.** oxytocin
**e.** insulin **f.** cortisol **g.** myxedema **h.** norepinephrine **i.** prostaglandin

## CHAPTER TEST

**1.** b **2.** c **3.** a **4.** a **5.** d **6.** b **7.** d **8.** b **9.** a
**10.** e **11.** a **12.** a **13.** d **14.** d **15.** c **16.** a
**17.** b **18.** b **19.** b **20.** a **21.** c **22.** a **23.** b
**24.** b **25.** c **26.** When an individual does not receive
enough iodine in the diet, the thyroid is unable to produce
thyroxine. The lack of thyroxine in the blood causes the
anterior pituitary to produce more TSH, and this hormone promotes increase in the size of the thyroid.
**27.** Renin leads to the formation of angiotensin II, which
stimulates the adrenal cortex to secrete aldosterone. Aldosterone causes sodium to be reabsorbed by the kidneys, and this leads to an increase in blood volume and
blood pressure.

# 21

# REPRODUCTIVE SYSTEM

## CHAPTER REVIEW

In males, **spermatogenesis** occurring in **seminiferous tubules** of the **testes** produces **sperm** that mature in the **epididymides** and may be stored in the **vasa deferentia.**

**Semen,** which contains mature sperm as well as secretions produced by **seminal vesicles,** the **prostate gland,** and **bulbourethral glands,** enters the **urethra** and is ejaculated during male orgasm when the **penis** is erect.

Hormonal regulation, involving secretions from the hypothalamus, the anterior pituitary, and the testes, maintains **testosterone** produced by the **interstitial cells** of the testes at a fairly constant level.

In females, an **egg** produced by an **ovary** enters an **oviduct,** which leads to the **uterus.** The uterus opens into the **vagina.** The external genital area includes the vaginal opening, the clitoris, the labia minora, and the labia majora.

In the nonpregnant female, the **ovarian** and **uterine cycles,** are under hormonal control of the hypothalamus, the anterior pituitary, and the female sex hormones **estrogen** and **progesterone.**

If fertilization occurs, the **corpus luteum** is maintained because of HCG production. Progesterone production does not cease, and the embryo implants itself in the thick uterine lining.

Estrogen and progesterone maintain the secondary sex characteristics of females, including less body hair than males, a wider pelvic girdle, a more rounded appearance, and development of breasts.

Infertile couples are increasingly resorting to assisted reproductive technologies. Numerous birth-control methods and devices are available for those who wish to prevent pregnancy.

Sexually transmitted diseases include **AIDS; herpes,** which repeatedly flares up; **genital warts,** which lead to cancer of the **cervix; gonorrhea** and **chlamydia,** which cause **pelvic inflammatory disease (PID); and syphilis,** which has cardiovascular and neurological complications if untreated.

## STUDY QUESTIONS

Study the text section by section. Answer the study questions so that you can fulfill the learning objectives for each section.

### 21.1 MALE REPRODUCTIVE SYSTEM (PP. 414–17)

The learning objectives for this section are:
- Describe the structure and function of the male reproductive system.
- State the path sperm take from the site of production until they exit the male.
- Name the glands that add secretions to semen.
- Discuss hormonal regulation of sperm production in the male.
- Name the actions of testosterone, including both primary and secondary sex characteristics.

1. Using the alphabetized list of terms and the blanks provided, identify and state a function for the parts of the human male reproductive system shown in the following diagram.

   bulbourethral gland    epididymis    penis    prostate gland    seminal vesicles
   testis    urethra    urinary bladder    vas deferens

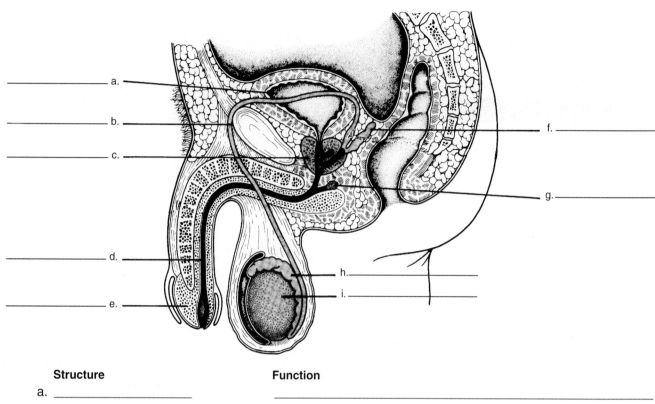

| Structure | Function |
|---|---|
| a. _____ | _____ |
| b. _____ | _____ |
| c. _____ | _____ |
| d. _____ | _____ |
| e. _____ | _____ |
| f. _____ | _____ |
| g. _____ | _____ |
| h. _____ | _____ |
| i. _____ | _____ |

2. Place the appropriate letters next to each statement.

   ST—seminiferous tubules         IC—interstitial cells

   a. _____ produce androgens
   b. _____ produce sperm
   c. _____ controlled by FSH
   d. _____ controlled by LH

In questions 3–7, fill in the blanks.

3. Trace the path of sperm through the male reproductive system.

   Testes to the <sup>a.</sup>_____ to the vas deferens to the <sup>b.</sup>_____.

4. What three organs add secretions to seminal fluid?

   a._____

   b._____

   c._____

5. What is the general function of these secretions? _____

   _____

6. The process of sperm production is called <sup>a.</sup>_____. This occurs inside

   <sup>b.</sup>_____ tubules inside each testis. Helper cells, known as <sup>c.</sup>_____ cells,

   nourish and regulate the developing sperm cells.

7. Mature sperm cells have three parts: <sup>a.</sup>_____, <sup>b.</sup>_____ and

   <sup>c.</sup>_____. A cap called the <sup>d.</sup>_____ contains enzymes that allow a sperm to enter

   an egg. What section of a sperm cell contains the mitochondria that provide energy for motility?

   <sup>e.</sup>_____

8. Indicate whether these statements are true (T) or false (F). Rewrite the false statements to make true
   statements.

   a._____ Testosterone exerts negative feedback control over the anterior pituitary secretion of LH.

   Rewrite:_____

   b._____ Inhibin exerts negative feedback control over the anterior pituitary secretion of FSH.

   Rewrite: _____

9. What are some of the effects of testosterone on the development of secondary sex characteristics?

   _____

   _____

## 21.2 FEMALE REPRODUCTIVE SYSTEM (PP. 418–20)

The learning objectives for this section are:
- Describe the structure and function of the female reproductive system.
- Describe the structure of the ovaries.
- Label a diagram of the external female genitals.

10. Using the alphabetized list of terms and the blanks provided, identify and state a function for the human female reproductive structures and urinary structures shown in the following diagram.

cervix    ovary    oviduct    urethra    urinary bladder    uterus    vagina

**Structure**                                    **Function**

a. _____    _____

b. _____    _____

c. _____    _____

d. _____    _____

e. _____    _____

f. _____    _____

g. _____    _____

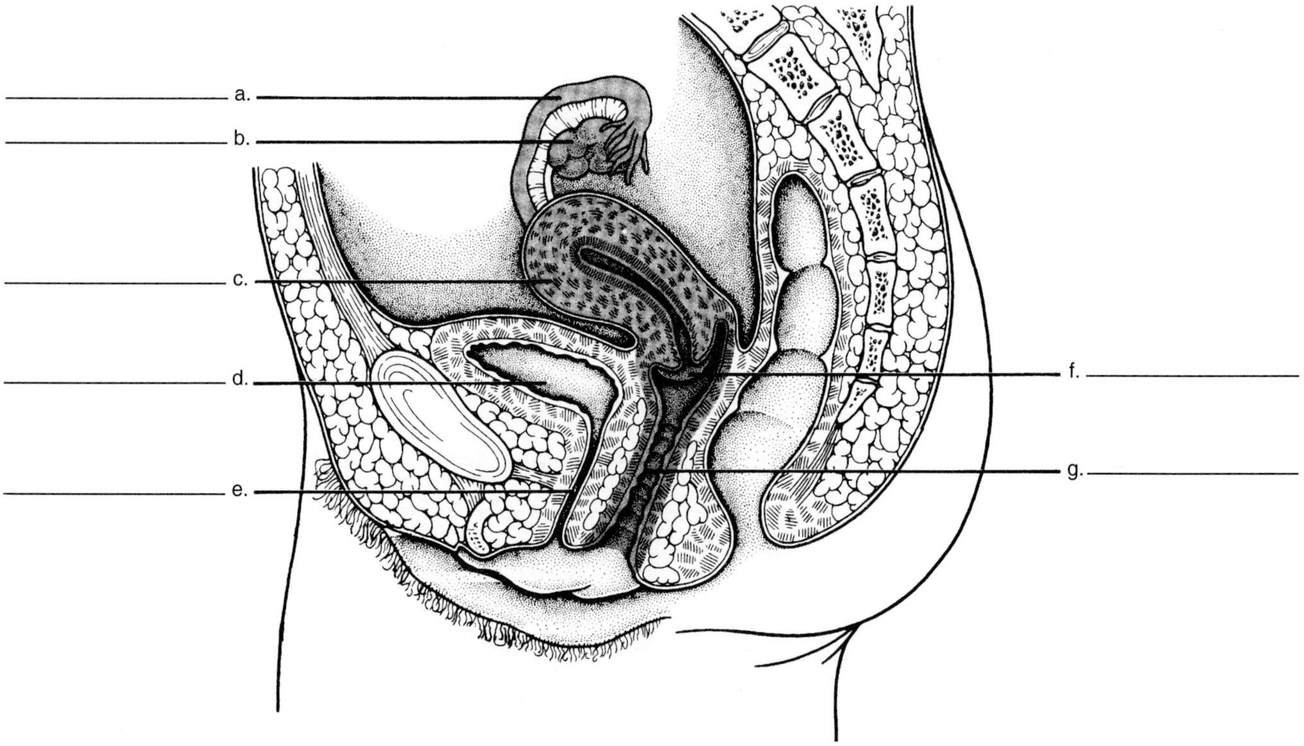

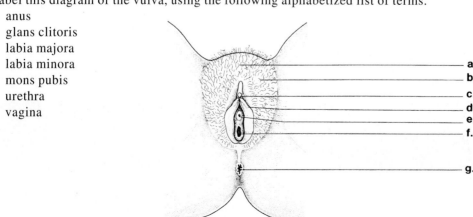

In question 11, fill in the blanks.

11. When sperm enter the female reproductive tract, they are deposited into the a._____. From there, they pass through the b._____ of the uterus. They swim up through the c._____ until they reach the egg cell.

12. Label this diagram of the vulva, using the following alphabetized list of terms.

anus
glans clitoris
labia majora
labia minora
mons pubis
urethra
vagina

a.
b.
c.
d.
e.
f.

g.

The learning objectives for this section are:
- Describe the ovarian and uterine cycles.
- Discuss hormonal regulation in the female, including feedback control.
- Name the actions of estrogen and progesterone, including the influence on secondary sex characteristics.

In question 13, fill in the blanks.

13. Each a._____ in the ovary contains an oocyte. A secondary follicle develops into a(n) b._____ follicle. c._____ is the release of the secondary oocyte (egg) from the ovary. Following ovulation, a follicle becomes a(n) d._____.

14. Fill in the following table to indicate the events in the ovarian and uterine cycles (simplified, and assuming a 28-day cycle).

| Ovarian Cycle | Events | Uterine Cycle | Events |
|---|---|---|---|
| Follicular phase—Days 1–13 | a. _____<br>Follicle maturation occurs.<br>Estrogen secretion is prominent. | b. _____—Days 1–5<br>d. _____—Days 6–13 | c. _____<br>e. _____ |
| Ovulation—Day 14 | LH spike occurs. | | |
| Luteal phase—Days 15–28 | LH secretion continues.<br>Corpus luteum forms.<br>h. _____ | f. _____—Days 15–28 | g. _____<br>_____ |

15. Match the definitions to these terms:
   estrogen     FSH     LH     progesterone

   a. _____ gonadotropic hormones

   b. _____ female sex hormones

   c. _____ primarily secreted by follicle

   d. _____ primarily secreted by corpus luteum

16. What are some of the effects of estrogen on the development of secondary sex characteristics?

   _____

   _____

17. Indicate whether the following statements are true (T) or false (F). Rewrite the false statements to make true statements.

   a. _____ Implantation occurs as soon as fertilization occurs. Rewrite: _____

   _____

   b. _____ HCG prevents degeneration of the corpus luteum. Rewrite: _____

   _____

   c. _____ During pregnancy, ovulation continues because estrogen and progesterone are still present. Rewrite:

   _____

## 21.4 CONTROL OF REPRODUCTION (PP. 425–28)

The learning objectives for this section are:
- Categorize birth control measures by the criteria used in the text.
- List the causes of infertility and the various assisted reproductive technologies.

18. Following are two groups of birth control measures. Rank the members of each group from the most effective (1) to the least effective (4).

|  | **A** |  | **B** |
|---|---|---|---|
| a. _____ | coitus interruptus | e. _____ | vasectomy |
| b. _____ | spermicidal jelly/cream | f. _____ | natural family planning |
| c. _____ | condom + spermicide | g. _____ | diaphragm + spermicide |
| d. _____ | natural family planning | h. _____ | IUD |

In questions 19–20, fill in the blanks.

19. The two common causes of infertility in females are <sup>a.</sup>_____ and
   <sup>b.</sup>_____.

20. The most common cause of infertility in males is <sup>a.</sup>_____ caused by <sup>b.</sup>_____.

21. In which assisted reproductive methods is the egg fertilized in laboratory glassware? _____
   _____

## 21.5 SEXUALLY TRANSMITTED DISEASES (PP. 429–35)

The learning objective for this section is:
* Identify several serious and prevalent sexually transmitted diseases.

In questions 22–25, fill in the blanks.

22. AIDS, genital herpes, and genital warts are <sup>a.</sup>_____ sexually transmitted diseases that do not respond to antibiotics. Gonorrhea and chlamydia are treatable with <sup>b.</sup>_____ therapy but are not always promptly diagnosed. AIDS (acquired immunodeficiency syndrome) is caused by a group of retroviruses known as <sup>c.</sup>_____ (human immunodeficiency viruses). HIV binds to the <sup>d.</sup>_____ . Once inside the host cell, HIV uses an enzyme called <sup>e.</sup>_____ to make a DNA copy of viral RNA. The viral RNA integrates into a host chromosome and makes more viral RNA. During the asymptomatic <sup>f.</sup>_____ phase, there are usually no symptoms, yet the person is highly infectious. Symptoms of a Category B infection begin to appear several months to several years after infection. The final stage of HIV infection is called <sup>g.</sup>_____, in which a person has a severe depletion of helper T lymphocytes and/or has an opportunistic infection.

23. Highly active antiretroviral therapy (HAART) therapy uses two drugs that inhibit <sup>a.</sup>_____ and one that inhibits protease needed for <sup>b.</sup>_____. The largest proportion of people with AIDS are <sup>c.</sup>_____ men, but the proportions attributed to intravenous drug users and heterosexuals are rising. Essentially, HIV is spread by passing virus-infected T lymphocytes found in <sup>d.</sup>_____ or in blood from one person to another. Genital herpes is caused by the <sup>e.</sup>_____ viruses. After the ulcers heal, the disease is only dormant, and blisters can reoccur repeatedly.

24. Genital warts are caused by the human <sup>a.</sup>_____ (HPVs), which are sexually transmitted and are now associated with cancer of the cervix and other tumors. Gonorrhea is caused by the bacterium <sup>b.</sup>_____ . In the male, a typical symptom of gonorrhea is a thick, greenish-yellow urethral discharge 3–5 days after contact. In females, it may spread to the oviducts, causing <sup>c.</sup>_____ disease (PID).

25. *Chlamydia* is named for the bacterium <sup>a.</sup>_____. Chlamydial infections of the genitals are the most common cause of <sup>b.</sup>_____ urethritis (NGU). Syphilis is caused by a bacterium called <sup>c.</sup>_____ and can be treated with penicillin. During the <sup>d.</sup>_____ stage of syphilis, a hard chancre (ulcerated sore with hard edges) indicates the site of infection. During the secondary stage, the victim breaks out in a <sup>e.</sup>_____ that does not itch. During the <sup>f.</sup>_____ stage, syphilis may affect the cardiovascular and nervous systems.

Review key terms by using the following alphabetized list of terms to fill in the blanks below. Then complete the wordsearch.

```
E T A T S O R P A P T E S T R
N B V C X X S A M U T O R C S
T E S T R O G E N T X V V V I
L K J H T R E W Q E C A V B T
P O I C E R V I X R F R C A S
B E S T A V W S X U H Y Z S E
V F I M B R I A S S A D F B T
G Y E F S M E N O P A U S E L
```

*cervix*
*estrogen*
*fimbria*
*menopause*
*ovary*
*Pap test*
*prostate*
*scrotum*
*testis*
*uterus*

a. _____ Narrow base of the uterus leading to vagina.
b. _____ Female sex hormone responsible for secondary sex characteristics.
c. _____ Muscular organ in which fetus develops.
d. _____ Pouch of skin that encloses testes.
e. _____ Fingerlike extension of oviduct.
f. _____ Termination of menstrual cycle in older women.
g. _____ Organ that produces sperm.
h. _____ Organ that produces eggs.
i. _____ Doughnut-shaped gland around male urethra.
j. _____ Clinical test to detect cervical cancer.

# CHAPTER TEST

## OBJECTIVE QUESTIONS

Do not refer to the text when taking this test.

1. The vas deferens
   a. becomes erect.
   b. carries sperm.
   c. is surrounded by the prostate gland.
   d. runs through bulbourethral glands.

____ 2. The prostate gland
   a. is removed when a vasectomy is performed.
   b. is not needed to maintain the secondary sex characteristics.
   c. receives urine from the bladder.
   d. almost never becomes cancerous.

____ 3. Which gland or organ secretes hormones?
   a. seminal vesicles
   b. prostate gland
   c. bulbourethral gland
   d. testes

____ 4. FSH
   a. stimulates sperm production in males.
   b. stimulates development of the follicle in females.
   c. is produced by the anterior pituitary.
   d. All of these are correct.

____ 5. Gonadotropic hormones are produced by the
   a. testes.
   b. ovaries.
   c. anterior pituitary.
   d. uterus.

____ 6. Which hormone stimulates the production of testosterone?
   a. LH
   b. FSH
   c. estrogen
   d. inhibin

____ 7. Which hormone regulates the production of testosterone?
   a. LH
   b. FSH
   c. estrogen
   d. inhibin

____ 8. The urethra is part of the reproductive tract in
   a. the female.
   b. the male.
   c. both the male and female.
   d. invertebrates.

____ 9. Which chromosome contains genes that determine whether a developing embryo develops into a female or a male?
   a. X chromosome
   b. Y chromosome
   c. chromosome #3
   d. chromosome #21

____ 10. The endometrium
   a. lines the vagina.
   b. along with the chorion is involved in the formation of the placenta.
   c. produces estrogen.
   d. None of these is correct.

____ 11. The uterus
   a. is connected to both the oviducts and the vagina.
   b. is not an endocrine gland.
   c. contributes to the development of the placenta.
   d. All of these are correct.

____ 12. Which structure is present after ovulation?
   a. primary follicle
   b. secondary follicle
   c. vesicular follicle
   d. corpus luteum

____ 13. Ovulation occurs
   a. due to hormonal changes.
   b. always on day 14.
   c. in postmenopausal women.
   d. as a result of intercourse.

____ 14. Which of these secretes hormones involved in the ovarian cycle?
   a. hypothalamus
   b. anterior pituitary gland
   c. ovary
   d. All of these are correct.

____ 15. FSH stimulates the
   a. release of an egg cell from the follicle.
   b. development of a follicle.
   c. development of the endometrium.
   d. beginning of menstrual flow.

____ 16. Secretions from which of the following structures are required before implantation can occur?
   a. the ovarian follicle
   b. the pituitary gland
   c. the corpus luteum
   d. All of these are correct.

____ 17. Human chorionic gonadotropin (HCG) is different from other gonadotropic hormones because it
   a. is produced by the hypothalamus.
   b. is not produced by a female endocrine gland.
   c. does not stimulate any tissue in the body.
   d. does not enter the bloodstream.

____ 18. Pregnancy is present
   a. when an egg develops in the corpus luteum.
   b. when ovulation occurs successfully.
   c. following fertilization and implantation.
   d. during the follicular phase only.

____ 19. Menstruation begins in response to
   a. increasing estrogen levels.
   b. decreasing progesterone levels.
   c. changes in blood chemistry.
   d. secretion of FSH.

____ 20. What do all the birth control methods have in common?
   a. They all use some device.
   b. They all interrupt intercourse.
   c. They are all terribly expensive and uncomfortable.
   d. None of these is correct.

____ 21. A vasectomy
   a. prevents the egg from reaching the oviduct.
   b. prevents sperm from reaching seminal fluid.
   c. prevents release of seminal fluid.
   d. inhibits sperm production.

____ 22. Which of these means of birth control prevents implantation?
   a. diaphragm
   b. IUD
   c. cervical cap
   d. vaginal sponge

____ 23. In vitro fertilization occurs in
   a. the vagina.
   b. a surrogate mother.
   c. laboratory glassware.
   d. the uterus.

____ 24. The _____ system provides nutrients for the growth of a developing fetus.
   a. digestive
   b. respiratory
   c. muscular
   d. lymphatic

____ 25. Androgens from the reproductive system stimulate the growth of skeletal muscle in which system?
   a. digestive system
   b. muscular system
   c. urinary system
   d. respiratory system

Answer in complete sentences.

26. How do the parts of a sperm assist its function?

27. Why do you expect to find sex hormones from the ovaries in pregnant women but not in menopausal women?

**Test Results:** _____ number correct ÷ 27 = _____ × 100 = _____%

## ANSWER KEY

### STUDY QUESTIONS

**1. a.** urinary bladder; stores urine **b.** vas deferens; conducts and stores sperm **c.** prostate gland; contributes to semen **d.** urethra; conducts both urine and sperm **e.** penis; organ of sexual intercourse **f.** seminal vesicles; contribute to semen **g.** bulbourethral gland; contributes nutrients and fluid to semen **h.** epididymis; stores sperm as they mature **i.** testis; production of sperm and male sex hormones **2. a.** IC **b.** ST **c.** ST **d.** IC **3. a.** epididymis **b.** urethra **4. a.** seminal vesicles **b.** prostate gland **c.** bulbourethral glands **5.** To nourish sperm cells, to increase the motility of sperm cells, and for lubrication. **6. a.** spermatogenesis **b.** seminiferous **c.** sustentacular (Sertoli) **7. a.** head **b.** middle piece **c.** tail **d.** acrosome **e.** middle piece **8. a.** T **b.** T **9.** Testosterone deepens the voice, promotes the development of muscles and body and facial hair, increases secretions from oil glands, and promotes the development of the sex organs. **10. a.** oviduct; conduction of egg **b.** ovary; production of eggs and sex hormones **c.** uterus; houses developing fetus **d.** urinary bladder; storage of urine **e.** urethra; conduction of urine **f.** cervix; opening of uterus **g.** vagina; receives penis during sexual intercourse and serves as birth canal **11. a.** vagina **b.** cervix **c.** oviduct **12. a.** mons pubis **b.** labia majora **c.** glans clitoris **d.** labia minora **e.** urethra **f.** vagina **g.** anus **13. a.** follicle **b.** vesicular **c.** Ovulation **d.** corpus luteum **14. a.** FSH secretion begins **b.** menstruation **c.** endometrium breaks down **d.** proliferative phase **e.** endometrium rebuilds **f.** secretory phase **g.** endometrium thickens and glands are secretory **h.** progesterone secretion is prominent **15. a.** FSH and LH **b.** progesterone and estrogen **c.** estrogen **d.** progesterone **16.** Estrogen promotes the deposition of body fat, the maturation and maintenance of the sex organs, and breast development. **17. a.** F, Implantation occurs several days after fertilization. **b.** T **c.** F, During pregnancy, ovulation discontinues because estrogen and progesterone secreted by the corpus luteum and the placenta exert feedback control over the hypothalamus and the anterior pituitary. **18. a.** 3 **b.** 2 **c.** 1 **d.** 4 **e.** 1 **f.** 4 **g.** 3 **h.** 2 **19. a.** blocked oviducts **b.** endometriosis **20. a.** low sperm count or large proportion of abnormal sperm **b.** environmental factors **21.** IVF and GIFT **22. a.** viral **b.** antibiotic **c.** HIV **d.** plasma membrane **e.** reverse transcriptase **f.** Category A acute **g.** AIDS **23. a.** reverse transcriptase **b.** viral assembly **c.** homosexual **d.** semen **e.** herpes simplex **24. a.** papillomaviruses **b.** *Neisseria gonorrhoeae* **c.** pelvic inflammatory **25. a.** *Chlamydia trachomatis* **b.** nongonococcal **c.** *Treponema pallidum* **d.** primary **e.** rash **f.** tertiary

### DEFINITIONS WORDSEARCH

```
E T A T S O R P A P T E S T
              M U T O R C S
E S T R O G E N T   V     I
              E     A     T
        C E R V I X R   R S
                  U   Y E
  F I M B R I A   S       T
      M E N O P A U S E
```

**a.** cervix **b.** estrogen **c.** uterus **d.** scrotum **e.** fimbria **f.** menopause **g.** testis **h.** ovary **i.** prostate **j.** Pap test

### CHAPTER TEST

**1.** b **2.** b **3.** d **4.** d **5.** c **6.** a **7.** d **8.** b **9.** b **10.** b **11.** d **12.** d **13.** a **14.** d **15.** b **16.** d **17.** b **18.** c **19.** b **20.** d **21.** b **22.** b **23.** c **24.** a **25.** b **26.** A sperm functions to fertilize an egg, and its various parts are specialized; for example, the head is capped by the acrosome, which releases enzymes that allow the sperm to penetrate the egg; the head contains 23 chromosomes; the middle piece contains mitochondria that provide energy; and the tail is a flagellum that allows the sperm cell to swim. **27.** The ovaries do not secrete female sex hormones in pregnant women or in menopausal women. During pregnancy, however, ovarian hormones are replaced by hormones secreted by the placenta. The placental hormones help to maintain the uterine lining, and thus, the pregnancy.

# 22

# DEVELOPMENT AND AGING

During **fertilization,** the acrosome of a sperm releases enzymes that digest a hole in the jelly coat around the egg and then extrudes a filament that attaches to a receptor on the vitelline membrane. The sperm nucleus enters the egg and fuses with the egg nucleus.

During the early developmental stages, **cleavage** leads to a **morula,** which becomes the **blastula** when an internal cavity (the **blastocoel**) appears. Then, at the **gastrula** stage, invagination of cells into the blastocoel results in formation of the germ layers: **ectoderm, mesoderm,** and **endoderm.** During neurulation, the nervous system develops from midline ectoderm, just above the **notochord.** At this point, it is possible to draw a typical cross section of a vertebrate embryo.

**Differentiation** begins with **cleavage,** when the egg's cytoplasm is partitioned among the numerous cells. In a frog embryo, only a daughter cell that receives a portion of the gray crescent is able to develop into a complete embryo. **Morphogenesis** involves the process of **induction,** as when the notochord induces the formation of the neural tube in frog embryos. Induction occurs because the inducing cells give off chemical signals that influence their neighbors.

Human development can be divided into embryonic development (months 1 and 2) and fetal development (months 3–9). The **extraembryonic membranes,** including the **chorion, amnion, yolk sac,** and **allantois,** appear early in human development. Following fertilization in an oviduct, cleavage occurs as the embryo moves toward the uterus. The **morula** becomes the **blastocyst** before implanting in the uterine lining.

Organ development begins with **neural tube** and heart formation. There follows a steady progression of organ formation during embryonic development. During fetal development, refinement of features occurs, and the fetus adds weight. Birth occurs about 280 days after the start of the mother's last menstruation.

Development after birth consists of infancy, childhood, adolescence, and adulthood. Young adults are at their prime, and then the aging process begins. Aging may be due to cellular repair changes, which are genetic in origin. Other factors that may affect aging are changes in body processes and certain extrinsic factors.

Study the text section by section. Answer the study questions so that you can fulfill the learning objectives for each section.

## 22.1 EARLY DEVELOPMENTAL STAGES (PP. 440—43)

The learning objectives for this section are:
- List the events that result in the fertilization of an egg.
- Describe the events in the early developmental stages of animals.
- Compare the development of three types of animals—lancelet, frog, and chick.

1. What is the proper sequence for these events, which describe fertilization? Indicate by letters. _____
    a. The plasma membrane depolarizes, which prevents the binding of any other sperm.
    b. When released, these enzymes digest a pathway for the sperm through the zona pellucida. The sperm binds to the plasma membrane of the egg.
    c. The head of a sperm has a membrane-bounded acrosome filled with digestive enzymes.
    d. The sperm enters the egg, and the sperm nucleus fuses with the egg nucleus, and fertilization is complete.

2. Label this diagram of early development, using the following alphabetized list of terms.
   archenteron
   blastocoel (used twice)
   blastopore
   blastula
   ectoderm
   endoderm
   gastrula
   morula

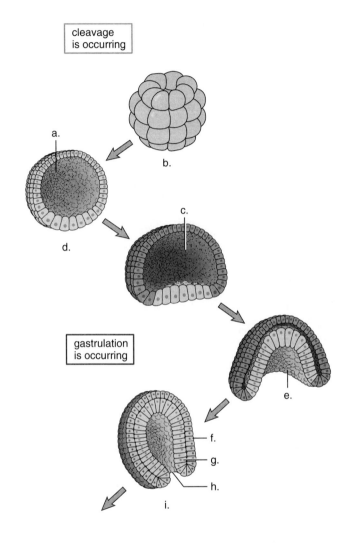

3. Indicate whether these statements are true (T) or false (F).
   a. _____ Cell division during cleavage does not produce growth.
   b. _____ The blastula is a solid ball of cells.
   c. _____ The ectoderm and endoderm form after the mesoderm in the gastrula.
   d. _____ The germ layer theory states that the development of later structures can be related to germ layers.
4. Indicate the germ layer (ectoderm, endoderm, mesoderm) of the vertebrate gastrula stage that is the source of the following:

   a. _____epidermis of the skin

   b. _____nervous tissue

   c. _____lining of the stomach

   d. _____muscles of the upper limb

   e. _____blood

In question 5, fill in the blanks.

5. Complete these sentences by using the terms *lancelet, frog,* or *chick.* In the <sup>a.</sup>_____ embryo, the cells have little yolk, and cleavage is equal. In the <sup>b.</sup>_____, the cells at the animal pole are smaller than those at the vegetal pole because those at the vegetal pole contain yolk. In the <sup>c.</sup>_____, the cells with yolk cleave more slowly than those without yolk. Still, in both the <sup>d.</sup>_____ and the <sup>e.</sup>_____, the blastula is a hollow ball of cells. In the <sup>f.</sup>_____, there is so much yolk that the embryo forms on top of the yolk, and the blastocoel is created when the cells lift up from the yolk.

6. Indicate which of these describes formation of mesoderm and the coelom in the *lancelet, frog,* or *chick.*

   a. _____ Invagination of cells along the edges of primitive streak is followed by a splitting of the mesoderm.

   b. _____ Migration of cells from the dorsal lip of the blastopore is followed by a splitting of the mesoderm.

   c. _____ Outpocketings of the primitive gut form two layers of mesoderm and the coelom.

7. Label this diagram of a vertebrate embryo, using the following alphabetized list of terms.
   coelom
   ectoderm
   endoderm
   gut
   mesoderm
   neural tube
   notochord
   somite

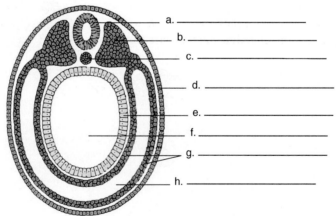

a. _____

b. _____

c. _____

d. _____

e. _____

f. _____

g. _____

h. _____

8. The diagram in question 7 shows the neural tube above the notochord. Explain the significance of this tissue relationship. _____

_____

The learning objectives for this section are:
- Explain how cytoplasmic segregation and induction bring about cellular differentiation and morphogenesis.
- Describe the benefits of research into the development of *C. elegans* and *Drosophila*.
- Understand how homeotic genes are involved in shaping the outward appearance of animals.

9. Place the appropriate letter(s) next to each statement.

   C—cytoplasmic segregation     I—induction     C, I—both

   a. _____ the parceling out of maternal determinants during cleavage and thereafter
   b. _____ embryos that receive a portion of the gray crescent develop normally
   c. _____ ability of one embryonic tissue to influence the development of another tissue
   d. _____ the nervous system develops above the notochord
   e. _____ the reciprocal development of the lens and optic vesicle
   f. _____ its importance is dependent on chemical signals that influence developing

10. Match the definitions to the terms. Answers can be used more than once, and there can be more than one answer per phrase.

   fate map     apoptosis     homeotic genes     morphogen gradients

   a. _____ *C. elegans*
   b. _____ induction is an ongoing process
   c. _____ during development, some cells die
   d. _____ proteins coded for by master genes
   e. _____ genes important in pattern formation
   f. _____ *Drosophila*
   g. _____ transcription factors; proteins that bind to DNA

11. Indicate whether these statements are true (T) or false (F).
   a. _____ Work with model organisms has shown that morphogen gradients coded for by master genes turn on the next set of master genes and so forth.
   b. _____ During induction, chemical signals pass from one tissue to the next.
   c. _____ Homeotic genes are restricted to *Drosophila*.
   d. _____ If development proceeds normally, each new cell is expected to have a particular fate—be a part of a particular organ and perform a particular function.

## 22.3 HUMAN EMBRYONIC AND FETAL DEVELOPMENT (PP. 448—58)

The learning objectives for this section are:
- Describe the stages of embryonic and fetal development.
- Name the parts of the extraembryonic membrane.
- Explain the role of the placenta in fetal development.

12. Label this diagram of the extraembryonic membranes of the human embryo, using the following alphabetized list of terms.

allantois
amnion
chorion
embryo
fetal portion of placenta
maternal portion of placenta
yolk sac

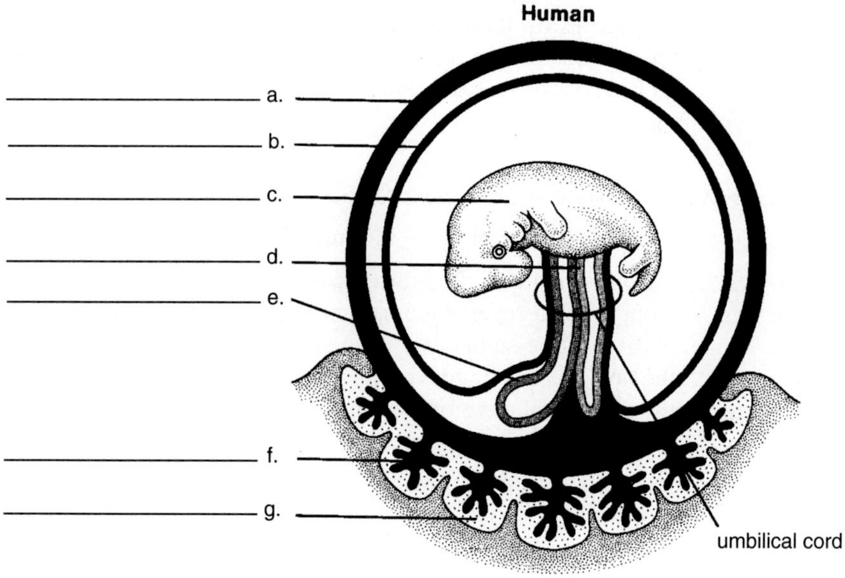

**Human**

a. _____
b. _____
c. _____
d. _____
e. _____
f. _____
g. _____

umbilical cord

13. Complete the following table:

| Membrane | Chick Function | Human Function |
|---|---|---|
| Chorion | | |
| Amnion | | |
| Allantois | | |
| Yolk sac | | |

14. To describe human embryonic development, complete the following table with the number of the event that occurs at each time indicated.

1  all internal organs formed; limbs and digits well formed; recognizable as human although still quite small
2  fertilization; cell division begins
3  limb buds begin; heart is beating; embryo has a tail
4  implantation; embryo has tissues; first two extraembryonic membranes
5  fingers and toes are present; cartilaginous skeleton
6  nervous system begins; heart development begins
7  head enlarges; sense organs prominent

| Time | Events |
|---|---|
| a. First week | |
| b. Second week | |
| c. Third week | |
| d. Fourth week | |
| e. Fifth week | |
| f. Sixth week | |
| g. Two months | |

15. Indicate whether these statements about fetal development (months 3–9) are true (T) or false (F).
    a. _____ It is possible to distinguish sex.
    b. _____ The notochord is replaced by the spinal column.
    c. _____ Limb buds are still present.
    d. _____ Fingernails and eyelashes appear.

In question 16, fill in the blanks.

16. The placenta produces the hormones progesterone and a. _____, which have two effects: to inhibit the hypothalamus and anterior pituitary from causing new follicles from maturing and to maintain the lining of the b. _____ so that the corpus luteum is not needed. The chorionic c. _____ are surrounded by maternal blood sinuses; yet maternal and fetal d. _____ never mix. The umbilical cord is the lifeline of the fetus, because the umbilical arteries take e. _____ and urea wastes to the placenta for disposal, and the umbilical f. _____ takes oxygen and nutrients to the fetus.

17. What events are associated with the three stages of parturition?
    a. first stage: _____
    b. second stage: _____
    c. third stage: _____

## 22.4 HUMAN DEVELOPMENT AFTER BIRTH (PP. 459–61)

The learning objectives for this section are:
- Describe the different stages of human development.
- Discuss underlying causes in the aging process.

In questions 18–19, fill in the blanks.

18. Three theories of aging: Some believe that aging is a. _____ in origin, meaning that our b. _____ inheritance causes us to age. Others maintain that c. _____ processes are involved; for example, the hormonal system and the immune system decrease in efficiency as we age. Still others believe that d. _____ factors influence aging more than we realize; for example, diet and exercise keep us healthy despite added years.

19. Aging affects body systems. Certain systems maintain the body; in regard to the cardiovascular system, a. _____ disease may be associated with b. _____ blood pressure, and reduced blood flow to the c. _____ may result in less efficiency at filtering wastes. In regard to those systems that integrate and coordinate the body, actually few d. _____ are lost from the brain, and the elderly can learn new material; loss of skeletal mass and osteoporosis can be controlled by e. _____. In regard to the reproductive system, there is a reduced level of f. _____ in both males and females, although males produce sperm until death. Young people should be aware that now is the time to begin the health habits that increase the life span.

Review key terms by using the following alphabetized list of terms to fill in the blanks below. Then complete the wordsearch.

```
D F E R G T H Y J U J Z P
C G E R O N T O L O G Y O
O U Y T R F G H A D F G A
L D S W E F R T N E R O M
O U M B I L I C U S D T N
S I O U K Y H R G S S E I
T M N B J O O P O T Y H O
R D P A R T U R I T I O N
U R E I Y T H R D C W S E
M H O E P I S I O T O M Y
V N I O Y R B M E A Q W S
```

*amnion*
*chorion*
*colostrum*
*embryo*
*episiotomy*
*gerontology*
*lanugo*
*parturition*
*zygote*

a. _____ Study of aging.

b. _____ Extraembryonic membrane that develops into placenta.

c. _____ Extraembryonic membrane that contains protective fluid.

d. _____ First milk.

e. _____ Diploid cell formed after union of two gametes.

f. _____ Fine down that covers the fetus.

g. _____ Process of giving birth.

h. _____ Incision to enlarge vagina while giving birth.

i. _____ Organism during first eight weeks of development.

# CHAPTER TEST

## OBJECTIVE QUESTIONS

Do not refer to the text when taking this test.

____ 1. The _____ develops first.
    a. morula
    b. blastula
    c. blastocoel
    d. gastrula

____ 2. The _____ is a hollow ball.
    a. morula
    b. blastula
    c. gastrula
    d. Both *a* and *b* are correct.

____ 3. The _____ contains germ layers.
    a. morula
    b. blastula
    c. gastrula
    d. All of these are correct.

____ 4. The _____ undergoes cleavage but lacks a morula.
    a. lancelet
    b. frog
    c. chick
    d. Both *a* and *b* are correct.

____ 5. The _____ has a notochord during development.
    a. lancelet
    b. frog
    c. human
    d. All of these are correct.

____ 6. The nervous system develops from the
    a. ectoderm.
    b. mesoderm.
    c. endoderm.
    d. notochord.

____ 7. Cellular differentiation is due to
   a. parceling out of cytoplasm.
   b. activation of particular genes.
   c. parceling out of genes.
   d. Both *a* and *c* are correct.
   e. Both *a* and *b* are correct.
____ 8. What induces the development of the nervous system?
   a. endoderm
   b. presumptive notochord
   c. ectoderm
   d. presumptive neural tube
____ 9. Homeodomain proteins
   a. occur in the nucleus.
   b. regulate transcription.
   c. regulate translation.
   d. Both *a* and *b* are correct.

For questions 10–14, match the descriptions to the extra-embryonic membranes.
   a. chorionic villi
   b. chorion
   c. amnion
   d. allantois
   e. yolk sac

____ 10. Placenta
____ 11. Umbilical blood vessels
____ 12. Watery sac
____ 13. First site of red blood cell formation
____ 14. Treelike extensions that penetrate the uterine lining

____ 15. The zygote begins to undergo cleavage in the
   a. cervix.
   b. ovary.
   c. oviduct.
   d. uterus.
____ 16. Which of these pairs is mismatched?
   a. cleavage—cell division
   b. morphogenesis—fertilization
   c. differentiation—specialization of cells
   d. growth—increase in size
____ 17. The placenta
   a. brings blood to the developing fetus.
   b. allows exchanges of substances between the mother's blood and fetal blood.
   c. forms the umbilical cord.
   d. Both *a* and *b* are correct.
   e. All of these are correct.
____ 18. When an embryo is clearly recognizable as a human being, it is called a
   a. developed embryo.
   b. fetus.
   c. newborn.
   d. blastocyst.
____ 19. During development, which system is the first to be visually evident?
   a. nervous
   b. respiratory
   c. digestive
   d. skeletal
____ 20. During which stage of parturition is the baby born?
   a. first
   b. second
   c. third

## THOUGHT QUESTIONS

Answer in complete sentences.
21. Should the chemicals functioning during induction be considered hormones?


22. What is the significance of the homeobox?


**Test Results:** _____ number correct ÷ 22 = _____ × 100 = _____ %

## STUDY QUESTIONS

**1.** c, b, d, a  **2. a.** blastocoel **b.** Morula **c.** Blastula **d.** blastocoel **e.** archenteron **f.** ectoderm **g.** endoderm **h.** blastopore **i.** Gastrula  **3. a.** T **b.** F **c.** F **d.** T  **4. a.** ectoderm **b.** ectoderm **c.** endoderm **d.** mesoderm **e.** mesoderm  **5. a.** lancelet **b.** frog **c.** frog **d.** lancelet **e.** frog **f.** chick  **6. a.** chick **b.** frog **c.** lancelet  **7. a.** neural tube **b.** somite **c.** notochord **d.** ectoderm **e.** endoderm **f.** gut **g.** mesoderm **h.** coelom  **8.** The notochord induces the formation of the neural tube. See Figure 22.5, p. 443, in text.  **9. a.** C **b.** C **c.** I **d.** I **e.** I **f.** C, I  **10. a.** fate map and apoptosis  **b.** fate map **c.** apoptosis  **d.** morphogen gradients  **e.** homeotic genes  **f.** homeotic genes and morphogen gradients **g.** morphogen gradients  **11. a.** T  **b.** T  **c.** F  **d.** T  **12. a.** chorion **b.** amnion **c.** embryo **d.** allantois **e.** yolk sac **f.** fetal portion of placenta **g.** maternal portion of placenta

**13.**

| Chick Function | Human Function |
|---|---|
| gas exchange | exchange with mother's blood |
| protection; prevention of desiccation and temperature changes | protection; prevention of temperature changes |
| collection of nitrogenous wastes | blood vessels become umbilical blood vessels |
| provision of nourishment | first site of blood cell formation |

**14. a.** 2 **b.** 4 **c.** 6 **d.** 3 **e.** 7 **f.** 5 **g.** 1  **15. a.** T **b.** F **c.** F **d.** T  **16. a.** estrogen **b.** uterus **c.** villi **d.** blood **e.** carbon dioxide **f.** vein  **17. a.** dilation of cervix **b.** The mother pushes as the baby moves down the birth canal. **c.** Afterbirth is expelled.  **18. a.** genetic **b.** genetic **c.** whole-body **d.** extrinsic  **19. a.** cardiovascular **b.** high **c.** kidneys **d.** neurons **e.** exercise **f.** hormones

## DEFINITIONS WORDSEARCH

```
                          Z
    C G E R O N T O L O G Y
    O           A     G A
    L           N     O M
    O           C U   T N
    S         H G E   E I
    T       O   O     O
    R   P A R T U R I T I O N
    U       I
    M   O E P I S I O T O M Y
      N   O Y R B M E
```

**a.** gerontology **b.** chorion **c.** amnion **d.** colostrum **e.** zygote **f.** lanugo **g.** parturition **h.** episiotomy **i.** embryo

## CHAPTER TEST

**1.** a  **2.** b  **3.** c  **4.** c  **5.** d  **6.** a  **7.** c  **8.** b  **9.** d  **10.** b  **11.** d  **12.** c  **13.** e  **14.** a  **15.** c  **16.** b  **17.** b  **18.** b  **19.** a  **20.** b  **21.** Traditionally, a hormone is considered to be a secretion of an endocrine gland that is carried in the bloodstream to a target organ. According to this definition, the chemicals that function during induction are not hormones. In recent years, some scientists have broadened the definition of a hormone to include all types of chemical signals. Therefore, in the broadest sense, these chemicals are hormones.  **22.** A particular sequence of DNA nucleotides, called the homeobox, occurs in homeotic genes in almost all eukaryotic organisms. This suggests that this sequence is important to development because it has been conserved for quite some time.

# 23

# PATTERNS OF GENE INHERITANCE

The genes are on the chromosomes; each gene has a minimum of two alternative forms, called **alleles.** Mendel's laws are consistent with the observation that each pair of alleles segregates independently of the other pairs during meiosis when the gametes form.

It is customary to use letters to represent the **genotype** of individuals. **Homozygous** dominant is indicated by two capital letters, and homozygous recessive is indicated by two lowercase letters. **Heterozygous** is indicated by a capital letter and a lowercase letter. In a one-trait cross, each heterozygous individual can form two types of gametes. In a two-trait cross, each heterozygous individual can form four types of gametes.

Use of the **Punnett square** allows us to make sure that all possible sperm have fertilized all possible eggs. These results tell us the chances of a child inheriting a particular phenotype. With regard to the **monohybrid** cross, there is a 25% chance of each child having the recessive **phenotype** and a 75% chance of each having the dominant phenotype.

**Testcrosses** are used to determine if an individual with the dominant phenotype is homozygous or heterozygous. If an individual expressing the **dominant allele** reproduces with an individual expressing the **recessive allele** and an offspring with the recessive phenotype results, we know that the individual is heterozygous.

Studies of human genetics have shown that many autosomal genetic disorders can be explained on the basis of simple Mendelian inheritance. When studying human genes, biologists often construct pedigree charts to show the pattern of inheritance of a characteristic within a family. The particular pattern indicates the manner in which a characteristic is inherited.

Tay-Sachs disease, cystic fibrosis, and PKU are autosomal recessive disorders that have been studied in detail. Neurofibromatosis and Huntington disease are autosomal dominant disorders that have been well studied.

There are many exceptions to Mendel's laws. These include **polygenic inheritance** (skin color), **multiple alleles** (ABO blood type), and degrees of dominance (curly hair).

For polygenic traits, several genes each contribute to the overall phenotype in equal, small degrees. The environment plays a role in the continuously varying expression that follows a bell-shaped curve. Several human disorders, such as cleft palate and human behaviors, are most likely controlled by polygenes. Sickle-cell disease is a human disorder that is controlled by incompletely dominant alleles.

Study the text section by section. Answer the study questions so that you can fulfill the learning objectives.

## 23.1 MENDEL'S LAWS (PP. 466–73)

The learning objectives for this section are:
- Explain the significance of the term alleles, and why the individual has two alleles and the gametes have one allele for each trait.
- Use Mendel's law of segregation to do one-trait genetics problems.
- Use Mendel's law of independent assortment to do two-trait genetics problems.

In question 1, fill in the blanks.

1. Letters on homologous chromosomes stand for genes that control a <sup>a.</sup> _____ , such as hair color. The genes are in a definite sequence and remain in their spots, or <sup>b.</sup> _____ , on the chromosomes. Alternate forms of a gene having the same position on a pair of homologous chromosomes and affecting the same trait are called <sup>c.</sup> _____ .

2. Indicate whether these statements about Mendel's law of segregation are true (T) or false (F).
   a. _____ Each individual has two alleles (factors) for each trait.
   b. _____ Alleles separate during fertilization.
   c. _____ Each gamete contains only one allele from each pair of factors.
   d. _____ Fertilization gives each new individual one factor for each trait.
   e. _____ Pairs of alleles must separate during meiosis.
   f. _____ Alleles are on chromosomes that are not homologous.

3. Complete the following table to distinguish between the terms genotype, phenotype, and gamete alleles. Assume that $W$ represents the dominant allele for widow's peak and $w$ represents the recessive allele for straight hairline.

| Genotype | Description of Genotype | Phenotype | Gamete(s) |
|---|---|---|---|
| a. | Homozygous dominant | b. | $W$ |
| $Ww$ | c. | d. | e. |
| f. | g. | Straight hairline | h. |

4. Using this key, $E$ = unattached, $e$ = attached, complete Punnett squares for cross 1 and cross 2.

Cross 1
heterozygous × homozygous recessive

Cross 2
heterozygous × heterozygous

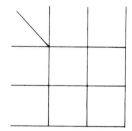

   a. What is the phenotypic ratio for cross 1?_____
   b. What are the chances of the recessive phenotype for cross 1?_____
   c. What is the phenotypic ratio for cross 2?_____
   d. What are the chances of the recessive phenotype for cross 2?_____

5. A man with a straight hairline reproduces with a woman with a widow's peak whose father has a straight hairline and whose mother had a widow's peak. The couple produce a child with a straight hairline. What are the genotypes of all the individuals involved?
   Use the alleles $W$ (widow's peak) and $w$ (straight hairline).

   a. man with a straight hairline _____ _____

   b. woman with a widow's peak _____ _____

   c. child _____ _____

   d. woman's father _____ _____

   e. woman's mother _____ _____

6. Indicate whether these statements are true (T) or false (F) about Mendel's law of independent assortment as it pertains to an individual with the genotype *AaBb*.

   a. _____ The alleles separate independently, so any possible combination (e.g., *AB, ab, Ab,* or *aB*) could be in the gametes.

   b. _____ The alleles separate independently, so even the combinations *Aa, Bb* could also be in the gametes.

   c. _____ The alleles do not separate independently, so only the combinations *AB* or *ab* could be in the gametes.

   d. _____ Each gamete contains one allele from each pair of alleles, and independent separation increases the variety of gametes for each individual.

In questions 7–8, fill in the blanks.

7. The cross *WwSs* × *WwSs* usually results in a phenotypic ratio close to 9:3:3:1. If *W* = widow's peak, *w* = straight hairline, *S* = short fingers, and *s* = long fingers, then out of 16 individuals:

   9 individuals are expected to have the phenotype ᵃ· _____.

   3 individuals are expected to have the phenotype ᵇ· _____.

   3 individuals are expected to have the phenotype ᶜ· _____.

   1 individual is expected to have the phenotype ᵈ· _____.

8. A parent with a widow's peak and short fingers has a child with a straight hairline and long fingers. What is the genotype of the parent? ᵃ· _____ If the parent were homozygous dominant for both traits, what would be the phenotype of the child? ᵇ· _____

9. a. Complete a Punnett square for the cross in horses *BbTt* × *Bbtt*, where:

   *B* = black
   *b* = brown
   *T* = trotter
   *t* = pacer

   b. What is the phenotypic ratio among offspring? _____ : _____ ;
   _____ : _____

## 23.2 GENETIC DISORDERS (PP. 474–76)

The learning objectives for this section are:
* Make use of a pedigree chart for both dominant and recessive genetic disorders.
* Describe the autosomal dominant and autosomal recessive disorders discussed in the text.
* Be able to solve genetics problems concerning autosomal dominant and autosomal recessive disorders.

In questions 10–11, fill in the blanks.

10. Following is a portion of a pedigree chart for a dominant disorder. Shaded individuals are affected.

Which of these individuals could be *AA*? ᵃ· _____ Explain: _____

_____

Which of these individuals is known to be *Aa*? ᵇ· _____ Explain: _____

_____

Which of these is *aa*? c. _____ Explain: _____

_____

Are heterozygotes affected or unaffected? d. _____

_____

11. Following is a portion of a pedigree chart for a recessive disorder.

Which of these individuals could be *AA*? a. _____ Explain: _____

_____

Which of these individuals is *Aa*? b. _____ Explain: _____

Which of these individuals is *aa*? c. _____ Explain:_____

Are homozygous dominant individuals and heterozygous individuals affected or unaffected? d._____

_____

For questions 12 and 13, study the pedigree charts. First, decide if the disorder (darkened shapes) is dominant or recessive. Then indicate the genotype of each person in the chart. Use the alleles *A* and *a* in each case.

12.

13.

14. Match the following disorders to each of the statements.
   1. neurofibromatosis
   2. Huntington disease
   3. Tay-Sachs disease
   4. cystic fibrosis
   5. phenylketonuria
   6. sickle-cell disease
   7. sickle-cell trait
   a. _____ nerve cell tumor growing under skin or in organs
   b. _____ accumulation of phenylketone in urine
   c. _____ allows better survival from malaria in Africa
   d. _____ brain cell degeneration causing muscle spasms
   e. _____ red blood cells are sickle shaped and clog arteries
   f. _____ mucus builds up in respiratory system
   g. _____ buildup of glycosphingolipid in lysosomes

15. A person heterozygous for Huntington disease reproduces with a person who is perfectly normal. What are the chances of an offspring developing Huntington disease when older? _____

16. Mary is 50 years old and has Huntington disease. Her father was killed accidentally at a young age; her mother is 80 years old and is normal. What is the most likely genotype of all persons involved? Use the alleles *A* and *a*.

    a. Mary's mother _____

    b. Mary's father _____

    c. Mary _____

17. Both parents appear to be normal, but their child has cystic fibrosis. What are the genotypes of all persons involved? Use the alleles *A* and *a*.

    a. parents _____ b. child _____

## 23.3 BEYOND SIMPLE INHERITANCE PATTERNS (PP. 477–79)

The learning objectives for this section are:
- Use skin color as an example of a polygenic trait.
- Use blood type as an example of a trait controlled by multiple alleles.
- Use wavy hair and sickle-cell trait as examples of incomplete dominance.

18. A woman has a genotype of *AABB* for skin color. She reproduces with a man whose genotype is *aabb*. What are the genotype and phenotype of their offspring?

    a. genotype _____ b. phenotype _____

19. Mrs. Doe and Mrs. Roe had babies at the same hospital. Mrs. Doe took home a girl, Nancy, and Mrs. Roe received a boy, Richard. However, Mrs. Roe was certain she gave birth to a girl and brought suit against the hospital. Blood tests showed that Mr. Roe was type O, Mrs. Roe was type AB, and Mr. and Mrs. Doe were both type B. Nancy was type A, and Richard was type O. List the possible genotypes of each individual, and then indicate the correct parents for each child. Genotypes:

    a. Mrs. Doe _____        d. Mrs. Roe _____

    b. Mr. Doe _____        e. Mr. Roe _____

    c. Nancy _____        f. Richard _____

    g. Whose baby is Nancy? _____

    h. Whose baby is Richard? _____

20. Curly and straight hair, when crossed, are an example of _____ dominance, because the resulting individuals have the intermediate character of wavy hair.

21. A curly-haired man has children with a wavy-haired woman. What are the genotypes and phenotypes of their children? Use the alleles *H* (curly hair) and *H'* (straight hair).

    a. man's genotype _____ b. woman's genotype _____
    Use this Punnett square to determine the outcome of this cross.

    c. Genotypes of children _____

    d. Possible phenotypes of children _____

    e. Could any of the children have straight hair? _____

22. A child has sickle-cell disease. What are the genotypes of the parents, who appear to be normal? _____

Review key terms by using the following alphabetized list of terms to fill in the blanks below. Then complete the wordsearch.

```
M E P O L Y G E N I C
U F H T H J K L O P A
L G E N O T Y P E U R
T C N F B T F D E S R
I O O X L I N K E D I
P P T I K Y H T G E E
L Y Y E D S R F G T R
E B P U N N E T T J U
F T E L E L L A P O L
```

*allele*
*carrier*
*genotype*
*multiple*
*phenotype*
*polygenic*
*Punnett*

a. _____ Alternative form of a gene on a chromosome.

b. _____ Outward expression of a gene.

c. _____ Pattern of inheritance in which many genes control one trait.

d. _____ Normal individual who carries a recessive allele.

e. _____ Genetic makeup of an individual.

f. _____ Type of square used to determine genetic outcome.

g. _____ More than two alleles for one trait are present in the population: _____ allele.

# CHAPTER TEST

## OBJECTIVE QUESTIONS

Do not refer to the text when taking this test.

____ 1. If 25% of the offspring of one set of parents show the recessive phenotype, the parents were probably
   a. both homozygous recessive.
   b. both homozygous dominant.
   c. both heterozygous.
   d. one homozygous dominant, one homozygous recessive.

____ 2. Alleles
   a. are alternate forms of a gene.
   b. have the same position on a pair of chromosomes.
   c. affect the same trait.
   d. All of these are correct.

____ 3. Which of these crosses could produce a blue-eyed child? ($B$ = brown, $b$ = blue)
   a. $BB \times bb$
   b. $Bb \times Bb$
   c. $bb \times Bb$
   d. $Bb \times BB$
   e. $Bb \times bb$

Questions 4 and 5 are concerned with the following pedigree chart. The shaded individuals are affected by a disorder, whereas the unshaded individuals are not affected.

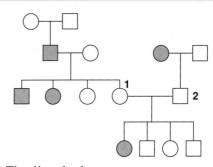

____ 4. The disorder is
   a. X-linked.
   b. dominant.
   c. recessive.
   d. not able to be determined.

____ 5. Individuals 1 and 2 are
   a. $AA \times AA$
   b. $aa \times aa$
   c. $Aa \times Aa$
   d. $Aa \times aa$
   e. $Aa \times AA$

____ 6. If a man is a carrier of Tay-Sachs disease, but a woman is homozygous normal, what are the chances of their child having Tay-Sachs?
   a. none
   b. 50%
   c. 25%
   d. 1:1

____ 7. Why is it that two normal parents could have a child with PKU?
  a. PKU is a dominant inherited disorder.
  b. PKU is a recessive inherited disorder.
  c. PKU results due to an error in gamete formation.
  d. There is no known explanation.

____ 8. If only one parent is a carrier for Huntington disease, what is the chance a child will have the condition?
  a. 25%
  b. 50%
  c. 75%
  d. no chance

____ 9. Polygenic inheritance can explain
  a. a range in phenotypes among the offspring.
  b. the occurrence of degrees of dominance.
  c. the inheritance of behavioral traits.
  d. Both a and c are correct.

____10. Sickle-cell disease illustrates
  a. dominance.
  b. recessiveness.
  c. incomplete dominance.
  d. multiple pairing.

____11. Two individuals with medium-brown skin color could have children who are both darker and lighter than they are.
  a. true
  b. false

____12. A female with light brown skin will be able to have a child with very dark skin if she reproduces with a very dark-skinned male, or a dark child if she reproduces with a light-skinned male.
  a. true
  b. false

____13. Children with which of the following blood types could not have parents who both have type A blood?
  a. type A
  b. type O
  c. type AB
  d. type B
  e. Both c and d are correct.

____14. Children with which of the following blood types could not have a parent with type AB blood?
  a. type A
  b. type B
  c. type AB
  d. type O

____15. Inheritance by multiple alleles is illustrated by the inheritance of
  a. skin color.
  b. blood type.
  c. sickle-cell disease.
  d. Both b and c are correct.

____16. When may complications arise regarding a pregnancy?
  a. $Rh^-$ woman and $Rh^+$ man
  b. $Rh^+$ woman and $Rh^-$ man
  c. $Rh^+$ woman and $Rh^+$ man
  d. $Rh^-$ woman and $Rh^-$ man

For questions 17–23, match the types of genetic crosses to each of these examples:

  a. multiple alleles        e. polygenic
  b. incomplete dominance     f. monohybrid cross
  c. disorder dominant        g. dihybrid cross
  d. disorder recessive

____17. $Rh^+ \times Rh^-$
____18. $AO \times AB$
____19. neurofibromatosis
____20. skin color
____21. Tay-Sachs disease
____22. $YySs \times YySs$
____23. wavy hair × straight hair

____24. In horses, $T$ = trotter, $t$ = pacer, $B$ = black, and $b$ = chestnut. If a dihybrid in both traits is mated with a chestnut pacer, the results will be
____  a. 1 black trotter : 1 chestnut pacer.
____  b. 3 black trotters : 1 chestnut pacer.
____  c. 1 black chestnut : 1 trotter pacer : 1 black pacer : 1 chestnut trotter.
____  d. 1 black trotter : 1 chestnut pacer : 1 black pacer : 1 chestnut trotter.

____25. If $Y$ = yellow seeds, $y$ = green seeds, $S$ = smooth seed coat, $s$ = rough seed coat, and a dihybrid is crossed with a dihybrid, then 9 out of 16 offspring will most likely be
  a. yellow, rough.
  b. yellow, smooth.
  c. green, rough.
  d. green, smooth.

Answer in complete sentences.

26. To test whether an animal has a homozygous dominant genotype or a heterozygous genotype, it is customary to mate them to the homozygous recessive animal rather than to a heterozygote. Why?

27. Discuss the concept that chance has no memory.

**Test Results:** _____ number correct ÷ 27 = _____ × 100 = _____%

# ANSWER KEY

## STUDY QUESTIONS

**1. a.** trait **b.** loci **c.** alleles **2. a.** T **b.** F **c.** T **d.** F **e.** T **f.** F **3. a.** *WW* **b.** widow's peak **c.** heterozygous **d.** widow's peak **e.** *W, w* **f.** *ww* **g.** homozygous recessive **h.** *w* **4. a.** 1:1 **b.** 50% **c.** 3:1 **d.** 25% **5. a.** *ww* **b.** *Ww* **c.** *ww* **d.** *ww* **e.** *Ww* **6. a.** T **b.** F **c.** F **d.** T **7. a.** widow's peak and short fingers **b.** widow's peak and long fingers **c.** straight hairline and short fingers **d.** straight hairline and long fingers **8. a.** *WwSs* **b.** widow's peak and short fingers
**9. a.**

|      | Bt   | bt   |
|------|------|------|
| BT   | BBTt | BbTt |
| Bt   | BBtt | Bbtt |
| bT   | BbTt | bbTt |
| bt   | Bbtt | bbtt |

**b.** 3 black trotters: 3 black pacers; 1 brown pacer: 1 brown trotter **10. a.** None. Both parents are affected, but they have an unaffected child. **b.** Both parents are *Aa*. Otherwise, they could not have an unaffected child. **c.** individual 3 because this individual is unaffected. **d.** affected **11. a.** None, Individuals 1 and 2 are not shaded, but they are heterozygous because they have an affected child. **b.** 1 and 2, They have to be heterozygous in order to have an affected child. **c.** Individual 3. Only the recessive can have parents that are unaffected. **d.** unaffected **12. top:** *aa, Aa* **middle:** *Aa, Aa; aa; aa; A? A?* **bottom:** *A? A? aa; aa, aa, aa; A?; A?, A?* **13. top:** *Aa, Aa* **middle:** *aa, Aa; aa, A?; A?, A?* **bottom:** *aa, Aa, Aa; Aa, Aa; A?, A?, A?* **14. a.** 1 **b.** 5 **c.** 7 **d.** 2 **e.** 6 **f.** 4 **g.** 3 **15.** 50% **16. a.** *aa* **b.** *Aa* **c.** *Aa* **17. a.** *Aa* **b.** *aa* **18. a.** *AaBb* **b.** medium brown skin color **19. a.** $I^B I^B$ or $I^B i$ **b.** $I^B I^B$ or $I^B i$ **c.** $I^A I^A$ or $I^A i$ **d.** $I^A I^B$ **e.** *ii* **f.** *ii* **g.** Nancy belongs to the Roes. **h.** Richard belongs to the Does. **20.** incomplete **21. a.** *HH* **b.** *HH'* **c.** *HH* or *HH'* **d.** curly or wavy hair **e.** no, because one parent must pass on an *H*. **22.** Both are heterozygous.

```
M   P O L Y G E N I C
U   H                 A
L G E N O T Y P E     R
T   N                 R
I   O                 I
P   T                 E
L   Y                 R
E   P U N N E T T
    E L E L L A
```

**a.** allele   **b.** phenotype   **c.** polygenic   **d.** carrier
**e.** genotype   **f.** Punnett   **g.** X-linked   **h.** multiple

**1.** c   **2.** d   **3.** b, c, e   **4.** c   **5.** c   **6.** a   **7.** b   **8.** b
**9.** d   **10.** c   **11.** a   **12.** b   **13.** e   **14.** d   **15.** b
**16.** a   **17.** f   **18.** a   **19.** c   **20.** e   **21.** d   **22.** g
**23.** b   **24.** d   **25.** b   **26.** If a heterozygote is mated to a heterozygote, there is a 25% chance for any offspring to be recessive. If a heterozygote is mated to a homozygous recessive, there is a 50% chance for any offspring to be recessive. It is only when the animal is mated to a homozygous recessive that its genotype can be determined.   **27.** The concept that chance has no memory refers to the idea that each pregnancy has the same probability as the previous one. If two heterozygous parents already have three children with a widow's peak and they are expecting a fourth child, this child still has a 75% chance of a widow's peak and a 25% chance of a continuous hairline.

**217**

# 24

# PATTERNS OF CHROMOSOME INHERITANCE

It is possible to karyotype the chromosomes of a cell during metaphase of mitosis. Then it is possible to observe that humans inherit 22 autosomes and one sex chromosome from each parent. Amniocentesis and chorionic villi sampling can provide fetal cells for the karyotyping of chromosomes when a medical disorder is suspected.

**Nondisjunction** during meiosis can result in an abnormal number of **autosomes** to be inherited. Down syndrome results when an individual inherits three copies of chromosome 21. Also, **chromosome mutations** lead to phenotypic abnormalities; for example, in cri du chat syndrome, one copy of chromosome 5 has a deletion.

The father determines the sex of a child because the mother gives only an **X chromosome** while the father gives an X or a **Y chromosome.** Males who inherit a fragile X chromosome are subject to mental retardation. Nondisjunction of the **sex chromosomes** can also cause abnormal sex chromosome numbers in offspring. Females who are XO have Turner syndrome, and those who are XXX are poly-X females. Males with Klinefelter syndrome are XXY. There are also XYY males.

Females have an inactive condensed X chromosome in their nuclei called a Barr body. If heterozygous, their cells differ in which allele is active. Sometimes this allows them to be tested to see if they are a carrier for a genetic disease.

Because males normally receive only one X chromosome, they are subject to disorders caused by the inheritance of a recessive allele on the X chromosome. For example, in a cross between a normal male and a carrier female, only the male children could have the X-linked disorder color blindness. Other well-known X-linked disorders are hemophilia and Duchenne muscular dystrophy.

All the genes on one chromosome form a **linkage group,** which is broken only when **crossing-over** occurs. Genes that are linked tend to go together into the same gamete. If crossing-over occurs, a dihybrid cross gives all possible phenotypes among the offspring, but the expected ratio is greatly changed. Crossing-over data have not helped to any degree to map the human chromosomes.

Study the text section by section. Answer the study questions so that you can fulfill the learning objectives for each section.

## 24.1 VIEWING THE CHROMOSOMES (PP. 484–85)

The learning objectives for this section are:
- Describe a normal karyotpe of a human being.
- Explain the purpose of amniocentesis and chorionic villi sampling.

1. Place A for amniocentesis and C for chorionic villi sampling on the appropriate lines.
   a. _____ 14 to 17 weeks of pregnancy
   b. _____ 5 weeks of pregnancy
   c. _____ Cells are obtained by suction.
   d. _____ Cells are obtained by needle.
   e. _____ Cells are from a cavity about the embryo/fetus.
   f. _____ Cells are from the extraembryonic membrane itself.

2. Match the descriptions with these terms:

   X and Y chromosomes    XX    homologous chromosomes    autosomes    XY

   a. _____ sex chromosomes
   b. _____ all chromosomes but the sex chromosomes
   c. _____ pairs of chromosomes
   d. _____ female
   e. _____ male

3. Label the photograph, using these terms:

   autosomes    homologous pair    karyotype of a male    sex chromosomes

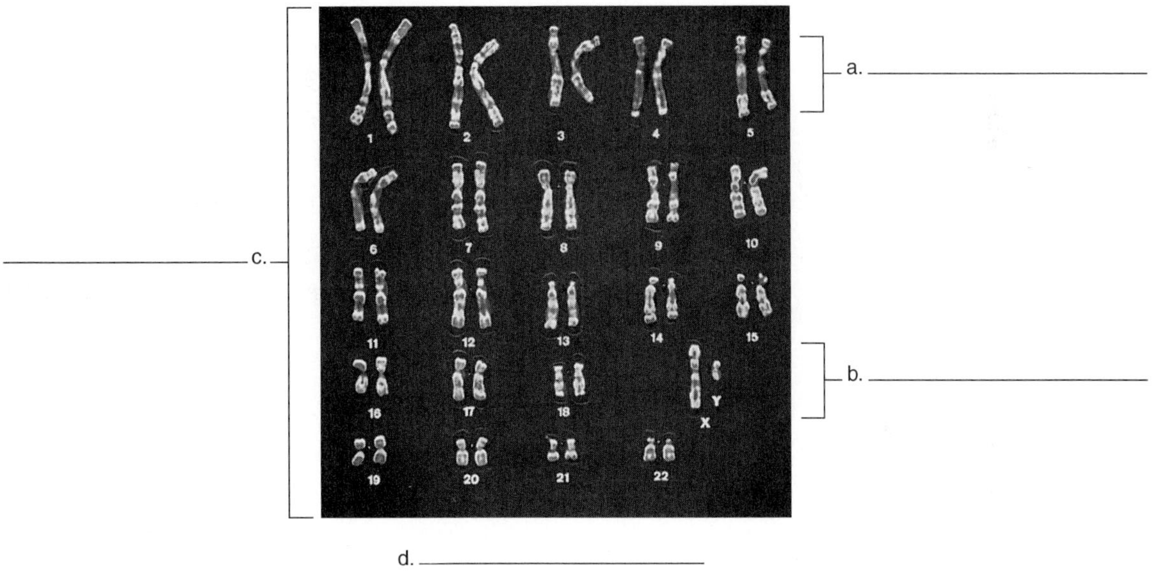

a. _____

c. _____

b. _____

d. _____

The learning objectives for this section are:
- Draw diagrams showing nondisjunction during meiosis I and meiosis II.
- Describe syndromes resulting from inheritance of an abnormal chromosome number.

4.  a. Down syndrome arises when the egg has two copies of chromosome 21. Complete this diagram to illustrate the occurrence of nondisjunction during meiosis II (left-hand side) and nondisjunction during meiosis I (right-hand side) to produce an individual with Down syndrome.

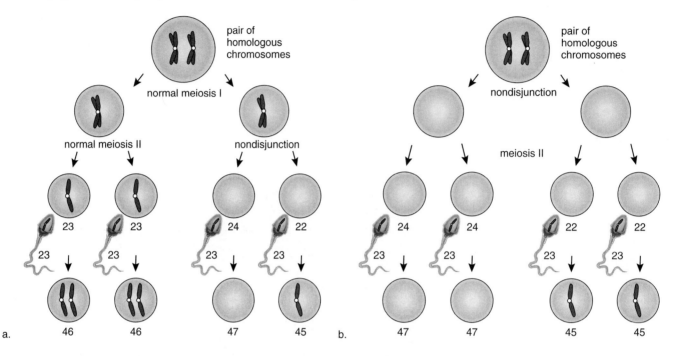

Following fertilization, if the zygote has one more chromosome than usual, a ᵇ· _____ has occurred. On the other hand, if the zygote has one less chromosome than usual, a ᶜ· _____ has occurred.

The learning objectives for this section are:
- Describe chromosome mutations, including deletions, duplications, translocations, and inversions.
- Describe the various syndromes associated with chromosome mutations.

5. Identify the types of chromosome mutations shown in the following illustration.

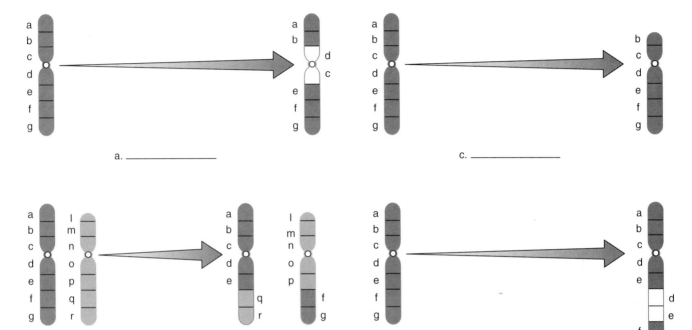

a. _____

c. _____

b. _____

d. _____

Cri du chat syndrome is due to which of these mutations? e. _____

6. Since males are XY and females are XX, the gender of the individual is normally determined by the

   a. _____ parent, depending on whether the offspring receives an b. _____ or a c. _____

   chromosome.

7. Match the following conditions to each of the descriptions below. Terms can be used more than once.
   1. Turner syndrome
   2. Klinefelter syndrome
   3. poly X individual
   4. Jacobs syndrome
   5. fragile X syndrome
   a. _____ female with no apparent physical abnormalities
   b. _____ hyperactive as children, protruding ears as adults, mentally retarded
   c. _____ male with some breast development, large hands
   d. _____ XYY male
   e. _____ XXY male
   f. _____ XXX female
   g. _____ XO female
   h. _____ X chromosome is nearly broken, leaving its tip hanging
   i. _____ female with no Barr body

The learning objectives for this section are:
• Describe traits that are controlled by genes located on the sex chromosomes.
• Explain why males always express X-linked disorders.

8. If $X^B$ = normal vision and $X^b$ = color blindness, state the sex and the phenotype of each of these genotypes:

$X^B X^B$    a. _____

$X^B X^b$    b. _____

$X^b X^b$    c. _____

$X^B Y$    d. _____

$X^b Y$    e. _____

9. a. Why do more males than females have X-linked genetic disorders? _____

   b. Why do sons inherit the disorder from their mothers? _____

10. a. Indicate the genotype of each person in the following pedigree chart. Use alleles *A* or *a* attached to an X chromosome in each case.

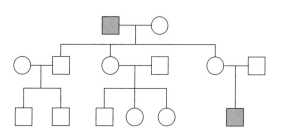

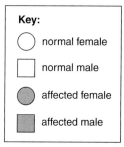

**Key:**
◯ normal female
▢ normal male
⬤ affected female
◼ affected male

   b. How do you know that this is a pedigree chart for an X-linked recessive trait? _____

11. Match the descriptions to these disorders:

     hemophilia    muscular dystrophy    color blindness

   a. _____ muscle weakness

   b. _____ can't see reds and greens

   c. _____ bleeder's disease

12. Use the Punnett square at right to show the expected outcome if a color-blind woman reproduces with a man who has normal vision.

   a. What are the chances of a color-blind daughter? _____

   b. What are the chances of a color-blind son? _____

13. A son is color blind, but his mother and father are not color blind. Give the genotype of all persons involved.

   son _____

   mother _____

   father _____

14. What is the genotype of a woman who is homozygous for widow's peak and a carrier for color blindness?

_____

## 24.5 LINKED GENES (PP. 496–97)

The learning objectives for this section are:
- Define linkage groups.
- Explain how linkage groups change the expected results of genetic crosses.

15. All the genes on one chromosome form a linkage group and tend to be inherited together. Mendel's law of

independent assortment _____ (does/does not) hold for linked genes.

In questions 16–18, consider that, in humans, arched eyebrow (*E*) is dominant over curved eyebrow (*e*), and hitchhiker thumb (*T*) is dominant over normal thumb (*t*). Imagine that these two genes are linked and that two dihybrids having these gametes reproduce.

16. From the diagram, indicate the phenotype for the following offspring:

    a._____

    b._____

    c._____

    d._____

17. a. What is the phenotypic ratio among the

    offspring? _____

    b. What would the ratio have been if the genes
were on nonhomologous chromosomes,
according to Mendel? _____

18. What are the chances (percent) that the offspring will have the following?

    a. _____ curved eyebrows and normal thumbs

    b. _____ arched eyebrows and hitchhiker thumbs

    c. _____ arched eyebrows and normal thumbs

    d. _____ curved eyebrows and hitchhiker thumbs

    e. _____ Is it correct to say that linkage cuts down on the possible number of phenotypes?

If crossing-over between the alleles occurs during meiosis, which of the genotypes would occur but in limited

number? f. _____ and g. _____

## DEFINITIONS WORDMATCH

Review key terms by completing this matching exercise, selecting from the following alphabetized list of terms:

*amniocentesis*
*chromosome mutation*
*karyotype*
*linkage group*
*nondisjunction*
*sex chromosomes*
*sex-linked trait*

a. _____ Alleles on the same chromosome are linked in the sense that they tend to move together to the same gamete; crossing-over interferes with linkage.

b. _____ Failure of homologues or sister chromatids to separate during the formation of gametes.

c. _____ Variation in regard to the normal number of chromosomes inherited or in regard to the normal sequence of alleles on a chromosome; the sequence can be inverted, translocated from a nonhomologous chromosome, deleted, or duplicated.

d. _____ Arrangement of all the chromosomes within a cell by pairs in a fixed order.

e. _____ Procedure for removing amniotic fluid surrounding the developing fetus for the testing of the fluid or cells within the fluid.

f. _____ Chromosome that determines the sex of an individual; in humans, females have two X chromosomes and males have an X and Y chromosome.

## OBJECTIVE QUESTIONS

Do not refer to the text when taking this test.

_____ 1. Which phrase best describes the human karyotype?
   a. 46 pairs of autosomes
   b. one pair of sex chromosomes and 23 pairs of autosomes
   c. X and Y chromosomes and 22 pairs of autosomes
   d. one pair of sex chromosomes and 22 pairs of autosomes

_____ 2. The gene arrangement on a chromosome changes from *ABCDEFG* to *ABCDEDEFG*. This is an example of
   a. deletion.
   b. duplication.
   c. inversion.
   d. linkage.

_____ 3. Which of the following conditions is NOT an example of a chromosome mutation?
   a. inversion
   b. translocation
   c. deletion
   d. duplication
   e. linkage

_____ 4. Which chromosome mutation does NOT require the presence of another chromosome?
   a. translocation
   b. duplication
   c. inversion
   d. All of these are correct.

_____ 5. Which type of chromosome mutation occurs when two simultaneous breaks in a chromosome lead to the loss of a segment?
   a. inversion
   b. translocation
   c. deletion
   d. duplication

_____ 6. Which of these is an autosomal abnormality?
   a. Turner syndrome
   b. Down syndrome
   c. Klinefelter syndrome
   d. poly-X syndrome

_____ 7. XYY (Jacob syndrome) males occur, due to nondisjunction during
   a. oogenesis.
   b. spermatogenesis.
   c. fertilization.
   d. mitosis.

_____ 8. Which condition is more likely to occur when the mother is over age 40?
   a. Turner syndrome
   b. poly-X syndrome
   c. Down syndrome
   d. Klinefelter syndrome

Questions 9 and 10 pertain to this pedigree chart.

_____ 9. The allele for this disorder is
   a. dominant.
   b. recessive.
   c. X-linked and recessive.
   d. None of these is correct.

_____ 10. The genotype of the starred individual is
   a. *Aa*.
   b. *aa*.
   c. $X^A X^a$.
   d. $X^A Y^a$.

_____ 11. A woman who is a carrier for color blindness reproduces with a man who has normal color vision. What is the chance they will have a color-blind daughter?
   a. 50%
   b. 25%
   c. 100%
   d. no chance

_____ 12. A color-blind woman reproduces with a man who has normal color vision. Their sons will
   a. be like the father because the trait is X-linked.
   b. be like the mother because the trait is X-linked.
   c. all have normal color vision.

_____ 13. A girl is color blind.
   a. She received a color-blind allele from her mother.
   b. She received a color-blind allele from her father.
   c. All her sons will be color blind.
   d. Her father is color blind.
   e. All of the above are correct.

_____ 14. Which chromosome has genes to determine male genital development?
   a. chromosome 5
   b. chromosome 10
   c. chromosome 21
   d. X chromosome
   e. Y chromosome

_____15. The _____ genotype indicates carrier female for color blindness.
 a. $X^B X^B$
 b. $X^B X^b$
 c. $X^b X^b$
 d. $X^B Y$
 e. $X^b Y$

_____16. The cri du chat syndrome in which an infant's cry resembles a cat's cry is due to a (an)
 a. deletion.
 b. duplication.
 c. inversion.
 d. translocation.
 e. invocation.

_____17. Which of the following statements is NOT correct?
 a. Females have an XX genotype.
 b. Males have an XY genotype.
 c. An egg always bears an X chromosome.
 d. Each sperm cell has an X and Y chromosome.
 e. The sex of the newborn child is determined by the father.

_____18. Which of the following genotypes is not paired with the correct number of Barr bodies?
 a. XY — none
 b. XO — one
 c. XX — one
 d. XXY — one
 e. XXX — two

_____19. Which of the following is NOT a characteristic of an X-linked recessive disorder?
 a. More males than females are affected.
 b. An affected son can have parents who have the normal phenotype.
 c. For a female to have the characteristic, her father must also have it.
 d. The characteristic often skips a generation from the grandmother to the granddaughter.
 e. If a woman has the characteristic, all of her sons will have it.

_____20. Which of the following is NOT considered an X-linked recessive disorder?
 a. Tay-Sachs disease
 b. color blindness
 c. hemophilia
 d. muscular dystrophy (some forms)
 e. agammaglobulinemia

_____21. Which of the following statements is NOT true about a linkage group?
 a. It includes all alleles on one chromosome.
 b. Traits controlled by linked genes tend to be inherited together.
 c. If linkage is complete, a dihybrid produces only two types of gametes in equal proportions.
 d. Incomplete linkage can be due to crossing-over between nonsister chromatids.

_____22. The X-linked disease prevalent among royal families of Europe at the turn of the century was
 a. muscular dystrophy.
 b. color blindness.
 c. hemophilia.
 d. fragile-X syndrome.
 e. cri du chat syndrome.

## THOUGHT QUESTIONS

Answer in complete sentences.

23. Why is it evident that a gene for maleness exists on the Y chromosome?

24. Why are color-blind women rare?

**Test Results:** _____ number correct ÷ 24 = _____ × 100 = _____ %

## STUDY QUESTIONS

**1. a.** A **b.** C **c.** C **d.** A **e.** A **f.** C   **2. a.** X and Y chromosomes **b.** autosomes **c.** homologous chromosomes **d.** XX **e.** XY   **3. a.** homologous pair **b.** sex chromosomes **c.** autosomes **d.** karyotype of a male   **4. a.** See Figure 24.2 in text. **b.** trisomy **c.** monosomy   **5. a.** inversion **b.** translocation **c.** deletion **d.** duplication **e.** deletion   **6. a.** male **b.** X **c.** Y   **7. a.** 3 **b.** 5 **c.** 2 **d.** 4 **e.** 2 **f.** 3 **g.** 1 **h.** 5 **i.** 1   **8. a.** female with normal vision **b.** female who is a carrier **c.** female who is color blind **d.** male with normal vision **e.** male who is color blind **9. a.** If a male inherits the recessive gene, he always has the disorder. **b.** Only mothers pass on an X chromosome to their sons.   **10. a. top:** $X^aY$, $X^AX^A$ **middle:** $X^AX^A$, $X^AY$, $X^AX^A$, $X^AY$, $X^AX^a$, $X^AY$ **bottom:** $X^AY$, $X^AY$, $X^AY$, $X^AX^A$, $X^AX^A$, $X^aY$ **b.** Only males have the disorder, and it passes from grandfather to grandson by way of a female.   **11. a.** muscular dystrophy **b.** color blindness **c.** hemophilia   **12. a.** none **b.** 100%   **13.** son $X^bY$, mother $X^BX^b$, father $X^BY$   **14.** $WWX^BX^b$   **15.** does not **16. a.** arched eyebrow and hitchhiker thumb **b.** curved eyebrow and normal thumb **c.** arched eyebrow and hitchhiker thumb **d.** arched eyebrow and hitchhiker thumb **17. a.** 3:1 **b.** 9:3:3:1   **18. a.** 25% **b.** 75% **c.** 0% **d.** 0% **e.** yes **f.** curved eyebrow and hitchhiker thumb **g.** arched eyebrow and normal thumb

## DEFINITIONS WORDMATCH

**a.** linkage group **b.** nondisjunction **c.** chromosome mutation **d.** karyotype **e.** amniocentesis **f.** sex chromosomes

## CHAPTER TEST

**1.** d **2.** b **3.** e **4.** c **5.** c **6.** b **7.** b **8.** c **9.** c **10.** c **11.** d **12.** b **13.** e **14.** e **15.** b **16.** a **17.** d **18.** b **19.** d **20.** e **21.** d **22.** c **23.** Any individual receiving a Y chromosome is male.   **24.** A color-blind woman has to receive an allele for color blindness from both parents. If she receives only one allele for color blindness and one normal allele, she will not be color blind.

# 25

# MOLECULAR BASIS OF INHERITANCE

DNA, the genetic material, is a double helix containing the nitrogen-containing bases A (adenine) paired with T (thymine), and G (guanine) paired with C (cytosine). During **replication,** DNA "unzips," and then a complementary strand forms opposite to each original strand.

RNA is a single-stranded nucleic acid in which A pairs with U (uracil) while G still pairs with C.

DNA specifies the synthesis of proteins because it contains a **triplet code:** Every three bases stand for one amino acid. During **transcription, mRNA** is made complementary to one of the DNA strands. mRNA, bearing codons, moves to the cytoplasm, where it becomes associated with the ribosomes. During **translation, tRNA molecules,** attached to their own particular amino acids, travel to a ribosome, and through **complementary base pairing** between **anticodons** and **codons,** the tRNAs and therefore the amino acids in a polypeptide are sequenced in a predetermined way.

The prokaryote **operon** model explains how one **regulator gene** controls the transcription of several **structural genes,** genes that code for proteins. The following levels of control of gene expression are possible in eukaryotes: transcriptional control, posttranscriptional control, translational control, and posttranslational control. In eukaryotic cells, the chromosome has to decompact before transcription can begin. Transcription factors attach to DNA and turn on particular genes.

In molecular terms, a **gene** is a segment of DNA, and a mutation is a change in the normal sequence of nucleotides of this segment. Frameshift mutations result when a base is added or deleted and the result is a nonfunctioning protein. Point mutations can range in effect, depending on the particular codon change. Gene mutation rates are rather low, because DNA polymerase proofreads the new strand during replication and because there are repair enzymes that constantly monitor the DNA.

**Cancer** is characterized by a lack of control: The cells grow uncontrollably and **metastasize.** Cancer development is a multistep process involving the mutation of genes. **Proto-oncogenes** and **tumor-suppressor genes** are normal genes that bring on cancer when they mutate because they code for factors involved in cell growth.

Study the text section by section. Answer the study questions so that you can fulfill the learning objectives for each section.

## 25.1 DNA STRUCTURE AND REPLICATION (PP. 502–5)

The learning objectives for this section are:
- Name the parts of the DNA double helix.
- Explain the process of DNA replication, and why it is termed semiconservative.

In question 1, fill in the blanks.

1. The following diagram shows that the two separate experiments used $^{32}$P to label a._____ and $^{35}$S to label b._____ of viruses. In each experiment, the viruses were allowed to infect bacteria, and then a blender was used to separate the capsids from the bacteria. Radioactivity was found inside the cell only when c._____ was labeled. Since replication of viruses followed, the hypothesis that d._____ is the genetic material was supported.

Label *e–i* on the diagram to tell the location of radioactivity.

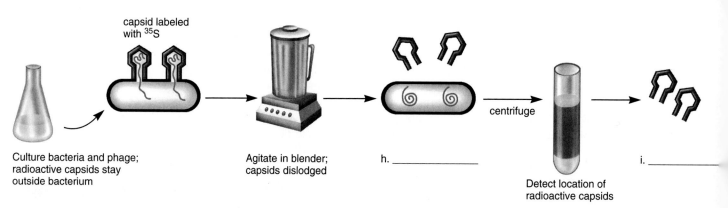

Culture bacteria and phage; e. _____

Agitate in blender; capsids dislodged

Empty capsids outside; f. _____

Detect location of radioactive DNA

g. _____

**a. Viral DNA is labeled** (red).

Culture bacteria and phage; radioactive capsids stay outside bacterium

Agitate in blender; capsids dislodged

h. _____

Detect location of radioactive capsids

i. _____

**b. Viral capsid is labeled** (red).

2. Four different nucleotides are found in DNA. Check the way(s) these nucleotides differ.
   a. _____ They differ in their sugar content.
   b. _____ They differ in their phosphate content.
   c. _____ They differ in their base content.
3. What are the four different nucleotide bases in DNA? _____

4. Examine this diagram, which shows the ladder structure of DNA.

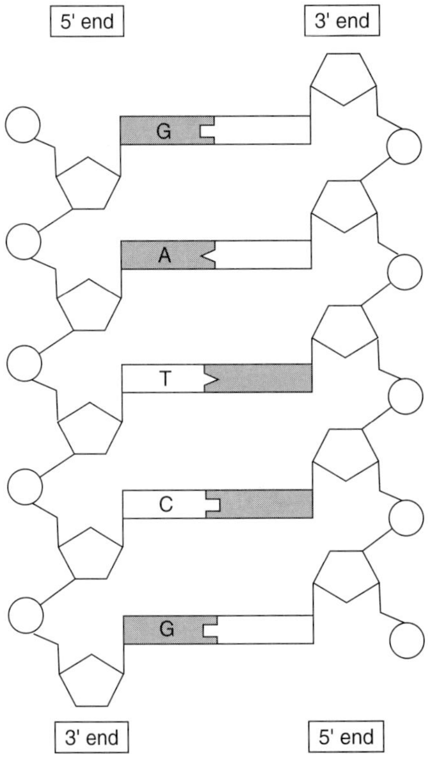

5' end        3' end

a. DNA is a polymer of _____.

b. Draw a box around one nucleotide.

c. The molecules making up the sides of the ladder are

   _____.

d. Label a sugar and a phosphate.

e. What is meant by the phrase *complementary base pairing?*

   _____

f. Add the bases that are complementary to those on the left.

g. What do you have to do to the ladder structure to have it match the

   Watson and Crick model? _____

h. Explain what is meant by *double-stranded helix.* _____

   _____

i. Explain what is meant by *antiparallel strands.* _____

   _____

3' end        5' end

5. Study this diagram of replication.

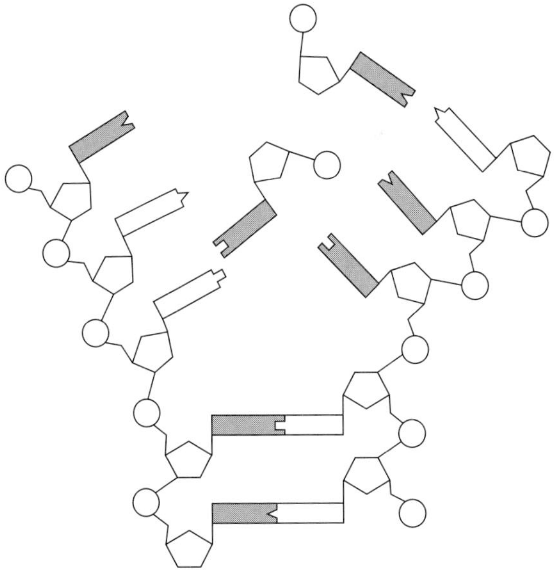

a. The bases in parental DNA are held together by what

   type of bond (not shown)? _____

b. What happens to these bonds for replication to take

   place? _____

c. During replication, new nucleotides move into proper

   position by what methodology? _____

d. Elongation of DNA is catalyzed by an enzyme called

   _____. When replication is finished there

   will be two DNA molecules.

e. Each double helix consists of an _____ strand and a _____ strand. Therefore, the

   process is called _____.

f. Each double helix has _____ (the same, a different) sequence of complementary paired bases.

The learning objective for this section is:
• Define the two steps of gene expression—translation and transcription.

6. Indicate whether these statements are true (T) or false (F).
   _____ a. Metabolic disorders are due to the inheritance of faulty enzymes.
   _____ b. Metabolic disorders are due to the inheritance of faulty proteins.
   _____ c. A gene is a segment of DNA that specifies the sequence of amino acids in a protein.
   _____ d. A gene is a segment of DNA that specifies the sequence of amino acids in a polypeptide.

7. Since genes (DNA) reside in the a._____ of the cell and polypeptide synthesis occurs in the

   b._____, they must have a go-between. The most likely molecule to fill this role is c._____.

8. Indicate whether these statements about differences between DNA and RNA are true (T) or false (F).
   a. _____ DNA is double stranded; RNA is single stranded.
   b. _____ DNA is a polymer; RNA is a building block of that polymer.
   c. _____ DNA occurs in three forms; RNA occurs in only one form.
   d. _____ The sugar of DNA is ribose, which is absent in RNA.
   e. _____ Uracil, in RNA, replaces the base thymine, found in DNA.

9. Complete this table to describe the function of the various types of RNA involved in protein synthesis.

| RNA | Function |
|---|---|
| Messenger RNA (mRNA) | a. _____ |
| Ribosomal RNA (rRNA) | b. _____ |
| Transfer RNA (tRNA) | c. _____ |

10. Label this diagram, which pertains to the central concept of genetics.

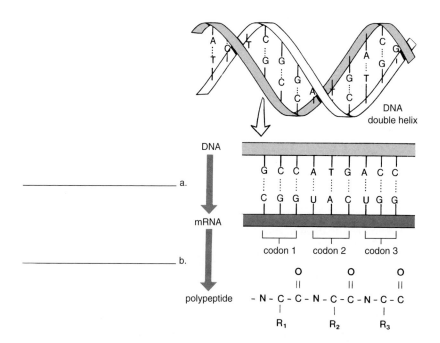

11. Study this figure, which lists the mRNA codons.

| First Base | Second Base | | | | Third Base |
|---|---|---|---|---|---|
| | U | C | A | G | |
| U | UUU phenylalanine | UCU serine | UAU tyrosine | UGU cysteine | U |
| | UUC phenylalanine | UCC serine | UAC tyrosine | UGC cysteine | C |
| | UUA leucine | UCA serine | UAA stop | UGA stop | A |
| | UUG leucine | UCG serine | UAG stop | UGG tryptophan | G |
| C | CUU leucine | CCU proline | CAU histidine | CGU arginine | U |
| | CUC leucine | CCC proline | CAC histidine | CGC arginine | C |
| | CUA leucine | CCA proline | CAA glutamine | CGA arginine | A |
| | CUG leucine | CCG proline | CAG glutamine | CGG arginine | G |
| A | AUU isoleucine | ACU threonine | AAU asparagine | AGU serine | U |
| | AUC isoleucine | ACC threonine | AAC asparagine | AGC serine | C |
| | AUA isoleucine | ACA threonine | AAA lysine | AGA arginine | A |
| | AUG (start) methionine | ACG threonine | AAG lysine | AGG arginine | G |
| G | GUU valine | GCU alanine | GAU aspartic acid | GGU glycine | U |
| | GUC valine | GCC alanine | GAC aspartic acid | GGC glycine | C |
| | GUA valine | GCA alanine | GAA glutamic acid | GGA glycine | A |
| | GUG valine | GCG alanine | GAG glutamic acid | GGG glycine | G |

a. What does it mean to say that the genetic code is a triplet code? _____

b. What are the mRNA codons for leucine? _____

_____

12. Complete this paragraph to describe transcription.

During transcription, an RNA molecule is formed that has a sequence of bases ᵃ·_____ to a

portion of one DNA strand. The bases pair in this manner: A in DNA pairs with ᵇ·_____, and G

pairs with ᶜ·_____

(and vice versa) in the mRNA being formed. If the sequence of bases in DNA is CGA AGC TCT, then the

sequence in mRNA is ᵈ· _____

Why is there a space between every three bases? ᵉ· _____

_____

_____

13. Which one, exons or introns, is spliced out when primary RNA is processed? ᵃ·_____

Organic catalysts called ᵇ·_____ do the splicing.

14. Label this diagram to describe translation.

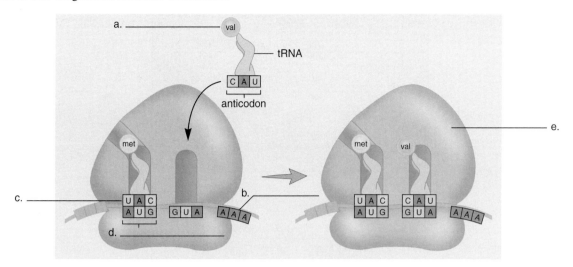

In questions 15–17, fill in the blanks.

15. Two types of RNA are seen in the previous diagram. Ribosomal RNA (rRNA) plus proteins make up the ribosomes. Each ribosome is composed of a ᵃ· _____ subunit and a ᵇ· _____ subunit. Transfer RNA is the second type of RNA in the diagram. At one end, an ᶜ· _____ attaches, and at the other end there is an ᵈ· _____, which is complementary to a codon in mRNA.

16. The three steps in protein synthesis are ᵃ· _____, when the ribosomal subunits ᵇ· _____; ᶜ· _____, when a polypeptide is ᵈ· _____; and ᵉ· _____, when the last tRNA, the mRNA, and the ribosome ᶠ· _____.

17. During elongation, the sequence of ᵃ· _____ in mRNA dictates the order of ᵇ· _____ in the polypeptide. For example, if the sequence of bases in mRNA is UUU UUA AUU GUC CCA, the sequence of amino acids in the polypeptide according to the figure in question 11 will be ᶜ· _____. Because several ribosomes, called a ᵈ· _____, can move along one mRNA molecule, ᵉ· _____ (one/many) polypeptides of the same type can be synthesized at a time.

The learning objective for this section is:
• Understand the four levels at which gene expression occurs in eukaryotic cells.

18. Label this diagram of a generalized operon, using the following alphabetized list of terms.

active repressor
DNA
mRNA
operator
promoter
regulator gene
structural genes

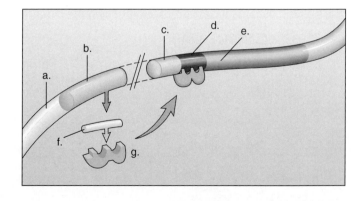

19. Complete the following table:

| Levels of Control of Gene Activity | Affects the Activity Of |
|---|---|
| | |
| | |
| | |
| | |

20. a. What are transcription factors? _____

_____

    b. What type of cell has transcription factors? _____

## 25.4 GENE MUTATIONS (P. 516)

The learning objective for this section is:
- Define gene mutations and give different examples.

21. Match the statements to these terms:
    frameshift    point    mutagen    transposons
    a. _____ type of mutation (requires two)
    b. _____ environmental influence that causes mutations
    c. _____ DNA sequence that can move between chromosomes
    d. _____ The codons AAU GUA CCU GGU become AUG UAC CUG GU
    e. _____ The codons AAU GUA CCU GGU become AUU CUA CCU GGU

## 25.5 CANCER: A FAILURE OF GENETIC CONTROL (PP. 518–21)

The learning objective for this section is:
- Describe characteristics of cancer cells.

22. Complete the following table:

| Characteristics of Normal Cells | Characteristics of Cancer Cells |
|---|---|
| Controlled growth | a. |
| Contact inhibition | b. |
| One organized layer in tissue culture | c. |
| Differentiated cells | d. |
| Normal nuclei | e. |

In question 23, fill in the blanks.

23. Instead of growing in a._____ layer(s), as normal cells do, cancer cells grow in b._____
    layer(s), losing the property of c._____ inhibition. Cancer cells divide to form a growth, or
    d._____. The cells of e._____ tumors remain in one place. The cells of f._____
    tumors wander, a characteristic called g._____.

24. Put a check next to the items that pertain to a regulatory pathway that controls the cell cycle in cells.
    a. _____ growth factor receptors in plasma membrane
    b. _____ signaling proteins within cytoplasm
    c. _____ various genes in nucleus
    d. _____ proteins that directly control the cell cycle
    e. _____ oncogenes and tumor-suppressor genes

25. Place the appropriate lette(s) next to each statement.

    P—proto-oncogenes   O—oncogenes   T—tumor-suppressor genes   MT—mutated tumor-suppressor gene

    a. _____ cell division is always promoted (requires two)
    b. _____ normal genes in cells that regulate cell division (requires two)
    c. _____ mutated genes that cause cancer (requires two)

## Gene Expression Maze

Can you find your way through the maze to a polypeptide by identifying each of the components depicted?

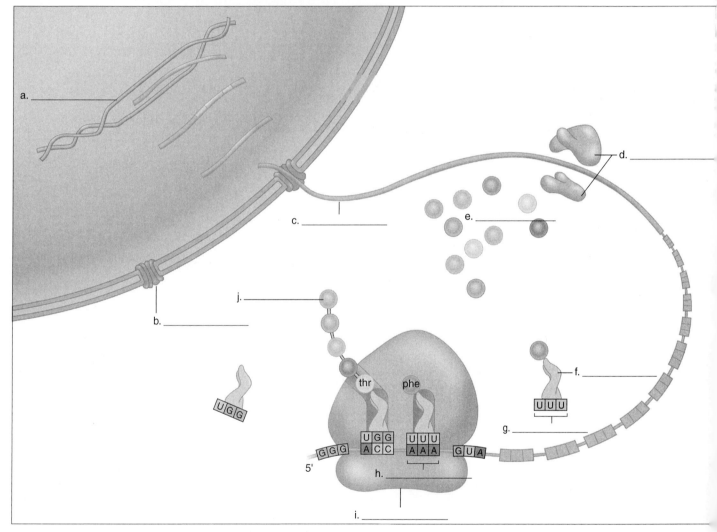

a. _____     g. _____

b. _____     h. _____

c. _____     i. _____

d. _____     j. _____

e. _____

f. _____     If you identified all correctly, you have found
                                         your way out.

Review key terms by completing this matching exercise, selecting from the following alphabetized list of terms:

anticodon
codon
complementary base
   pairing
double helix
mutagen
purine
pyrimidine
replication
RNA polymerase
template
transcription
translation

a. _____ Agent, such as radiation or a chemical, that brings about a mutation.

b. _____ Enzyme that speeds the formation of RNA from a DNA template.

c. _____ Process whereby a DNA strand serves as a template for the formation of mRNA.

d. _____ Process whereby the sequence of codons in mRNA determines (is translated into) the sequence of amino acids in a polypeptide.

e. _____ Three nucleotides on a tRNA molecule attracted to a complementary codon on mRNA.

f. _____ Three-base sequence in messenger RNA that causes the insertion of a particular amino acid into a protein or termination of translation.

g. _____ Double spiral; describes the three-dimensional shape of DNA.

## CHAPTER TEST

### OBJECTIVE QUESTIONS

Do not refer to the text when taking this test.

____ 1. In the experiment that used radioactive labeling to show that DNA is the genetic material,
  a. the entire virus entered bacteria, so determining whether the capsid or the DNA controls replication of viruses was difficult.
  b. just the capsid entered bacteria and controlled replication of viruses.
  c. just the DNA entered bacteria and controlled replication of viruses.
  d. the capsid must have been digested for DNA to control replication of viruses.

____ 2. In a DNA molecule, the sugar
  a. bonds covalently to phosphate groups.
  b. bonds covalently to nitrogen-containing bases.
  c. is deoxyribose.
  d. All of these are correct.

____ 3. If the structure of DNA is compared to a ladder, then the
  a. sides of the ladder consist of phosphate and sugar.
  b. rungs of the ladder are hydrogen-bonded bases.
  c. ladder is twisted.
  d. All of these are correct.

____ 4. Semiconservative replication means that
  a. sometimes DNA can replicate and sometimes it cannot—this accounts for aging.
  b. sometimes daughter DNA molecules are exact copies of parental molecules and sometimes they are not, so that genetic variability may occur.
  c. the DNA molecule consists of an old strand and a new strand.
  d. All of these are correct.

____ 5. Which is(are) correct regarding DNA?
  a. C is paired with G.
  b. The sugar is deoxyribose.
  c. Hydrogen bonds exist between the bases.
  d. All of these are correct.

____ 6. Before replication begins,
  a. enzymes must be present.
  b. the parental strands must unzip.
  c. "free" nucleotides must be present.
  d. All of these are correct.

____ 7. Select the incorrect association.
  a. mRNA—takes DNA message to the ribosome
  b. mRNA—takes amino acids to the ribosome
  c. rRNA—combines with protein in ribosomal subunits
  d. tRNA—has an anticodon

8. The base sequence of DNA is ATAGCATCC. The sequence of RNA transcribed from this strand is
   a. ATAGCATCC.
   b. CCTACGATA.
   c. CCUACGAUA.
   d. UAUCGUAGG.

9. An mRNA base sequence is UUAGCA. The two anticodons complementary to this are
   a. AAT CGT.
   b. AAU CGU.
   c. TTA GCA.
   d. UUA GCA.

10. A DNA base sequence changes from ATGCGG to ATGCGC. This type of mutation is
   a. deletion.
   b. frameshift.
   c. point.
   d. translocation.

11. Which of the following pairs is NOT a valid comparison of DNA and RNA?

   | | DNA | RNA |
   |---|---|---|
   | a. | double helix | single stranded |
   | b. | replicates | replicates |
   | c. | deoxyribose | ribose |
   | d. | thymine | uracil |

12. Which of these is true of an anticodon but is not true of a codon?
   a. part of an RNA molecule
   b. sequence of three bases
   c. part of a tRNA molecule
   d. part of a mRNA molecule

13. RNA nucleotides are joined during transcription by
   a. helicase.
   b. DNA polymerase.
   c. RNA polymerase.
   d. ribozymes.

14. In the DNA double helix, if 20% of the bases are A, then _____ of the bases are G.
   a. 10%
   b. 20%
   c. 30%
   d. 80%

15. Which of these is happening when translation takes place?
   a. mRNA is still in the nucleus.
   b. tRNAs are bringing amino acids to the ribosomes.
   c. rRNA is exposing its anticodons.
   d. DNA is being replicated.
   e. All of these are correct.

16. If the triplet code in DNA is TAG, what is the anticodon?
   a. UTC
   b. AUG
   c. UAG
   d. ATG

17. A drug prevents the exit of mRNA from the nucleus. This control is
   a. transcriptional.
   b. posttranscriptional.
   c. translational.
   d. posttranslational.

18. Which of the following does NOT describe the behavior of cells in a malignant tumor?
   a. carry out metastasis
   b. lose the ability of contact inhibition
   c. multiply rapidly
   d. remain in one site

19. Which of the following is NOT a suggested measure to prevent cancer?
   a. Avoid foods of the cabbage family.
   b. Cut down on salt-cured foods.
   c. Eat more high-fiber foods.
   d. Increase the intake of vitamins A and C.

20. A promoter
   a. turns on and off the transcription of a set of structural genes.
   b. binds to RNA polymerase.
   c. codes for the enzymes necessary for the transcription of polypeptides.
   d. is an intron that breaks up a structural gene.

Answer in complete sentences.

21. Compare the Mendelian concept of a gene to the biochemical concept of a gene.

22. Cancer research shows that there is communication from cytoplasm to the nucleus and from the nucleus to the cytoplasm. Explain.

**Test Results:** _____ number correct ÷ 22 = _____ × 100 = _____ %

# ANSWER KEY

## STUDY QUESTIONS

**1. a.** DNA **b.** capsids **c.** DNA **d.** DNA **e.** Radioactive DNA enters bacterium. **f.** Radioactivity stays within bacteria. **g.** radioactivity in bacteria **h.** Radioactivity stays within capsids **i.** radioactivity in liquid medium (See also Figure 25.1, p. 502 in text.) **2.** c **3.** adenine (A), guanine (G), cytosine (C), thymine (T) **4. a.** nucleotides **b.** See figure at right. **c.** sugar (deoxyribose) and phosphate **d.** See figure at right. **e.** A binds with T, and G binds with C. **f.** See figure at right. **g.** twist it **h.** Each nucleotide polymer is a strand; when the ladder twists, a helix results. **i.** The strands run opposite to one another. **5. a.** hydrogen bond **b.** They become unzipped. **c.** complementary base pairing **d.** DNA polymerase **e.** old, new, semiconservative **f.** the same **6. a.** T **b.** T **c.** T **d.** T **7. a.** nucleus **b.** cytoplasm **c.** RNA. **8. a.** T **b.** F **c.** F **d.** F **e.** T **9. a.** takes a message from DNA in the nucleus to the ribosomes in the cytoplasm **b.** is found in ribosomes, where proteins are synthesized **c.** transfers amino acids to the ribosomes **10. a.** transcription **b.** translation **11. a.** Every three bases stands for an amino acid. **b.** UUA, UUG, CUU, CUC, CUA, CUG **12. a.** complementary **b.** U **c.** C **d.** GCU UCG ACA **e.** The code is a triplet code and each codon contains three bases. **13. a.** introns **b.** ribozymes **14. a.** amino acid **b.** mRNA **c.** anticodon **d.** codon **e.** ribosome **15. a.** large **b.** small **c.** amino acid **d.** anticodon **16. a.** initiation **b.** associate **c.** elongation **d.** lengthened **e.** termination **f.** dissociate **17. a.** bases **b.** amino acids **c.** phenylalanine, leucine, isoleucine, valine, proline **d.** polyribosome **e.** many **18. a.** DNA **b.** regulator gene **c.** promoter **d.** operator **e.** structural genes **f.** mRNA **g.** active repressor

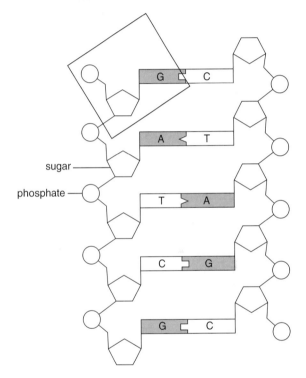

sugar

phosphate

**19.**

| Levels of Control of Gene Activity | Affects the Activity of |
|---|---|
| transcriptional | DNA |
| posttranscriptional | mRNA during formation and processing |
| translational | mRNA life span during protein synthesis |
| posttranslational | protein |

20. **a.** factors that must bind to DNA before transcription can begin. **b.** eukaryotes  **21. a.** frameshift, point **b.** mutagen **c.** transposons **d.** frameshift **e.** point **22. a.** uncontrolled growth **b.** no contact inhibition **c.** disorganized, multilayered **d.** nondifferentiated cells **e.** abnormal nuclei  **23. a.** one **b.** multiple **c.** contact **d.** tumor **e.** benign **f.** malignant **g.** metastasis  **24.** a, b, c, d, e **25. a.** O, MT **b.** P, T **c.** O, MT

## GAME: GENE EXPRESSION MAZE

**a.** DNA  **b.** nuclear pore  **c.** mRNA  **d.** ribosomal subunits  **e.** amino acids  **f.** tRNA  **g.** anticodon **h.** codon  **i.** ribosome  **j.** peptide

## DEFINITIONS WORDMATCH

**a.** mutagen  **b.** RNA polymerase  **c.** transcription **d.** translation  **e.** anticodon  **f.** codon  **g.** double helix

## CHAPTER TEST

**1.** c  **2.** d  **3.** d  **4.** c  **5.** d  **6.** d  **7.** b  **8.** d  **9.** b **10.** c  **11.** b  **12.** c  **13.** c  **14.** d  **15.** b  **16.** c **17.** b  **18.** d  **19.** a  **20.** b  **21.** According to the Mendelian concept, genes are portions of a chromosome that are passed from one generation to the next. According to the biochemical concept, a gene is a portion of a DNA molecule that specifies the sequence of amino acids in a protein. The biochemical concept explains how genes control metabolism.  **22.** Growth factors received by plasma membrane receptors set in motion a series of events in the cytoplasm that ends when certain genes are turned on in the nucleus. These genes in turn control proteins in the cytoplasm, some of which are involved in the cell cycle.

# 26

# BIOTECHNOLOGY

To clone a gene, a vector is first prepared. To genetically engineer a **plasmid** or virus, **restriction enzymes** are used to cleave plasmid DNA and to cleave foreign DNA. The "sticky ends" produced facilitate the insertion of foreign DNA into **vector** DNA. The foreign gene is sealed into the vector DNA by **DNA ligase.** When the plasmid replicates or the virus reproduces, the foreign gene is **cloned.**

The **polymerase chain reaction (PCR)** uses the enzyme DNA polymerase to make multiple copies of target DNA. Then the base sequence of this DNA can be determined, or it can be subjected to DNA fingerprinting. During **DNA fingerprinting,** restriction enzymes are used to fragment DNA. Gel electrophoresis separates the fragments according to size and charge. Analysis of the resulting pattern identifies the DNA as belonging to a particular human or other organism.

**Transgenic organisms** have also been made. Genetically engineered bacteria, agricultural plants, and farm animals now produce commercial products of interest to humans, such as hormones and vaccines. Transgenic bacteria also perform bioremediation, extract minerals, and produce chemicals. Transgenic agricultural plants have been engineered to resist herbicides and pests. Transgenic animals have been given bovine growth hormone. Pigs have been genetically altered to serve as a source of organs for transplant patients. Cloning of animals is now possible.

Constructing a genetic map of the chromosomes is still in progress. Sequencing of all the bases in the human genome is now complete. This information is expected to facilitate testing for genetic disorders, developing treatments that lead to a longer life and making gene therapy commonplace.

Gene therapy is used to correct the genotype of humans and to cure various human ills. Ex vivo therapy involves withdrawing cells from the patient, inserting a functioning gene, usually via a retrovirus, and then returning the treated cells to the patient. Many investigators are trying to develop in vivo therapy, in which viruses, laboratory-grown cells, or synthetic chemicals will be used to carry healthy genes into the patient.

Study the text section by section. Answer the study questions so that you can fulfill the learning objectives for each section.

## 26.1 CLONING OF A GENE (PP. 526–27)

The learning objectives for this section are:
• List and describe the means by which genes are cloned.
• Describe the polymerase chain reaction.

1. In the diagram, write the numbers of the following descriptions in the appropriate blanks.
   1  Cloning occurs when host cell reproduces.
   2  Host cell takes up recombined plasmid.
   3  DNA ligase seals human gene and plasmid.
   4  Restriction enzyme cleaves DNA.

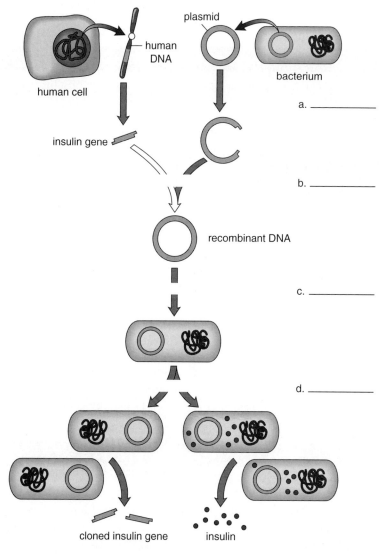

2. What is meant by the expression that restriction enzymes produce "sticky ends"? _____

_____

3. Change the following false statements to true statements:

   a. Plasmids are used as vectors in genetic engineering experiments involving humans. Rewrite: _____

   _____

   b. Recombinant DNA contains two types of bacterial DNA recombined together. Rewrite: _____

   _____

   c. Genetic engineering usually means that an organism receives genes from a member of its own species.
      Rewrite: _____

   _____

   d. Gene cloning occurs when a gene produces many copies of various genes. Rewrite: _____

   _____

In questions 4–6, fill in the blanks.

4. For human gene expression to take place in bacteria, the genes must be accompanied by a._____ regions that are necessary for the expression of mammalian genes. If genes have been made using reverse transcriptase, then the cDNA contains the b._____ but not the introns.

5. Explain the polymerase chain reaction by telling what *polymerase* refers to: a._____ ; and what chain reaction means: b._____. At the beginning of the reaction, very little DNA may be available, but at the end of the reaction, c._____ copies of a segment of DNA are available.

6. In DNA fingerprinting, a._____ enzymes digest the two samples to be compared. b._____ separates the fragments, and their different lengths are compared. If the pattern is similar, the samples are from c._____.

## 26.2 BIOTECHNOLOGY PRODUCTS (PP. 528–29)

The learning objectives for this section are:
- Give examples of biotechnology products produced by bacteria, plants, and animals.
- List and describe the means by which animals are cloned.

7. Complete the following table on transgenic organisms:

| Type of Organism | Engineered for What Purpose |
| --- | --- |
|  |  |
|  |  |
|  |  |

8. a. The advantage of using bacteria to make a product is that _____.

   b. The advantage of using plants to make a product is that _____.

   c. The advantage of using farm animals to make a product is that _____.

9. Place a check to indicate which of these is true of xenotransplantation.
   _____ a. Xenotransplantation uses humans as a source of organs for transplants.
   _____ b. Xenotransplantation uses other species, such as the pig, as a source of organs for transplants.
   _____ c. Pigs can be genetically altered to prevent rejection of their organs by humans.
   _____ d. Other species can pass new and different viruses to humans.

10. Put these statements in the proper sequence (1–5) to describe the making of a transgenic female goat that will produce a medicine needed by humans in its milk.
   _____ a. Development within a host animal
   _____ b. Remove egg from donor animal
   _____ c. Isolate a human gene
   _____ d. Microinject the human gene into the egg of the donor animal
   _____ e. Trangenic goat is born

11. Put these statements in the proper sequence (1–6) to describe the cloning of the transgenic goat produced in question 10.
   _____ a. Birth of cloned transgenic goats
   _____ b. Remove nuclei from adult cells of transgenic goat
   _____ c. Collect the milk that contains the medicine of interest
   _____ d. Development within host goats
   _____ e. Microinject 2n nuclei into the donor eggs
   _____ f. Remove eggs from donor animal

## 26.3 THE HUMAN GENOME PROJECT (P. 531)

The learning objective for this section is:
- Discuss what we now know about the human genome and how this knowledge may benefit humans.

12. Place a check beside those statements that are true.
    _____ a. We now know the sequence of base pairs in the human genome.
    _____ b. We now know the sequence of all the genes on all the chromosomes.
    _____ c. Humans have hundreds of thousands more genes than do the other animals.
    _____ d. The sequence of the bases in our DNA differs greatly from that of bacteria.

13. Place a check beside the statements that are true. Knowing the sequence of bases in our DNA holds great promise for which of these?
    _____ a. Helping locate genes on the chromosomes
    _____ b. Helping determine why some people get sick
    _____ c. Helping develop drugs that are specific for the various causes of the same type of disorder

## 26.4 GENE THERAPY (P. 534)

The learning objective for this section is:
- Describe two possible methods for gene therapy in humans.

14. Change these false statements to true statements.

    a. Ex vivo methods of gene therapy require that the therapeutic gene be placed in the body either directly or by using a viral vector. Rewrite: _____
    _____

    b. A common ex vivo method is to microinject normal genes into bone marrow stem cells removed from the patient. Then the stem cells are returned to the patient. Rewrite: _____
    _____

    c. Gene therapy is currently restricted to curing genetic diseases and is not used to treat illnesses like cystic fibrosis or cardiovascular diseases. Rewrite: _____
    _____

## DEFINITIONS WORDMATCH

Review key terms by completing this matching exercise, selecting from the following alphabetized list of terms:

clone
DNA ligase
gene cloning
plasmid
polymerase chain
  reaction
recombinant DNA
restriction enzyme
transgenic organism
vector
xenotransplantation

a. _____ Bacterial enzyme that stops viral reproduction by cleaving viral DNA; used to cut DNA at specific points during production of recombinant DNA.

b. _____ Free-living organisms in the environment that have had a foreign gene inserted into them.

c. _____ Production of identical copies; in genetic engineering, the production of many identical copies of a gene.

d. _____ Self-duplicating ring of accessory DNA in the cytoplasm of bacteria.

e. _____ Technique that uses the enzyme DNA polymerase to produce copies of a particular piece of DNA within a test tube.

f. _____ DNA that contains genes from more than one source.

## OBJECTIVE QUESTIONS

Do not refer to the text when taking this test.

_____ 1. Select the incorrect description of a plasmid.
   a. used as vector
   b. consists of chromosomal DNA
   c. found in some bacteria
   d. small, ringlike structure

_____ 2. Restriction enzymes
   a. cleave DNA into small fragments.
   b. restrict the growth of eukaryotic cells.
   c. seal pieces of DNA together.
   d. serve as introns in cells.

_____ 3. The problem with using pigs as a source of human organs is
   a. it's not possible to genetically alter pig cells to avoid rejection.
   b. their organs are too small in comparison to human organs.
   c. pig organs might carry viruses that would be new to humans.
   d. transplant patients would rather die than receive organs from a pig.

_____ 4. To clone a transgenic animal, you have to have
   a. an egg donor.
   b. special food to feed them.
   c. a way to protect human beings from coming in contact with transgenic animals.
   d. a transgenic animal.
   e. Both _a_ and _d_ are correct.

_____ 5. A final step in the use of a plasmid to clone a gene is
   a. to insert a foreign gene into a bacterium.
   b. to introduce a plasmid into a treated cell.
   c. to remove a plasmid from a bacterium.
   d. reproduction of plasmid in a cell.

_____ 6. The polymerase chain reaction does NOT typically
   a. take place in vats called bioreactors.
   b. occur at a high temperature.
   c. copy the entire human genome at one time.
   d. work if the target DNA is limited in size.
   e. Both _a_ and _c_ are not typical.

_____ 7. A transgenic organism is
   a. free-living and receives a foreign gene.
   b. free-living and transmits a foreign gene.
   c. parasitic and receives a foreign gene.
   d. parasitic and transmits a foreign gene.

_____ 8. The human genome
   a. will probably be completely sequenced in 50 years.
   b. has no usefulness to humans.
   c. contains only about 30,000 genes.
   d. is known only to government employees because it is top secret.

_____ 9. Gene therapy is
   a. on the back burner because human genes are so different from one another.
   b. going full speed ahead despite its being still investigative.
   c. the use of foreign genes to cure a human ill.
   d. Both _a_ and _c_ are correct.
   e. Both _b_ and _c_ are correct.

_____10. Genetically engineered plants have been or will be used to
   a. resist insects.
   b. resist herbicides.
   c. produce protein-enhanced beans, corn, and wheat.
   d. produce animal neuropeptides, blood factors, and growth hormones.
   e. All of these are correct.

_____11. A DNA fingerprint is the type of
   a. plasmid chosen for recombinant DNA.
   b. restriction enzyme used to fragment DNA.
   c. fragment pattern that results following gel electrophoresis.
   d. sequence pattern discovered in a section of DNA.

_____12. Genetically engineered bacteria can be used to
   a. protect plants from frost.
   b. clean up oil spills on beaches.
   c. produce organic chemicals.
   d. extract copper and gold from low-grade sources.
   e. All of these are correct.

_____13. Which statement(s) are true?
   a. A biotechnology product is a protein made by a transgenic organism that is desired by humans.
   b. Transgenic organisms have had a foreign gene inserted into their cells.
   c. If a transgenic organism can be cloned, it would only be necessary to make the first organism transgenic.
   d. Most likely, it will never be possible to insert foreign genes into the human genome, and therefore there can never be transgenic humans.
   e. All but _d_ are true statements.

_____14. Which of these would you NOT expect to be a biotechnology product produced by a bacterium?
   a. steroid sex hormones
   b. hormones of any kind
   c. nucleic acids instead of proteins
   d. Both _b_ and _c_ are correct.

_____15. If a cell is altered while outside the human body for gene therapy, it is considered _____ therapy.
   a. ex vivo
   b. in vivo
   c. in vitro
   d. extraneous
   e. intravenous

_____16. What does the Human Genome Project have to do with gene therapy?
   a. Once we know the location of all the genes on the chromosomes, it will make it easier to isolate particular genes to cure humans.
   b. Both the Human Genome Project and gene therapy are unethical, and therefore will never be completed.
   c. Both the Human Genome Project and gene therapy require the use of restriction enzymes and gel electrophoresis.
   d. Both the Human Genome Project and gene therapy require the prior use of the polymerase chain reaction.
   e. All but *b* are correct.

_____17. The polymerase chain reaction
   a. produces many copies of different segments of DNA.
   b. produces many copies of the same segment of DNA.
   c. requires denaturing double-stranded DNA into single strands by heating.
   d. Both *b* and *c* are correct.

_____18. Which of these is NOT a step to prepare recombinant DNA?
   a. remove plasmid from bacterial cell
   b. use restriction enzyme to acquire foreign gene and cut open vector
   c. use ligase to seal foreign gene into vector
   d. use a virus to carry recombinant DNA into a plasmid

_____19. Which of these is a benefit to having insulin produced by biotechnology?
   a. It can be mass-produced.
   b. It is nonallergenic.
   c. It is less expensive.
   d. All of these are correct.

_____20. Vaccines produced by biotechnology could be
   a. pathogens treated to be nonvirulent.
   b. proteins produced by a pathogen's gene.
   c. only enzymes taken from a pathogen.
   d. Both *a* and *b* are correct.

## THOUGHT QUESTIONS

Answer in complete sentences.

21. How do studies of genetic engineering prove that the genetic code is nearly universal?


22. What do you think are some objections our society may have regarding genetic engineering?


**Test Results:** _____ number correct ÷ 23 = _____ × 100 = _____ %

## STUDY QUESTIONS

**1. a.** 4 **b.** 3 **c.** 2 **d.** 1 **2.** Cleavage results in unpaired bases. **3. a.** . . . involving bacteria **b.** contains DNA from two different sources **c.** . . . from a member of a different species **d.** . . . many copies of the same gene **4. a.** regulatory **b.** exons **5. a.** DNA polymerase, the enzyme involved in DNA replication **b.** the reaction occurs over and over again **c.** many **6. a.** restriction **b.** gel electrophoresis **c.** from the same individual **7.**

| Type of Organism | Engineered for What Purpose |
| --- | --- |
| bacteria | to make products to protect plants, for bioremediation, to produce chemicals, and to mine metals |
| plants | to resist insects, pesticides, and herbicides, and to make products |
| animals | to have improved qualities and to make products |

**8. a.** They will take up plasmids. **b.** They will grow from single cells (protoplasts). **c.** The product is easily obtainable in milk. **9.** b, c, d **10.** c, b, d, a, e **11.** b, f, e, d, a, c **12.** a **13.** a, b, c **14. a.** In vivo **b.** . . . use a viral vector to carry normal genes . . . **c.** . . . is not restricted to curing genetic disease and is used to treat illnesses . . .

## DEFINITIONS WORDMATCH

**a.** restriction enzyme **b.** transgenic organism **c.** clone **d.** plasmid **e.** polymerase chain reaction **f.** recombinant DNA

## CHAPTER TEST

**1.** b **2.** a **3.** c **4.** e **5.** d **6.** e **7.** a **8.** c **9.** b **10.** e **11.** c **12.** e **13.** e **14.** a **15.** a **16.** a **17.** d **18.** d **19.** d **20.** b **21.** Genes transmitted to new cells through vectors and other means are still transcribed and translated by the same process and with the same accuracy. **22.** Some people may object to genetic engineering on religious grounds because we are now able to change the inherited characteristics of organisms, including human beings. Some scientists have concerns because a disease-causing transgenic bacterium may be produced for which humans have no immunity or a transgenic bacterium, plant, or animal may be produced that could wreak havoc in the environment.

# 27

# EVOLUTION OF LIFE

## CHAPTER REVIEW

The fossil record and biogeography, as well as comparative anatomy, development, and biochemistry, all give evidence of evolution. The **fossil record** gives us the history of life in general and allows us to trace the descent of a particular group. **Biogeography** shows that the distribution of organisms on earth is explainable by assuming organisms evolved in one locale. Comparing the anatomy and the development of organisms reveals a unity of plan among those that are closely related. All organisms have certain biochemical molecules in common, and similarities indicate the degree of relatedness.

A chemical evolution is believed to have resulted in the first cell(s). Inorganic chemicals, probably derived from the primitive atmosphere, reacted to form small organic molecules. These reactions occurred in the ocean, either on the surface or in the region of hydrothermal vents, deep within the ocean.

Next, macromolecules evolved and interacted. The RNA-first hypothesis is supported by the discovery of ribozymes, RNA enzymes. The protein-first hypothesis is supported by the observation that amino acids polymerize abiotically when exposed to dry heat. The Cairns-Smith hypothesis suggests that macromolecules could have originated in clay. The **protocell** must have been a **heterotrophic** fermenter living on the preformed organic molecules in the organic soup. Eventually the DNA —> RNA —> protein self-replicating system evolved, and a true cell came into being.

**Evolution** is described as a process that involves a change in gene frequencies within the **gene pool** of a sexually reproducing **population.** The Hardy-Weinberg law states that the gene pool frequencies arrive at an equilibrium that is maintained generation after generation unless disrupted by **mutations, genetic drift, gene flow, nonrandom mating,** or **natural selection.** Any change from the initial allele frequencies in the gene pool of a population signifies that evolution has occurred.

Speciation is the origin of **species.** This usually requires geographic isolation, followed by reproductive isolation.

The evolution of several species of finches on the Galápagos Islands is an example of adaptive radiation, because each species has a different way of life, but all species came from one common ancestor.

Currently, there are two hypotheses about the pace of speciation. Traditionalists support **phyletic gradualism**—slow, steady change leading to speciation. A new model, **punctuated equilibrium,** says that a long period of stasis is interrupted by speciation.

Classification involves the assignment of species to a hierarchy of categories: species, genus, family, order, class, phylum, kingdom, and in this text, domain. Each higher category is more inclusive; members of the same domain share general characteristics, and members of the same species share quite specific characteristics. The five-kingdom system of classification recognizes these kingdoms: Monera (the bacteria), Protista (e.g., algae, protozoans), Fungi, Plantae, and Animalia. The three-domain system used by this text is based on molecular data and recognizes three domains: Bacteria, Archaea, and Eukarya. The domain Eukarya contains the kingdoms Protista, Fungi, Plantae, and Animalia.

## STUDY QUESTIONS

Study the text section by section. Answer the study questions so that you can fulfill the learning objectives for each section.

The learning objective for this section is:
• Give examples of the evidence for common descent.

In questions 1–5, fill in the blanks.

1. Earth is about 4.6 billion years old, life evolved in the form of prokaryotes about a. _____ billion years ago, and eukaryotic cells arose about 2.1 billion years ago. Evolution explains the unity and the b. _____ of life. All living things share the same fundamental characteristics because they descended from a common c. _____. d. _____ are the remains and traces of past life and are found embedded in e. _____ rock. f. _____ are biologists who discover and study the fossil record. g. _____ is a transitional link between reptiles and birds. Fossils allow us to deduce that h. _____ preceded amphibians, which preceded reptiles, which preceded both birds and i. _____ .

2. Fossils can be dated by the a. _____ dating method, which determines the relative order of fossils and strata based on their position, or by the b. _____ dating method, which relies on radioactive dating techniques to assign an actual date to a fossil. Mass c. _____ are times when a large percentage of existing species becomes extinct within a relatively short period of time. The Cretaceous extinction of the dinosaurs may be due to an asteroid falling to earth, leaving high levels of d. _____ .

3. a. _____ evidence for evolution refers to the study of the distribution of plants and animals throughout the world. Such distributions are consistent with the hypothesis that related forms evolve in one locale and then b. _____ out into other regions. Continental c. _____ is a hypothesis that states that the continents are not fixed but change over time. The d. _____ biological diversity of today's world is the result of isolated evolution on separate continents.

4. A common descent hypothesis offers a plausible explanation for a. _____ similarities among organisms. b. _____ structures are similar in structure (not function) because they were inherited from a common ancestor, such as the forelimbs in birds, whales, horses, and monkeys. The wing of a bird and insect are c. _____ structures; they are all adaptations for flying that are structurally unrelated. Anatomical features that are fully developed in one group of organisms but are reduced and may have no function in similar groups are called d. _____ structures. During embryological development, all vertebrates have a e. _____ and paired pharyngeal pouches. In fishes, the pouches develop into gills, but in humans, other structures are formed. Organisms share a f. _____ of plan when they are closely related because of common descent.

5. Almost all living organisms use the same basic a. _____ molecules, including DNA, ATP, amino acid sequences of proteins, etc. Their similarity can be explained by b. _____ from a common ancestor.

In questions 6–10, match each of the statements to the type of evidence supporting evolution.
   a. The succession of life-forms is revealed through preserved remnants.
   b. Organisms have similarities and differences in structures.
   c. Closely related organisms have high correlations in DNA base sequences.
   d. Organisms arise and disperse from place of origin.
   e. In their early phases of development, vertebrates show a unity of plan.
   6. _____ biogeography
   7. _____ comparative anatomy
   8. _____ comparative biochemistry
   9. _____ comparative embryology
   10. _____ fossil record

## 27.2 ORIGIN OF LIFE (PP. 546–48)

The learning objectives for this section are:
• Describe the chemical evolution that may have led to the origin of the protocell.
• Explain different hypotheses for the origin of life.

11. In the top half of the diagram, place these labels next to the correct arrows:
    cooling     energy capture     polymerization
    Then place these labels in the boxes:
       inorganic chemicals     macromolecules     plasma membrane     primitive earth     small organic molecules

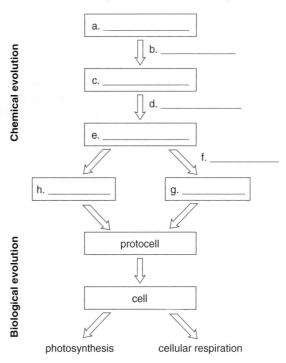

12. A student decides to reproduce Miller's experiment, so she assembles all the necessary equipment and adds the proper gases. What is still needed and why? _____

    _____

13. A student decides to reproduce Fox's experiment, so he puts a solution of amino acids on heated rocks. What is still needed and why? _____

    _____

14. What is the evidence for suggesting that it was an "RNA world" some 4 billion years ago? _____

    _____

15. Indicate whether these statements about the protocell are true (T) or false (F).
    a. _____ carried on cellular respiration
    b. _____ was a heterotrophic fermenter
    c. _____ contained a self-replication system that allowed it to reproduce
    d. _____ had a plasma membrane
16. To be a true cell, which statement, labeled as false in question 15, must be fulfilled? a._____ Add the
    label *self-replication system* to the diagram in question 11. b._____

17. a. Why does it seem logical that the protocell was heterotrophic? _____

    b. Why would the protocell have been a fermenter? _____

18. What is the proper sequence of events in the evolution of a self-replication system, assuming that it was an

    "RNA world" at the time? Indicate by letters. _____
    a. replication of RNA
    b. reverse transcription of DNA
    c. RNA → proteins
    d. DNA → RNA → proteins

19. Place the appropriate letter next to each statement about the earth's atmosphere to describe it as either:

P—primitive    C—current

a. _____ It exists without the ozone shield.

b. _____ It favors the polymerization of organic molecules.

c. _____ It is a reducing atmosphere.

d. _____ It is an oxidizing atmosphere.

e. _____ It tends to break down organic molecules.

f. _____ Oxygen-producing autotrophs made it.

## 27.3 PROCESS OF EVOLUTION (PP. 549–56)

The learning objectives for this section are:
- Explain why allele frequencies in the gene pool of a population change from one generation to the next.
- Explain Darwin's mechanisms for natural selection, which result in adaptation to the environment.

20. When the gene frequency within the gene pool of a population changes, then a. _____ is said to have occurred. The Hardy-Weinberg law states that the frequencies of alleles in a sexually reproducing population will remain the same in each succeeding generation as long as there is

b. _____    e. _____

c. _____    f. _____

d. _____

21. The Hardy-Weinberg law uses the binomial expression $p^2 + 2pq + q^2$ to calculate the genotypic and a. _____ frequencies of a population. $p^2$ represents the genotype in a population that is the b. _____ phenotype, $2pq$ represents the c. _____ phenotype, and $q^2$ represents the d. _____.

22. Assuming a Hardy-Weinberg equilibrium, 1% of a population of fruit flies had the homozygous recessive trait of vestigial wings ($v$). Answer the following questions.

a. What is the frequency of the recessive allele $v$? _____

b. What is the frequency of the dominant normal allele $V$? _____

Fill in this Punnett square using the allele frequencies you calculated to determine the genotypic frequencies in the offspring of this population:

|         | _____ V | _____ v |
|---------|---------|---------|
| _____ V | c.      | d.      |
| _____ v | e.      | f.      |

g. The percentage of homozygous dominant (normal) fruit flies is _____ .

h. The percentage of heterozygous dominant (normal) fruit flies is _____ .

i. The percentage of homozygous recessive (vestigial) fruit flies is _____ .

23. Label the statements with the correct agents of evolutionary change:

gene flow    genetic drift    mutations    natural selection    nonrandom mating

a. _____ Investigators have discovered that multiple alleles are common in a population.

b. _____ Populations are subject to new alleles entering by the migration of organisms between populations.

c. _____ Female birds of paradise choose mates with the most splendid feathers.

d. _____ Investigators discovered that if they randomly picked out a few flies from each generation to start the next generation, gene pool frequency changes appeared.

e. _____ Giraffes with longer necks get a larger share of resources and tend to have more offspring.

24. Match these descriptions to one of the agents of evolutionary change listed in question 23. (Some agents are used more than once.)

a. _____ Dwarfism is common among the Amish of Lancaster County, Pennsylvania.

b. _____ Cheetahs are homozygous for a larger proportion of their genes.

c. _____ This agent of change tends to make the members of a population dissimilar to one another.

d. _____ This agent of change tends to make the members of a population similar to one another.

e. _____ Certain members of a population are more fit than other members.

f. _____ Bacteria and insects become resistant to agents that formerly killed them.

25. Natural selection can now be understood in terms of genetics. Many of the variations that exist between members of a population are due to differences in $^{a.}$_____. Some of these genotypes result in $^{b.}$_____ that are better adapted to the environment. Individuals that are better adapted to the environment reproduce to a(n) $^{c.}$_____ extent, and therefore, these genotypes and phenotypes become more prevalent in the population.

26. Match the observations to the correct type of natural selection at work:
   directional selection      disruptive selection      stabilizing selection

a. _____ Trees in a windy area tend to remain the same size each year.

b. _____ The brain size of hominids steadily increases.

c. _____ The same species of moths tends to have blue stripes in open areas and orange stripes in forested areas.

27. Match the types of natural selection listed in question 26 with the following diagrams:

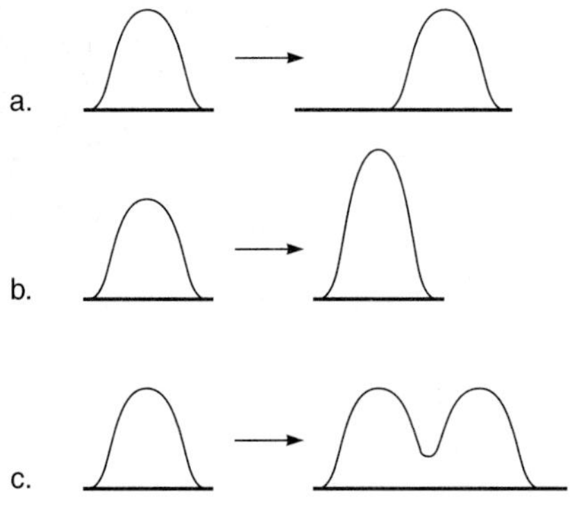

In question 28, fill in the blanks.

28. Variation is maintained in a population despite directional and stabilizing selection when members have $^{a.}$_____ alleles for every trait. That's because the $^{b.}$_____ allele is hidden by the $^{c.}$_____ allele. In instances such as sickle-cell disease, in which the $^{d.}$_____ genotype is more fit, the two $^{e.}$_____ genotypes are maintained due to the reproductive process, which involves meiosis and fertilization.

## 27.4 SPECIATION (PP. 557–560)

The learning objectives for this section are:
• Distinguish between premating and postmating isolating mechanisms and explain the process of speciation.
• Relate adaptive radiation to the occurrence of many related species.
• Contrast the concepts of phyletic gradualism and punctuated equilibrium.

29. Bush babies (a type of primate) living higher in the tropical canopy are a different species from those living lower in the canopy. Answer the following *yes* or *no:*

    a. Would the two species of bush babies reproduce with each other? _____

    b. Would a premating mechanism separate the two species of bush babies? _____

    c. Could the two species of bush babies eat the same food? _____

    d. Would the two species of bush babies have to look dissimilar enough to be distinguishable by the naked eye? _____

30. Match the numbered statements to the letters in the diagram.
    1 A newly formed barrier comes between the populations.
    2 The barrier is removed.
    3 A species contains one population or several interbreeding populations.
    4 Reproductive isolation has occurred.
    5 Two species now exist.
    6 Divergent evolution occurs.
    7 Allopatric speciation occurs.

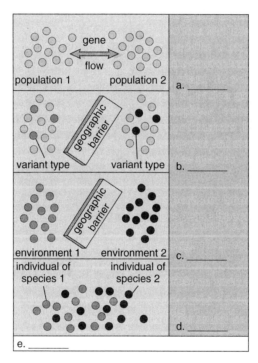

31. Label each of these statements with *pre* for premating isolating mechanism or *post* for postmating isolating mechanism.
    a. _____ Zygote mortality exists.
    b. _____ Species reproduce at different times.
    c. _____ Species have genitalia that are unsuitable to each other.

32. Explain the adaptive radiation of Darwin's finches on the Galápagos Islands. _____

_____

_____

33. Indicate whether these comparisons between the phyletic gradualism and punctuated equilibrium models of evolutionary change are true (T) or false (F):

|  Phyletic Gradualism | Punctuated Equilibrium |
|---|---|

_____ a. Speciation occurs gradually.      Speciation occurs rapidly.

_____ b. Transitional links are expected.      Transitional links may not be found.

_____ c. New species are easily recognizable in the fossil record.      New species cannot be recognized.

_____ d. An ancestral species can gradually become a new species.      A subpopulation usually becomes a new species.

_____ e. Speciation is always occurring.      Speciation occurs sporadically.

_____ f. Stasis is rare.      Stasis is common.

## 27.5 CLASSIFICATION (PP. 560–61)

The learning objectives for this section are:
- Contrast the three domains of life and state how they are related.
- Contrast the four kingdoms within the domain Eukarya.

34. a. Create a saying that will help you remember the order of the classification categories. _____

_____

b. Which category is just below family? _____

c. Which category is just above class? _____

d. Which category is just below class? _____

e. Which two categories are used in the scientific name? _____ , _____

35. a. Prokaryotes are placed in what two domains? _____ , _____

b. Eukaryotes are placed in what domain? _____

36. Complete this table to show major differences between the domains.

|  | Archaea and Bacteria | Eukarya |
|---|---|---|
| Type of cell | a. | b. |
| Nucleus? | c. | d. |
| Complexity | e. | f. |

37. Match the descriptions to the four kingdoms with the domain Eukarya.
   1. Protista
   2. Fungi
   3. Plantae
   4. Animalia
   _____ a. multicellular, motile by contractile fibers, and ingest their food; diplontic life cycle
   _____ b. usually unicellular, nutrition various, life cycle various, motile by flagella or cilia
   _____ c. multicellular, photosynthetic, alternation of generations life cycle

38. Match the same listing as in question 37 to these organisms.
   _____ a. includes algae, protozoans, water molds, and slime molds
   _____ b. includes yeasts, mushrooms, and molds
   _____ c. includes trees, grasses, and vines
   _____ d. includes human beings, insects, clams, and birds

Review key terms by completing this matching exercise, selecting from the following alphabetized list of terms:

*biogeography*
*continental drift*
*directional selection*
*evolution*
*fossil*
*gene flow*
*homologous structure*
*microsphere*
*natural selection*
*proteinoid*
*protocell*
*punctuated equilibrium*

a. _____ Process by which populations become adapted to their environment.

b. _____ Natural selection in which an extreme phenotype is favored, usually in a changing environment.

c. _____ An evolutionary model that proposes there are periods of rapid change dependent on speciation followed by long periods of stasis.

d. _____ Structure that is similar in two or more species because of common ancestry.

e. _____ Movement of genes from one population to another via sexual reproduction between members of the populations.

f. _____ Any past evidence of an organism that has been preserved in the earth's crust.

g. _____ In biological evolution, a possible cell forerunner that became a cell once it could reproduce.

# CHAPTER TEST

## OBJECTIVE QUESTIONS

Do not refer to the text when taking this test.

____ 1. Each of the following was present in the primitive atmosphere of the earth EXCEPT
   a. carbon dioxide.
   b. carbon monoxide.
   c. molecular nitrogen.
   d. molecular oxygen.

____ 2. Miller's experiments produced
   a. coacervate droplets from macromolecules.
   b. inorganic substances from organic molecules.
   c. organic molecules from inorganic substances.
   d. protocell(s) from macromolecules.

____ 3. Select the correct sequence that occurred on the primitive earth.
   a. gases, small molecules, macromolecules, protocell(s)
   b. macromolecules, small molecules, protocell(s), gases
   c. protocell(s), macromolecules, small molecules, gases
   d. small molecules, gases, macromolecules, protocell(s)

____ 4. Microspheres formed from the polymerization of
   a. amino acids.
   b. nucleotides.
   c. sugars.
   d. water.

____ 5. Protocell(s) exhibited each of the following EXCEPT
   a. the ability to separate from water.
   b. a lipid-protein membrane.
   c. conduction of energy metabolism.
   d. means of self-replication.

____ 6. Vertebrate forelimbs are most likely to be studied in
   a. biogeography.
   b. comparative anatomy.
   c. comparative biochemistry.
   d. ecological physiology.

____ 7. Biochemical evidence supporting evolution would show that
   a. there are more base differences between yeasts and humans than between horses and humans.
   b. there are more base differences between apes and humans than between horses and humans.
   c. apes and humans have almost the same sequence of bases.
   d. Both *a* and *c* are correct.

____ 8. Comparative anatomy demonstrates that
   a. each species has its own structures, indicating no relationship with any other species.
   b. different vertebrates have widely different body plans.
   c. different species can have similar structures that are traceable to a common ancestor.
   d. fossils bear no anatomical similarities to modern-day species.

____ 9. The study of biogeography shows that
   a. the same species of plants and animals are found on different continents whenever the environment is the same.
   b. one species can spread out and give rise to many species, each adapted to varying environments.
   c. the structure and function of organisms bear no relationship to the environment.
   d. barriers do not prevent the same species from spreading around the world.

____ 10. Which is NOT true of fossils?
   a. They are evidences of life in the past.
   b. They look exactly like modern-day species, regardless of their age.
   c. In general, the older the fossil, the less it resembles modern-day species.
   d. They indicate that life has a history.

____ 11. Each is a condition of the Hardy-Weinberg law EXCEPT that
   a. gene flow is absent.
   b. genetic drift does not occur.
   c. mutations are lacking.
   d. random mating does not occur.

____ 12. Establishment of polydactylism among the Amish is an example of
   a. artificial selection.
   b. natural selection.
   c. the bottleneck effect.
   d. the founder effect.

____ 13. Industrial melanism is an example of selection that is
   a. directional.
   b. disruptive.
   c. sexual.
   d. stabilizing.

____ 14. Each factor contributes to the maintenance of variation from a recessive allele EXCEPT
   a. diploidy.
   b. heterozygosity.
   c. homozygosity.
   d. sexual reproduction.

____ 15. Select the premating isolating mechanism.
   a. $F_2$ fitness
   b. gamete mortality
   c. habitat type
   d. hybrid sterility

____ 16. Select the postmating isolating mechanism.
   a. behavior
   b. mechanical differences in genitalia
   c. temporal factors
   d. zygote mortality

____ 17. During the usual process of speciation, a species is first isolated
   a. behaviorally.
   b. geographically.
   c. reproductively.
   d. mechanically.
   e. genetically.

____ 18. Which of the following sequences is in correct order, starting from the most specific but fewest in number of species?
   a. class, family, kingdom, order, phylum
   b. family, order, class, phylum, kingdom
   c. order, family, kingdom, class, phylum
   d. phylum, order, kingdom, family, class

____ 19. In the scientific name *Elaphe obsoleta*, which is the genus name?
   a. *Elaphe*
   b. *obsoleta*
   c. *bairdi*
   d. *Elaphe obsoleta*

____ 20. In which kingdom are the members unicellular, but have a nucleus?
   a. Protista
   b. Fungi
   c. Plantae
   d. Animalia
   e. Both *b* and *e* are correct.

Answer in complete sentences.

21. In a sample of geological strata, where are the oldest life-forms most likely to be found? Where are the most recent life-forms?

22. A dominant allele produces a desirable coloration pattern in a fish species. A pond owner stocks a pond with a small number of heterozygous members showing this desirable trait, hoping to maintain it. Will this approach work?

**Test Results:** _____ number correct ÷ 22 = _____ × 100 = _____ %

# ANSWER KEY

## STUDY QUESTIONS

**1. a.** 3.5 **b.** diversity **c.** ancestor **d.** Fossils **e.** sedimentary **f.** Paleontologists **g.** *Archaeopteryx* **h.** fishes **i.** mammals **2. a.** relative **b.** absolute **c.** extinctions **d.** iridium **3. a.** Biogeographical **b.** spread **c.** drift **d.** mammalian **4. a.** anatomical **b.** Homologous **c.** analogous **d.** vestigial **e.** notochord **f.** unity **5. a.** biochemical **b.** descent **6.** d **7.** b **8.** c **9.** e **10.** a **11. a.** primitive earth **b.** cooling **c.** inorganic chemicals **d.** energy capture **e.** small organic molecules **f.** polymerization **g.** plasma membrane **h.** macromolecules **12.** an energy source, because amino acids do not react unless energy is provided **13.** To obtain microspheres, proteinoids must be placed in water. **14.** discovery of ribozymes, which are nucleotides with enzymatic properties **15. a.** F **b.** T **c.** F **d.** T **16. a.** must contain a self-replication system **b.** add *self-replication system* between protocell and cell **17. a.** The ocean contained organic molecules that could serve as food. **b.** The atmosphere did not contain any oxygen. **18.** a, c, b, d **19. a.** P **b.** P **c.** P **d.** C **e.** C **f.** C **20. a.** evolution **b.** no mutation **c.** no genetic drift **d.** no gene flow **e.** no natural selection **f.** random mating **21. a.** allele **b.** homozygous dominant **c.** heterozygous **d.** homozygous recessive **22. a.** 0.1 **b.** 0.9 ($p + q = 1$) **c.** 0.81 $VV$ **d.** 0.09 $Vv$ **e.** 0.09 $Vv$ **f.** 0.01 $vv$ **g.** 81% **h.** 18% (9% + 9%) **i.** 1.0% **23. a.** mutations **b.** gene flow **c.** nonrandom mating **d.** genetic drift **e.** natural selection **24. a.** genetic drift (founder effect) **b.** genetic drift (bottleneck effect) **c.** genetic drift **d.** gene flow **e.** natural selection **f.** mutations **25. a.** genotype **b.** phenotypes **c.** greater **26. a.** stabilizing **b.** directional **c.** disruptive **27. a.** directional **b.** stabilizing **c.** disruptive **28. a.** two **b.** recessive **c.** dominant **d.** heterozygous **e.** homozygous **29. a.** no **b.** yes **c.** yes **d.** no **30. a.** 3 **b.** 1 **c.** 6 **d.** 2, 4, 5 **e.** 7 **31. a.** post **b.** pre **c.** pre **32.** Populations of finches were geographically isolated from one another on islands where the environments were different. Each population of finches evolved to suit its particular environment. **33. a.** T **b.** T **c.** F **d.** T **e.** T **f.** T **34. a.** Example: Dumb Karen Pushed Cans Off Friendly Grandmother's Stove (domain, kingdom, phylum, class, order, family, genus, species) **b.** genus **c.** phylum **d.** order **e.** genus, species **35. a.** Archaea and Bacteria **b.** Eukarya **36. a.** prokaryotic **b.** eukaryotic **c.** no nucleus **d.** nucleus **e.** unicellular **f.** unicellular/multicellular **37. a.** 4 **b.** 1 **c.** 3 **38. a.** 1 **b.** 2 **c.** 3 **d.** 4

## DEFINITIONS WORDMATCH

**a.** natural selection **b.** directional selection **c.** punctuated equilibrium **d.** homologous structure **e.** gene flow **f.** fossil **g.** protocell

## CHAPTER TEST

**1.** d **2.** c **3.** a **4.** a **5.** d **6.** b **7.** d **8.** c **9.** b **10.** b **11.** d **12.** d **13.** a **14.** c **15.** c **16.** d **17.** b **18.** b **19.** a **20.** a **21.** The oldest forms are in the deepest strata. The most recent life-forms are in the more recently added strata, which are not as deep. **22.** It may not work because of genetic drift—the population may diverge to homozygous forms through random events.

# 28

# MICROBIOLOGY

**Viruses** are noncellular obligate parasites that have a protein coat called a capsid and a nucleic acid core. In addition, animal viruses have an envelope. Viral DNA must enter a host cell before reproduction is possible. In the **lytic cycle,** a **bacteriophage** immediately reproduces, and in the **lysogenic cycle,** viral DNA integrates into the host genome but may eventually reproduce.

The **archaea** (domain Archaea) and the **bacteria** (domain Bacteria) are prokaryotes. Prokaryotic cells lack a nucleus and most of the other cytoplasmic organelles that are found in eukaryotic cells. Most bacteria are **saprotrophs** (heterotrophic by absorption), and along with fungi, fulfill the role of decomposers in ecosystems. The **cyanobacteria** are photosynthetic in the same manner as plants. Reproduction of prokaryotes is by **binary fission,** but sexual exchange occasionally takes place in bacteria. Some bacteria form endospores, which can survive the harshest of treatment except sterilization.

**Protists** are members of the domain Eukarya and the kingdom Protista. Protists are unicellular organisms, but there are also some multicellular forms. Algae are aquatic autotrophs by photosynthesis; protozoans are aquatic heterotrophs by ingestion. Slime molds, which are terrestrial, and water molds, which are aquatic, have some characteristics of fungi.

**Algae** are classified according to their pigments (colors). Green algae are diverse: Some are unicellular or colonial flagellates, some are filamentous, and some are multicellular sheets. **Diatoms** and **dinoflagellates** are unicellular producers in oceans; brown and red algae are seaweeds; euglenoids are unicellular, and some have chloroplasts in addition to flagella. Every type of life cycle is seen among the algae.

**Protozoans** are classified according to the type of locomotor organelle. The amoebas phagocytize, the **ciliates** are very complex, and the sporozoans are all animal parasites. Malaria is a significant disease caused by a sporozoan.

Slime molds have an amoeboid stage and then form **fruiting bodies,** which produce **spores** that are dispersed by the wind. Water molds have threadlike bodies.

**Fungi** are members of the domain Eukarya and the kingdom Fungi. Fungi are multicellular saprotrophic organisms composed of **hyphae,** which form a **mycelium.** Along with heterotrophic bacteria, they are **decomposers.** The fungi produce windblown spores during both sexual and asexual reproduction. The major groups of fungi are distinguished by type of sexual spore and fruiting body. **Zygospore** fungi produce spores in **sporangia;** sac fungi produce ascospores in **asci;** and club fungi form basidiospores in **basidia.**

Fungi form two symbiotic associations of interest. **Lichens** contain both a fungus and an alga; **mycorrhizae** are symbiotic relationships of mutual benefit between soil fungi and the roots of plants.

Study the text section by section. Answer the study questions so that you can fulfill the learning objectives for each section.

## 28.1 VIRUSES (PP. 566—69)

The learning objectives for this section are:
- Explain why viruses are not classified with cellular organisms.
- Describe the replication of viruses.

1. Fill in the following table to describe viruses.

| | Viruses |
|---|---|
| Structure | |
| Life cycle occurs where? | |
| Parasitic? | |

2. a. Label this diagram, which describes how bacteriophages replicate, using the following alphabetized list of terms.

attachment
biosynthesis
integration
lysogenic cycle
lytic cycle
maturation
penetration
release

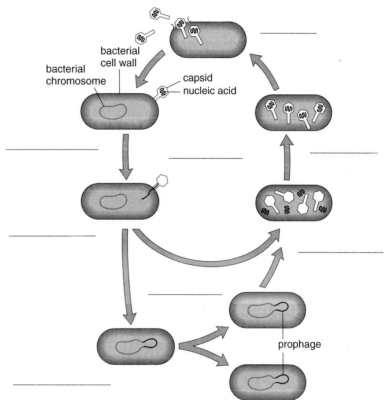

bacterial cell wall
bacterial chromosome
capsid
nucleic acid
prophage

b. Which cycle produces bacteriophages? _____

c. Which cycle is dormant? _____

d. Which cycle kills, or lyses, the host? _____

3. Match the statements to these portions of a retrovirus:

envelope    capsid    viral genome    transcriptase    cDNA

a. _____ integrated into host genome
b. _____ composed of RNA
c. _____ allows the virus to adhere to plasma membrane receptors
d. _____ carries out RNA
e. _____ enters the host cell (choose two)

4. Describe these additional steps needed for the virus to reproduce.

a. biosynthesis _____

b. maturation _____

c. release _____

257

5. Indicate whether these statements are true (T) or false (F).
   _____ a. Antibiotics are helpful for viral infections.
   _____ b. Antiviral drugs act by interfering with viral replication.
   _____ c. There are no vaccines for viral infections.
   _____ d. Prions are neither viruses nor bacteria; they are protein particles.

## 28.2  THE PROKARYOTES (PP. 569—74)

The learning objectives for this section are:
• Give different characteristics of prokaryotes, and describe their structure.
• Describe domain Bacteria and domain Archaea.

6. Label this diagram of a bacterial cell, using the following alphabetized list of terms.
   capsule
   cell wall
   cytoplasm
   fimbriae
   flagellum
   nucleoid
   plasma membrane
   plasmid
   ribosome

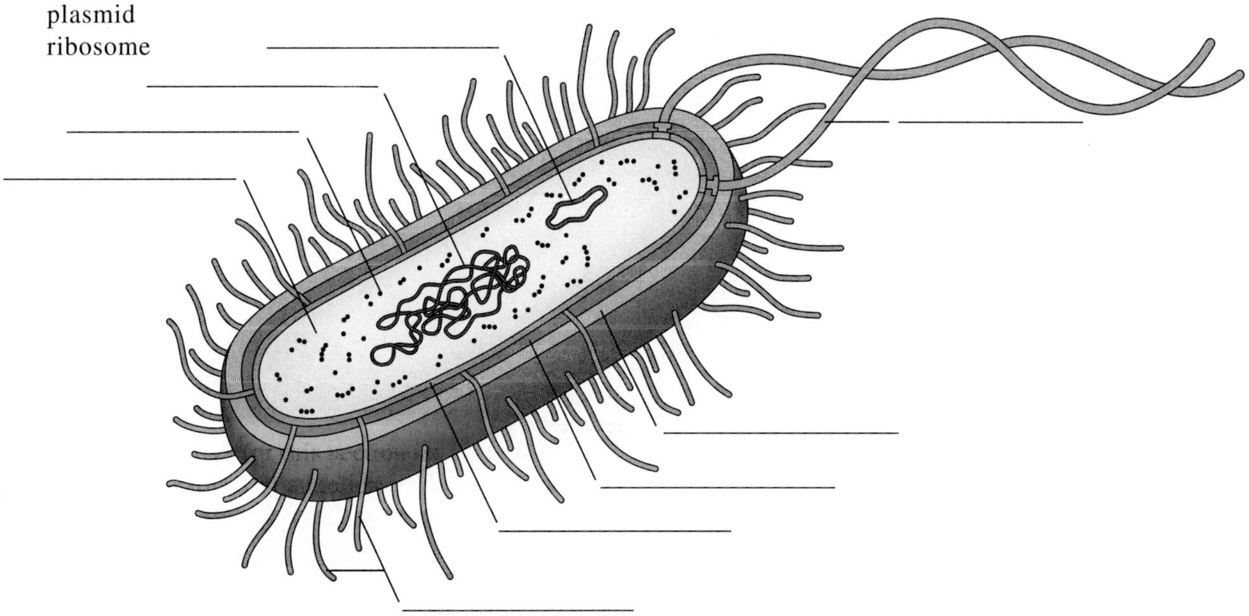

7. Based on the diagram you labeled in question 6, which of these structures are present in eukaryotic animal/plant cells but not in a bacterial cell? _____
   a. plasma membrane
   b. nucleus
   c. ribosomes
   d. mitochondria
   e. cell wall
   f. chloroplasts
   g. flagella

8. Based on the diagram you labeled in question 6, what four structures are present in a bacterial cell but absent from a eukaryotic cell? What are their functions?

   |  | Structure | Function |
   |---|---|---|
   | a. | _____ | _____ |
   | b. | _____ | _____ |
   | c. | _____ | _____ |
   | d. | _____ | _____ |

9. Match the descriptions to these organisms:
    1 chemosynthetic prokaryote
    2 cyanobacteria
    3 obligate anaerobes
    4 saprotrophic bacteria
    a. _____ decomposers
    b. _____ $O_2$ given off
    c. _____ $NH_3 \rightarrow NO_3^-$
    d. _____ cannot survive in $O_2$ environment
10. Match the relationships to these terms:
    1 commensalism
    2 mutualism
    3 parasitism
    4 symbiotic
    a. _____ includes all the others
    b. _____ bacteria living in nodules of legumes
    c. _____ bacteria living on your skin
    d. _____ bacteria that cause strep throat
11. Label the three shapes of bacteria in the following diagram:

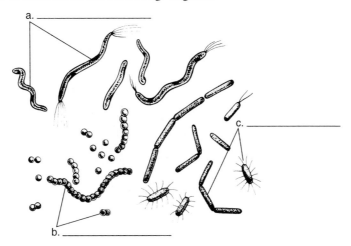

a. _____

c. _____

b. _____

12. Place a check next to all characteristics that are typical of cyanobacteria.
    a. _____ many forms of nutrition
    b. _____ always photosynthetic
    c. _____ have flagella
    d. _____ form lichens
    e. _____ associated with algal bloom
    f. _____ nitrogen fixing

13. Match the descriptions to these terms, which pertain to the reproduction of bacteria:
    1 binary fission
    2 conjugation
    3 transformation
    4 transduction
    5 endospores
    a. _____ bacteria pick up free pieces of DNA
    b. _____ a means of survival
    c. _____ asexual division
    d. _____ male passes DNA to female
    e. _____ bacteriophages carry DNA from one cell to the next

The learning objective for this section is:
* Classify organisms in the kingdom Protista.

14. Green algae are believed to be related to plants because they have a cell wall that

    contains <sup>a.</sup>_____, they possess chlorophylls <sup>b.</sup>_____

    and <sup>c.</sup>_____, and they store reserve food as <sup>d.</sup>_____.

15. Complete the table describing the algae by placing the terms in the appropriate columns. (Some terms are used more than once.)

    **I**
    unicellular
    filamentous
    colonial
    multicellular

    **II**
    isogametes
    heterogametes
    conjugation
    alternation of generations
    daughter colonies
    zoospores

| **Algae** | I. | II. |
|---|---|---|
| *Volvox* | | |
| *Chlamydomonas* | | |
| *Spirogyra* | | |
| *Ulva* | | |

16. Label this diagram of the life cycle of *Chlamydomonas*, using these terms:

    fertilization    meiosis    zoospores    zygote

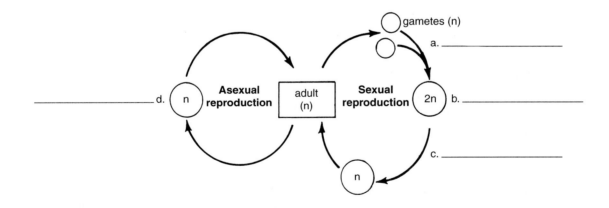

    e. Which portions of this life cycle are haploid? _____

    f. Which portion is diploid? _____

    g. What type of life cycle is this? _____

17. Label this diagram of the life cycle of *Ulva*, using the following alphabetized list of terms.

fertilization
gametophyte
meiosis
sporophyte

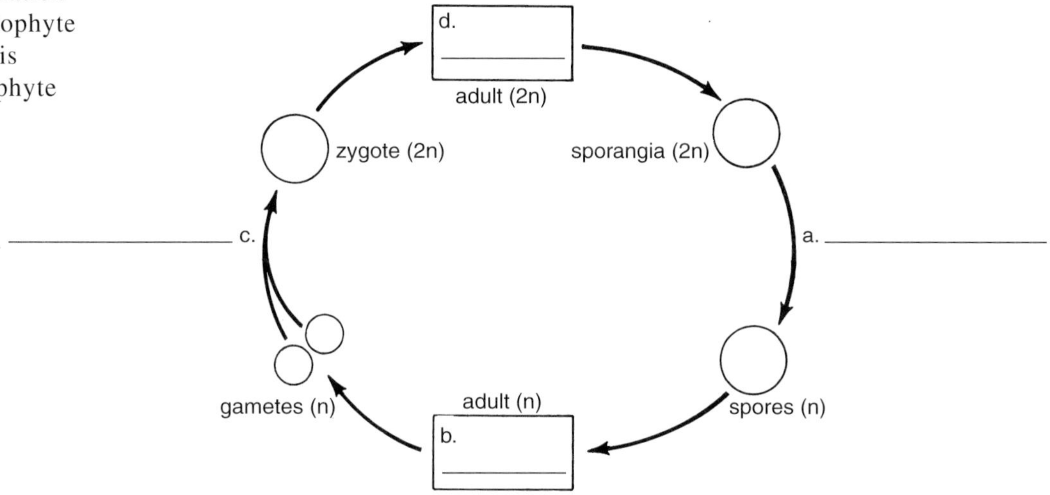

18. Match the traits with these algae (some numbers are used more than once):
    1 brown algae
    2 diatoms
    3 euglenoids
    4 dinoflagellates
    a. _____ are numerous photosynthesizers in the ocean
    b. _____ have animal-like and plantlike characteristics
    c. _____ have chlorophylls *a* and *c,* and carotenoid pigment
    d. _____ have silica-impregnated valves
    e. _____ are used as filtering agents and scouring powders
    f. _____ cause red tide
    g. _____ have a symbiotic relationship with corals
    h. _____ are seaweeds

19. Place the appropriate letter next to each description.

    B—brown algae     R—red algae

    a. _____ *Fucus,* a rockweed
    b. _____ *Laminaria,* a kelp
    c. _____ adapted to cold, rough water
    d. _____ adapted to warm, gentle water
    e. _____ economically important as source of agar

20. Protozoans are typically a._____, b._____, and c._____

    organisms. Some protozoans d._____ and engulf their food; others are e._____

    and absorb nutrients; others are f._____ and cause disease.

21. Complete this table, classifying the protozoans by means of locomotion and giving an example organism of each:

| Classes | Organelle of Locomotion | Example |
|---|---|---|
| Amoeboids | | |
| Ciliates | | |
| Zooflagellates | | |
| Sporozoans | | |

22. Label this diagram of *Paramecium*, using the following alphabetized list of terms.
    anal pore
    contractile vacuole
    food vacuole
    macronucleus
    micronucleus
    oral groove
    pellicle

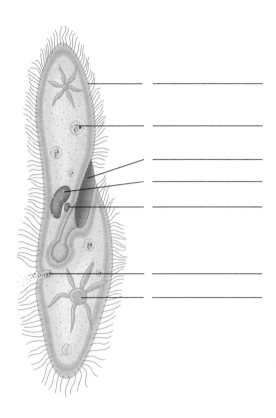

23. Ciliates, such as *Paramecium,* have hundreds of <sup>a.</sup>_____ that extend through a pellicle. Beneath the pellicle are numerous oval capsules that contain <sup>b.</sup>_____, which are used for defense. Food is swept down a(n) <sup>c.</sup>_____, at the end of which food vacuoles form. Ciliates have two nuclei: a large <sup>d.</sup>_____ that controls normal metabolism and one or more micronuclei used during conjugation.

24. Concerning the life cycle of *Plasmodium,*
    a. Where does sexual reproduction occur? _____
    b. Where does asexual reproduction occur? _____
    c. What causes the cycle of recurring chills and fever? _____
    _____

25. Complete this table to describe slime molds and water molds.

| Type of Mold | Body Organization | Nutrition | Reproduction |
|---|---|---|---|
| Plasmodial slime molds | | | |
| Cellular slime molds | | | |
| Water molds | | | |

## 28.4 THE FUNGI (PP. 583–87)

The learning objective for this section is:
- Classify organisms in the kingdom Fungi.

26. Indicate whether these statements about fungi are true (T) or false (F).
    a. _____ usually multicellular
    b. _____ usually unicellular
    c. _____ composed of hyphae
    d. _____ saprotrophic
    e. _____ can be parasitic
    f. _____ can be photosynthetic
    g. _____ cell wall contains cellulose
    h. _____ cell wall contains chitin
    i. _____ have flagella at some time in their life cycle
    j. _____ do not have flagella at any time in their life cycle
    k. _____ form spores only during asexual reproduction
    l. _____ form spores during both asexual and sexual reproduction

27. Fungi are mostly a._____ decomposers that assist in the recycling of nutrients in ecosystems.

    The bodies of most fungi are made up of filaments called b._____, a collection of which are

    called a(n) c._____. They reproduce in accordance with the d._____ life cycle.

    Classification is largely based on the mode of e._____.

28. Label this diagram of the life cycle of black bread mold, using the following alphabetized list of terms. (Some terms are used more than once.)

asexual reproduction
gametangia
meiosis
mycelium
nuclear fusion
sexual reproduction
sporangiophore
sporangium
spores
zygospore
zygospore germination
zygote

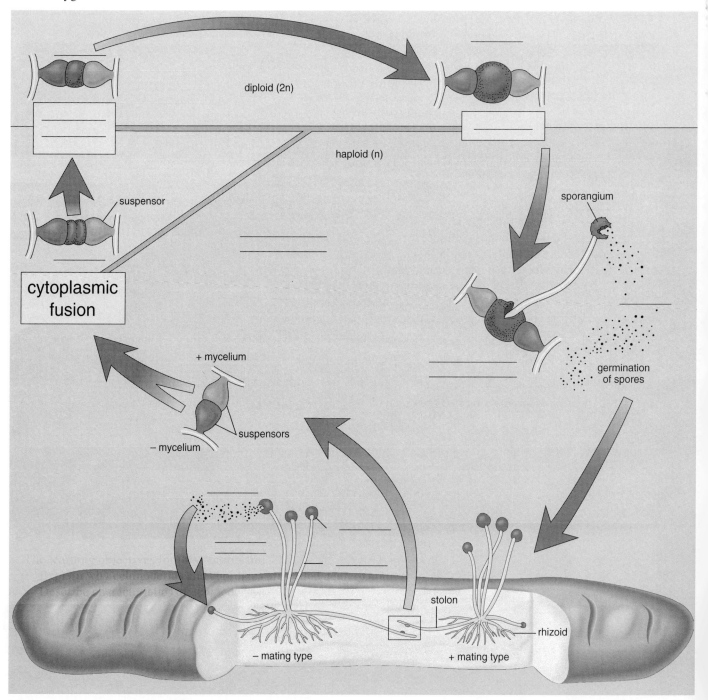

diploid (2n)

haploid (n)

suspensor

cytoplasmic fusion

sporangium

germination of spores

+ mycelium

suspensors

− mycelium

stolon

rhizoid

− mating type

+ mating type

29. Place the appropriate letter next to each description.

F—free-living sac fungi    P—parasitic sac fungi

a. _____ powdery mildew that grows on leaves
b. _____ red mold that grows on bread
c. _____ cup fungi that grow on the forest floor
d. _____ chestnut blight that grows on chestnut trees
e. _____ ergot that grows on rye plants
f. _____ unicellular yeasts

30. Why are all these fungi classified as sac fungi? _____
_____

31. Explain what is happening in each of the following sequential drawings of asci:

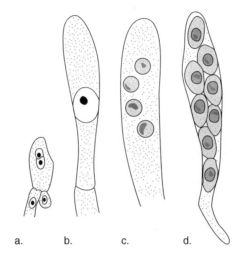

a.          b.          c.          d.

32. What do mushrooms, puffballs, bird's nest fungi, stinkhorn fungi, and bracket fungi have in common?
_____

33. Label this diagram of the life cycle of a mushroom, using the following alphabetized list of terms.

basidiospores
basidium
cap
cytoplasmic fusion
dikaryotic (n+n)
dikaryotic mycelium
diploid (2n)
fruiting body
gill

gill (portion of)
meiosis
monokaryotic
nuclear fusion
nuclei
spore germination
spore release
stalk
zygote

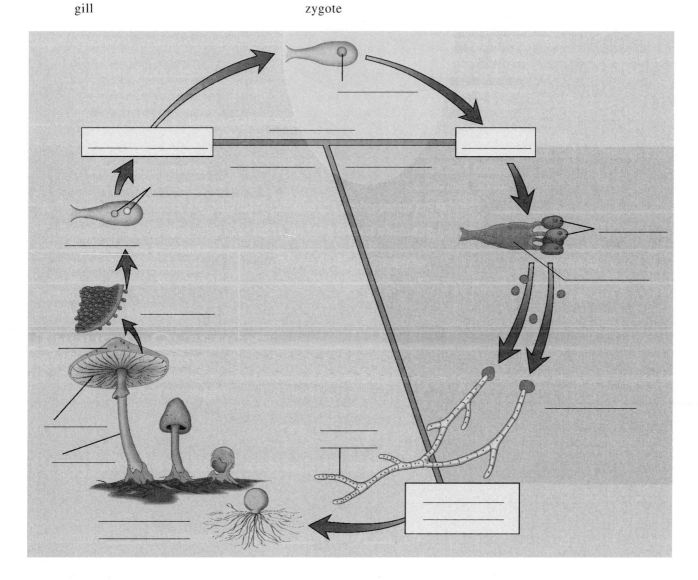

34. Complete the following table concerning imperfect fungi.

| Fungus | Significance | Associated Disease |
|---|---|---|
| *Penicillium* | | |
| *Aspergillus* | | |
| *Candida albicans* | | |

35. Like sac fungi, imperfect fungi reproduce asexually by producing spores called a._____. Unlike sac and club fungi, however, sexual reproduction b._____ in imperfect fungi.

36. Label this diagram of a lichen, using these terms:

cells of algae     mycelium of fungus

a. _____

b. _____

37. Match each description to the type of lichen, using these terms:

crustose     fruticose     foliose

a. _____ compact
b. _____ leaflike
c. _____ shrublike

38. a._____ (fungus roots), which are b._____ relationships between

a(n) c._____ and d._____ roots, help plants acquire e._____

nutrients.

# DEFINITIONS WORDMATCH

Review key terms by completing this matching exercise, selecting from the following alphabetized list of terms:

*bacteriophage*
*binary fission*
*decomposer*
*dinoflagellate*
*lichen*
*prion*
*protozoan*
*saprotroph*
*symbiotic*
*virus*

a. _____ Relationship that occurs when two different species live together in a unique way; it may be beneficial, neutral, or detrimental to one and/or the other species.

b. _____ Splitting of a parent cell into two daughter cells; serves as an asexual form of reproduction in bacteria.

c. _____ Organism that secretes digestive enzymes and absorbs the resulting nutrients back across the plasma membrane.

d. _____ Symbiotic relationship between certain fungi and algae, in which the fungi possibly provide inorganic food or water and the algae provide organic food.

e. _____ Virus that infects bacteria.

f. _____ Heterotrophic unicellular protist that moves by flagella, cilia, or pseudopodia, or is immobile.

g. _____ Noncellular obligate parasite of living cells consisting of an outer capsid and an inner core of nucleic acid.

Do not refer to the text when taking this test.

_____ 1. In the lytic cycle, the term *maturation* refers to the
   a. translation of RNA.
   b. integration of cDNA.
   c. assembly of parts into new viruses.
   d. All of these are correct.

_____ 2. Which of the following is NOT a form of genetic recombination in bacteria?
   a. binary fission
   b. conjugation
   c. transduction
   d. transformation

_____ 3. The function of the bacterial endospore is to
   a. increase the rate of anaerobic respiration.
   b. promote asexual reproduction.
   c. protect against attack from immune systems.
   d. withstand harsh environmental conditions.

_____ 4. A bacterium that can exist in the presence or absence of oxygen is a(n)
   a. autotroph.
   b. facultative anaerobe.
   c. obligate anaerobe.
   d. saprotroph.

_____ 5. Viruses are not in the classification system because
   a. they are obligate parasites.
   b. they are noncellular.
   c. they can integrate into the host genome.
   d. All of these are correct.

_____ 6. Chemosynthetic prokaryotes
   a. give off oxygen just like plants do.
   b. are exemplified by the nitrifying bacteria that oxidize ammonia ($NH_3$) to nitrites ($NO_2^-$).
   c. are decomposers like all bacteria.
   d. Both *b* and *c* are correct.

_____ 7. Why can't cyanobacteria be classified with the eukaryotic algae?
   a. They fix atmospheric nitrogen.
   b. They form a symbiotic relationship with fungi.
   c. They cause disease.
   d. They do not have a nucleus.

_____ 8. Classification of algae according to color
   a. can no longer be justified.
   b. is based on the type of pigments they contain.
   c. suggests that they do not have chlorophyll.
   d. means that some algae are colorless.

_____ 9. *Volvox* is a colonial alga that
   a. does not reproduce.
   b. produces heterogametes.
   c. produces daughter colonies.
   d. Both *b* and *c* are correct.

_____ 10. Diatoms
   a. reproduce sexually.
   b. have a cell wall impregnated with cellulose.
   c. are flagellated.
   d. resemble a pill box.

_____ 11. Which is (are) true of euglenoids?
   a. They have flagella.
   b. Some have chloroplasts.
   c. They reproduce asexually.
   d. All of these are correct.

_____ 12. Both red algae and brown algae
   a. have the same pigments.
   b. are delicate in appearance.
   c. are seaweeds.
   d. are economically unimportant.

_____ 13. Amoebas
   a. have pseudopods.
   b. never have a shell.
   c. always live in fresh water.
   d. All of these are correct.

_____ 14. Ciliates
   a. have a macronucleus and a micronucleus.
   b. do not move.
   c. are parasitic.
   d. are usually saprotrophic.

_____ 15. A trypanosome causes
   a. malaria.
   b. trichinosis.
   c. an intestinal infection.
   d. African sleeping sickness.

_____ 16. In the life cycle of *Plasmodium vivax,* a cause of malaria,
   a. sexual reproduction occurs in a mosquito.
   b. red blood cells burst, causing chills and fever.
   c. spores and gametes form.
   d. All of these are correct.

_____ 17. Slime molds
   a. are exactly like fungi.
   b. have a body composed of hyphae.
   c. produce spores.
   d. All of these are correct.

_____ 18. Club fungi
   a. include the mushrooms.
   b. have a basidiocarp that looks like a cup.
   c. include more parasites than all the other types of fungi.
   d. All of these are correct.

_____ 19. In fungi, the gametes are
   a. heterogametes.
   b. flagellated.
   c. the ends of hyphae.
   d. produced by meiosis.

_____20. Which of the following is NOT characteristic of lichens?
   a. soil formers
   b. algal cells and fungal hyphae
   c. form a type of moss
   d. can live in extreme conditions
_____21. A fruiting body is
   a. a special type of vacuole found in fungi.
   b. a symbiotic relationship between algae and bacteria.
   c. a reproductive structure found in fungi.
   d. always the same shape.
_____22. Sexual reproduction in a bread mold involves the production of
   a. a sperm and an egg.
   b. flagellated zoospores.
   c. zygospores.
   d. fruiting bodies.
_____23. In a mushroom, the _____ is (are) analogous to the asci of a sac fungus.
   a. stalk
   b. cap
   c. basidia
   d. spores

## THOUGHT QUESTIONS

Answer in complete sentences.
24. Algae and protozoans are in the same kingdom. Do they seem closely related? Why or why not?

25. How do you think the earth would change ecologically if fungi were not present?

**Test Results:** _____ number correct ÷ 25 = _____ × 100 = _____ %

## ANSWER KEY

### STUDY QUESTIONS

**1.**

| Viruses |
| --- |
| capsid plus nucleic acid core |
| in host cell |
| always |

**15.**

| I. | II. |
| --- | --- |
| colonial | heterogametes, daughter colonies |
| unicellular | isogametes, zoospores |
| filamentous | conjugation |
| multicellular | isogametes, alternation of generations, zoospores |

**2. a.** See Figure 28.2, page 567, in text. **b.** lytic **c.** lysogenic **d.** lytic **3. a.** cDNA **b.** viral genome **c.** envelope **d.** transcriptase **e.** capsid, viral genome **4. a.** Viral components are synthesized. **b.** assembly of viral components **c.** Budding gives virus an envelope **5. a.** F **b.** T **c.** F **d.** T **6.** See Figure 28.4, page 569, in text. **7.** b, d, f **8. a.** plasmid, accessory ring of DNA **b.** fimbriae, attachment to a substratum **c.** capsule, protection **d.** nucleoid, location of DNA **9. a.** 4 **b.** 2 **c.** 1 **d.** 3 **10. a.** 4 **b.** 2 **c.** 1 **d.** 3 **11. a.** spirillum **b.** coccus **c.** bacillus **12.** b, d, e, f **13. a.** 3 **b.** 5 **c.** 1 **d.** 2 **e.** 4 **14. a.** cellulose **b.** *a* **c.** *b* **d.** starch

**16. a.** fertilization **b.** zygote **c.** meiosis **d.** zoospores **e.** zoospores, adult, gametes **f.** zygote **g.** haplontic **17. a.** meiosis **b.** gametophyte **c.** fertilization **d.** sporophyte **18. a.** 2, 4 **b.** 3 **c.** 1, 2 **d.** 2 **e.** 2 **f.** 4 **g.** 4 **h.** 1 **19. a.** B **b.** B **c.** B **d.** R **e.** R **20. a.** heterotrophic **b.** unicellular **c.** motile **d.** capture **e.** saprotrophic **f.** parasitic **21.** See page 580 in text. **22.** See Figure 28.20, page 581, in text. **23. a.** cilia **b.** trichocysts **c.** gullet **d.** macronucleus **24. a.** in the mosquito **b.** in the human **c.** Toxins, or poisons, enter the blood when red blood cells release spores.

**25.**

| Body Organization | Nutrition | Reproduction |
|---|---|---|
| 2n plasmodium | phagocytosis | Sporangium produces spores by meiosis, which produce flagellated haploid cells that fuse. |
| individual amoeboid cells | phagocytosis | Sporangium produces spores. |
| 2n filamentous, cell walls are cellulose | parasitism | Meiosis produces haploid gametes; otherwise, asexual by zoospores. |

**26. a.** T **b.** F **c.** T **d.** T **e.** T **f.** F **g.** F **h.** T **i.** F **j.** T **k.** F **l.** T  **27. a.** saprotrophic **b.** hyphae **c.** mycelium **d.** haplontic **e.** sexual reproduction  **28.** See Figure 28.24, page 584, in text.  **29. a.** P **b.** F **c.** F **d.** P **e.** P **f.** F  **30.** because they form asci during sexual reproduction  **31. a.** Ascus with two nuclei is forming. **b.** Nuclei have fused, and zygote has formed. **c.** Meiosis has occurred. **d.** Mitotic divisions have resulted in eight ascospores.  **32.** They are all basidiomycetes.  **33.** See Figure 28.26, page 586, in text.

**34.**

| Significance | Associated Disease |
|---|---|
| makes penicillin | none |
| makes various chemicals | none |
| yeastlike | vaginal infections |

**35. a.** conidiospores **b.** has never been observed  **36. a.** cells of algae **b.** mycelium of fungus  **37. a.** crustose **b.** foliose **c.** fruticose  **38. a.** Mycorrhizae **b.** symbiotic **c.** fungus **d.** plant **e.** mineral

## DEFINITIONS WORDMATCH

**a.** symbiotic **b.** binary fission **c.** saprotroph **d.** lichen **e.** bacteriophage **f.** protozoan **g.** virus

## CHAPTER TEST

**1.** c **2.** a **3.** d **4.** b **5.** b **6.** b **7.** d **8.** b **9.** d **10.** d **11.** d **12.** c **13.** a **14.** a **15.** d **16.** d **17.** c **18.** a **19.** c **20.** c **21.** c **22.** c **23.** c **24.** They do not seem related in that algae are photosynthetic and protozoans are heterotrophic. They do seem related in that some of the algae are motile in the same way protozoans are. **25.** Recycling would be reduced and organic waste would accumulate. Without efficient recycling, the carrying capacities of ecosystems would diminish.

# 29

# PLANTS

Plants (domain Eukarya, kingdom Plantae) are multicellular, photosynthetic organisms adapted to a land existence. Among the various adaptations, all plants protect the developing embryo from drying out. Plants have an **alternation of generations** life cycle, but some have a dominant **gametophyte** (haploid generation) and others have a dominant **sporophyte** (diploid generation).

The **nonvascular plants,** which include the liverworts and the mosses, lack true roots, stems, and leaves. In the moss life cycle, antheridia produce swimming sperm that use external water to reach the eggs in the archegonia. Following fertilization, the moss sporophyte, which consists of a foot, a stalk, and a capsule, is dependent on the female gametophyte. Windblown spores are produced by meiosis in the capsule, and they disperse the species.

**Vascular plants** have vascular tissue—that is, xylem and phloem. In the life cycle of vascular plants, the sporophyte is dominant.

Whisk ferns, club mosses, horsetails, and ferns are the seedless vascular plants that were prominent in swamp forests during the Carboniferous period. In the life cycle of seedless plants, spores disperse the species, and the separate gametophyte produces flagellated sperm. Vegetative (asexual) reproduction is used to a degree to disperse ferns in dry habitats.

Seed plants have a life cycle in which there are **microgametophytes** (male gametophytes) and **megagametophytes** (female gametophytes). The microgametophyte is the **pollen grain,** which produces nonflagellated sperm, and the megagametophyte, which is located within an ovule, produces an egg. The pollen grain replaces the flagellated sperm of seedless vascular plants.

**Gymnosperms** produce **seeds** that are uncovered. In gymnosperms, the microgametophytes develop in pollen cones. The megagametophyte develops within an **ovule** located on the scales of seed cones. Following **pollination** and fertilization, the ovule becomes a winged seed that is dispersed by wind.

**Angiosperms** produce seeds that are covered by **fruits.** The petals of **flowers** attract pollinators, and the **ovary** develops into a fruit, which aids the dispersal of seeds. Angiosperms provide most of the food that sustains terrestrial animals, and they are the source of many products used by humans.

Study the text section by section. Answer the study questions so that you can fulfill the learning objectives for each section.

## 29.1 CHARACTERISTICS OF PLANTS (PP. 592–93)

The learning objectives for this section are:
- Name characteristics of plants that facilitate a land existence.
- Compare plants according to the presence of vascular tissue and their reproductive strategy.

1. Indicate whether these statements about plants are true (T) or false (F).
   a. _____ adapted to living on land
   b. _____ diploid sporophyte produces diploid spores
   c. _____ haploid gametophyte, which produces sex cells
   d. _____ photosynthetic organisms

2. Match the plants to the following plant categories. (The categories can be used more than once.)

    1 nonvascular plants         2 seedless vascular plants

    3 gymnosperms             4 angiosperms

    _____ a. club mosses

    _____ b. conifers

    _____ c. mosses

    _____ d. division Magnoliophyta

    _____ e. ferns

    _____ f. cycads

## 29.2 NONVASCULAR PLANTS (PP. 594–95)

The learning objective for this section is:

• Describe the three characteristics of nonvascular plants—hornworts, liverworts, and mosses.

3. Label this diagram of part of a moss life cycle, using these terms:

    antheridium    archegonium    egg    sperm

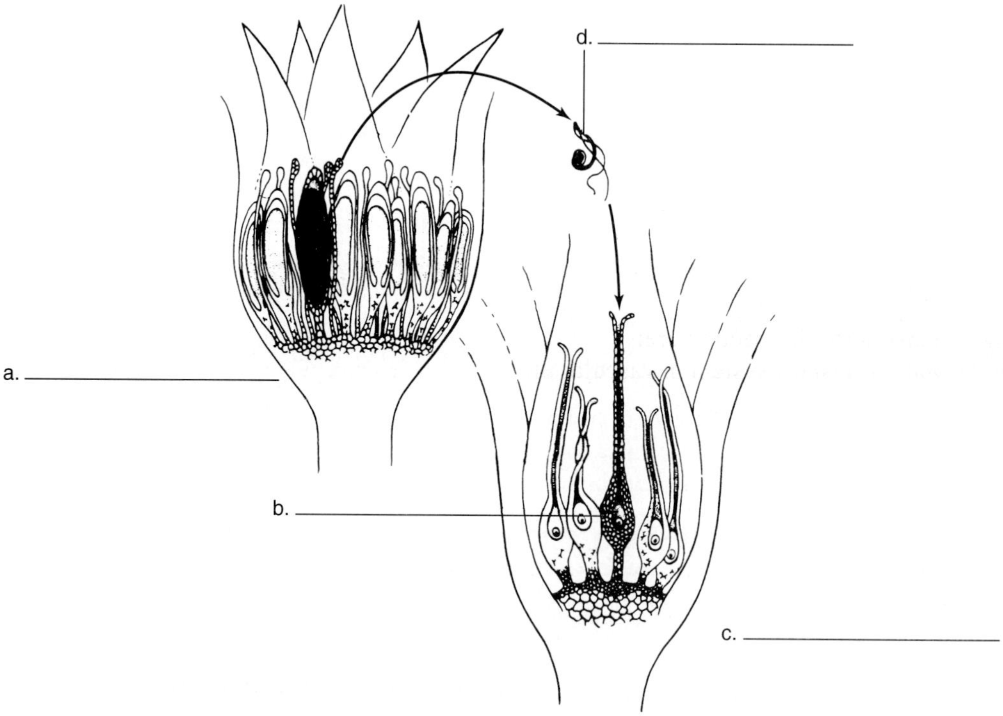

4. a. In the diagram in question 3, the antheridium and archegonium are part of what

    generation? _____

  b. Do mosses have flagellated sperm? _____

  c. Do mosses protect the zygote? _____

5. Is the sporophyte generation dependent on the gametophyte generation in the moss? a._____

Do either of these generations have vascular tissue? b._____ The capsule contains the

sporangium where the cellular process of c._____ occurs during the production

of d._____. The latter disperse the species. When they germinate,

a(n) e._____ forms to begin the f._____ generation.

6. The a._____ anchor the moss gametophyte plant in the soil while absorbing minerals and

water. There must be an external source of b._____ for the sperm to move from

the c._____ to the eggs found in the d._____.

7. Label this diagram of part of the moss life cycle, using the following alphabetized list of terms.

capsule
gametophyte generation
rhizoids
sporophyte generation

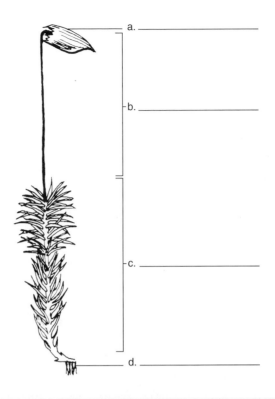

a. _____

b. _____

c. _____

d. _____

## 29.3 SEEDLESS VASCULAR PLANTS (PP. 596–99)

The learning objectives for this section are:
• Describe the water transport system in vascular plants, which also provides internal support.
• Describe different seedless vascular plants.

8. In all vascular plants, the a._____ generation is the dominant generation, and it

is b._____ (haploid/diploid). c._____ vascular tissue conducts water and

minerals up from the soil. d._____ vascular tissue transports organic nutrients from one part

of the body to another. An advantage of a diploid plant is that a functional gene can e._____ a

faulty gene.

9. Indicate whether these statements are true (T) or false (F). Rewrite the false statement(s) to be true.
   a. _____ All vascular plants produce pollen grains and seeds. Rewrite: _____

   b. _____ Spores disperse seedless vascular plants, while seeds disperse seed plants. Rewrite: _____

   c. _____ In all vascular plants, the gametophyte generation is dependent on the sporophyte. Rewrite: _____

10. Indicate whether these descriptions of rhyniophytes are true (T) or false (F).
    a. _____ had true stem, roots, and leaves
    b. _____ spores were produced without benefit of sporangia
    c. _____ had a forked stem, no roots or leaves

11. Match the descriptions to these plants:
    1 whisk ferns            2 club mosses
    3 horsetails             4 ferns
    a. _____ Scalelike leaves cover stems and branches; there are terminal strobili.
    b. _____ Whorls of slender branches form at the nodes of the stem where the leaves are; they are sometimes called scouring rushes.
    c. _____ These plants, alive today, best resemble *Rhynia*.
    d. _____ Large fronds subdivide into leaflets.
12. Label this diagram of part of the fern life cycle with these terms: antheridium      archegonium

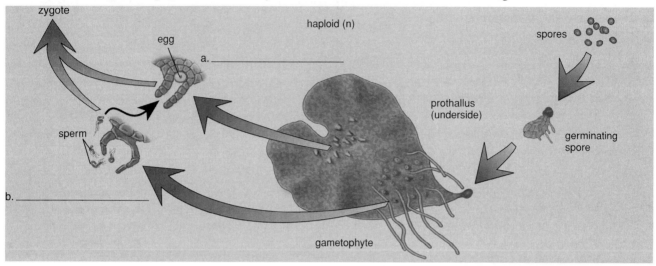

13. In the diagram in question 12, this entire structure is called the ᵃ·_____. This structure has

    a(n) ᵇ·_____ shape and represents the ᶜ·_____ generation.

    It ᵈ·_____ (does/does not) have vascular tissue and ᵉ·_____ (does/does not)

    have flagellated sperm. The sporophyte ᶠ·_____ (is/is not) the dominant

    generation. ᵍ·_____ disperse the species.
14. Label this diagram of part of the fern life cycle, using these terms:

    fiddlehead      fronds      rhizome      sori

15. The frond is part of the a._____ generation in the fern, and it b._____ (does/does not) have vascular tissue. This generation is c._____ (diploid/haploid).

16. Label the following as a bryophyte (B) or a fern (F):
    _____ a. peat moss
    _____ b. ornamental
    _____ c. Native Americans used this during childbirth

## 29.4 SEED PLANTS (PP. 600—606)

The learning objective for this section is:
• Describe gymnosperms and angiosperms and their life cycles.

17. Label this diagram, showing the alternation of generations in seed plants with the notation n or 2n.

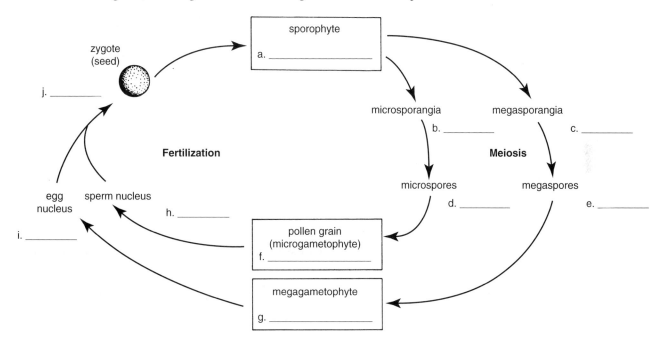

18. Is the gametophyte generation dependent on the sporophyte generation in seed plants? a._____ Does the sporophyte generation have vascular tissue? b._____ What structure in seed plants replaces flagellated sperm in nonseed plants? c._____ What structure disperses the species in seed plants? d._____

19. Match the descriptions with these types of gymnosperms:
    conifer
    cycad
    ginkgo
    a. _____ stout, unbranched stem with large, compound leaves
    b. _____ fan-shaped leaves shed in autumn; planted in parks
    c. _____ pine trees, hemlocks, spruces

20. Study this diagram of the life cycle of a pine tree, and then describe the numbered events.

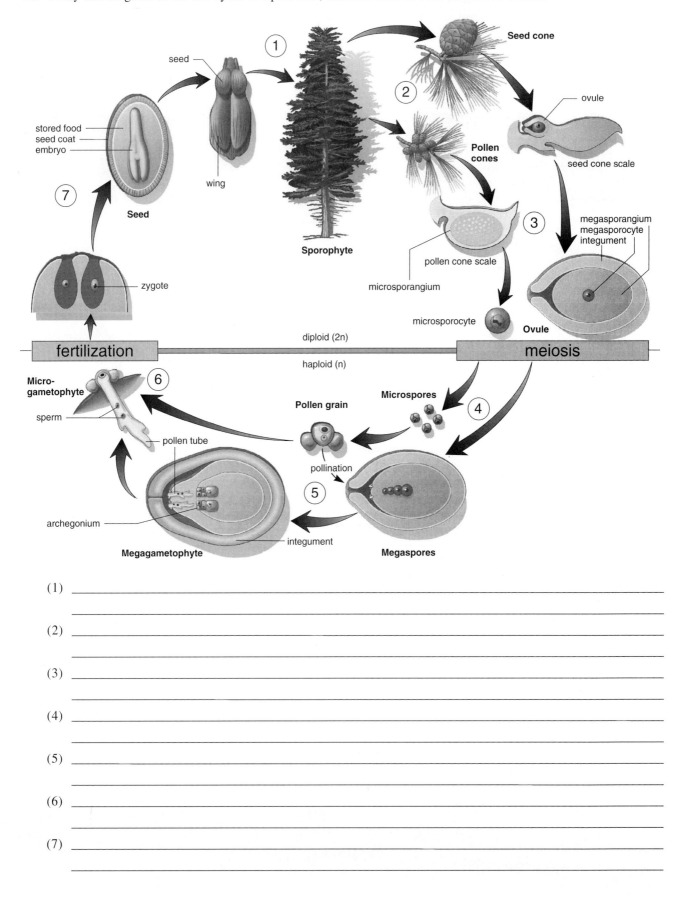

(1) _____
_____

(2) _____
_____

(3) _____

(4) _____
_____

(5) _____

(6) _____
_____

(7) _____
_____

21. Gymnosperms have an alternation of generations life cycle. Which structure in the diagram in question 20 is described in each of the following?

    a. is the sporophyte _____

    b. produces the microspore _____

    c. produces the megaspore _____

    d. is the microgametophyte _____

    e. contains the megagametophyte _____

    f. contains the sperm _____

    g. contains the egg _____

    h. contains the embryonic sporophyte _____

22. Place the appropriate letter next to each statement.

    M—monocot    D—dicot

    a. _____ almost always herbaceous
    b. _____ either woody or herbaceous
    c. _____ flower parts in fours or fives
    d. _____ flower parts in threes
    e. _____ net-veined leaves
    f. _____ parallel-veined leaves
    g. _____ vascular bundles arranged in a circle in the stem
    h. _____ vascular bundles scattered in the stem

23. Label this diagram of the structures of a flower, using the following alphabetized list of terms.

    anther
    ovary
    ovule

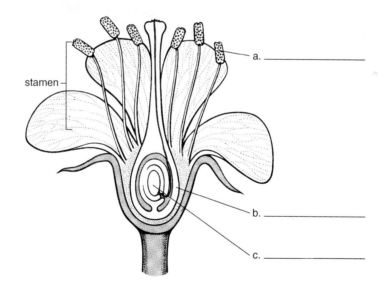

stamen

a. _____

b. _____

c. _____

24. Study this diagram of the life cycle of a flowering plant, and then describe the numbered events.

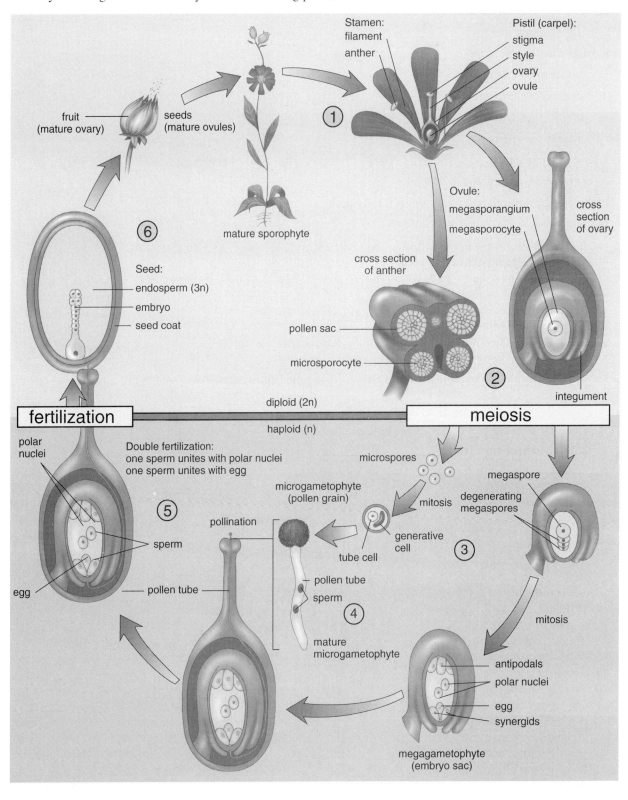

(1) _____
_____

(2) _____
_____

(3) _____
_____

(4) _____
_____

(5) _____
_____

(6) _____
_____

25. Flowering plants have an alternation of generations life cycle. Which structure in the diagram in question 24 is described in each of the following?

a. is the sporophyte _____

b. produces the microspore _____

c. produces the megaspore _____

d. is the microgametophyte _____

e. contains the megagametophyte _____

f. contains the sperm _____

g. contains the egg _____

h. contains the embryonic sporophyte _____

i. matures to become a fruit _____

26. The color and arrangement of flower parts are designed

to a._____.

A flower, which is necessary to the reproduction of flowering plants,

contains b._____

_____.

A flower produces seeds enclosed by c._____, which helps d._____.

27. Complete the following table to compare the different plants:

| Plant | Vascular Tissue (yes or no) | Dominant Generation | Spores or Seeds Disperse Species | Fruit (yes or no) |
|---|---|---|---|---|
| Mosses | | | | |
| Ferns | | | | |
| Gymnosperms | | | | |
| Angiosperms | | | | |

1. These cups are of different sizes. Sort the cups according to size of classification category. Not all cups are used.
   a. Domains: _____
   b. Kingdoms: _____
   c. Other Groups (A category that contains at least one of the types of organisms on the list): _____
   d. Specific type of organism:

   **Cups**

   plants                    mushroom
   cyanobacterium            nonvascular plant
   protozoan                 vascular plants
   alga                      seed plants
   protists                  fungi
   moss                      fern
   bacteria                  pine
   flowering plants          slime mold

2. The cups are of different colors. Put all the like colors, regardless of size, together.

   a. Any category pertaining to plants: _____

   b. Any category pertaining to protists: _____

   c. Any category pertaining to bacteria: _____

   d. Any category pertaining to fungi: _____

3. Players are given disks and required to flip them into the correct cup. Put the disk in the largest cup possible, but it must pertain to all the organisms of the group. Cups can contain more than one disk.

   **Disks**

   a. red, brown, golden brown, or green: _____

   b. single cell, motile, food vacuoles: _____

   c. live on the land: _____

   d. trees: _____

   e. lack nucleus: _____

   f. always protect the embryo: _____

   g. saprotrophic: _____

   h. photosynthesis: _____

   i. alternation of generations: _____

   j. haplontic cycle: _____

   k. malaria: _____

   l. *Amoeba, Paramecium, Euglena:* _____

   m. zygospore, sac, club, imperfect: _____

   n. heterospores: _____

   o. fruit: _____

   p. fruiting body: _____

   q. hyphae: _____

   r. rhizoids, shoot, leaflets: _____

   s. zoospores: _____

   t. naked seeds: _____

   u. pollen carried by insects: _____

   v. endospores: _____

   w. monocot and dicot: _____

   x. fronds: _____

   y. plasmodium: _____

   z. fruiting body: _____

# DEFINITIONS WORDMATCH

Review key terms by completing this matching exercise, selecting from the following alphabetized list of terms:

*alternation of generations*
*antheridium*
*archegonium*
*ginkgo*
*gymnosperm*
*megaspore*
*ovule*
*phloem*
*pollination*
*seed*
*sporophyte*
*xylem*

a. _____ In seed plants, a large spore that develops into the egg-producing megagametophyte.

b. _____ Seed plant with uncovered (naked) seeds; examples are conifers and cycads, which bear cones.

c. _____ In the life cycle of a plant, the diploid generation that produces spores by meiosis.

d. _____ Mature ovule that contains a sporophyte embryo with stored food enclosed by a protective coat.

e. _____ In seed plants, a structure where the megaspore becomes an egg-producing megagametophyte; develops into a seed following fertilization.

f. _____ Vascular tissue that conducts organic solutes in plants; contains sieve-tube elements and companion cells.

g. _____ In seed plants, the delivery of pollen to the vicinity of the egg-producing megagametophyte.

# CHAPTER TEST

## OBJECTIVE QUESTIONS

Do not refer to the text when taking this test.

_____ 1. Select the incorrect association.
   a. gametophyte—diploid generation
   b. gametophyte—produces sex cells
   c. sporophyte—diploid generation
   d. sporophyte—produces haploid spores

_____ 2. Select the nonvascular plant.
   a. bryophyte
   b. cycad
   c. fern
   d. rosebush

_____ 3. The antheridium is part of the
   a. megagametophyte.
   b. megasporophyte.
   c. microgametophyte.
   d. microsporophyte.

_____ 4. Rhyniophytes are significant because they
   a. are currently the most successful conifers.
   b. are currently the most successful flowering plants.
   c. were the first to evolve flowers.
   d. were the first to evolve vascular tissue.

_____ 5. Xylem and phloem are
   a. the covering tissues on roots, stems, and leaves.
   b. the male and female parts of a flower.
   c. two kinds of flowering plants.
   d. two types of vascular tissue.

_____ 6. Ferns are plants that are
   a. nonvascular with seeds.
   b. nonvascular without seeds.
   c. vascular with seeds.
   d. vascular without seeds.

_____ 7. Which structure develops into a pollen grain?
   a. antheridium
   b. archegonium
   c. megaspore
   d. microspore

_____ 8. Select the characteristic NOT descriptive of conifers.
   a. can withstand cold winters
   b. can withstand hot summers
   c. needlelike leaves
   d. reproduce through flowers

_____ 9. Select the incorrect statement about angiosperms.
   a. contain only tracheids in their vascular tissue
   b. did not diversify until the Cenozoic era
   c. include tiny plants living on pond surfaces
   d. the most successful group of plants

_____ 10. Select the incorrect association.
   a. dicot—woody or herbaceous
   b. dicot—vascular bundles arranged in a circle within the stem
   c. monocot—almost always herbaceous
   d. monocot—net-veined leaf

_____ 11. In comparing alternation of generations for non-seed and seed plants, it is apparent that seed plants have
   a. flowers only.
   b. heterospores.
   c. flagellated sperm.
   d. megagametophytes and microgametophytes.
   e. Both *b* and *d* are correct.

____ 12. The fern is a nonseed plant and
  a. is a bryophyte.
  b. has flagellated sperm.
  c. lacks vascular tissue.
  d. All of these are correct.

____ 13. The dominant generation in the seed plants is the
  a. sporophyte.
  b. gametophyte.
  c. green leafy shoot.
  d. flower only.

____ 14. Ferns are restricted to moist places because
  a. of the sporophyte generation called the frond.
  b. of a sensitive type of chlorophyll.
  c. of the water-dependent gametophyte generation.
  d. they never grow very tall.

____ 15. In the life cycle of seed plants, meiosis
  a. produces the gametes.
  b. produces microspores and megaspores.
  c. does not occur.
  d. produces spores.

____ 16. Double fertilization refers to the fact that in angiosperms
  a. two egg cells are fertilized within an ovule.
  b. a sperm nucleus fuses with an egg cell and with polar nuclei.
  c. two sperm are required for fertilization of one egg cell.
  d. a flower can engage in both self-pollination and cross-pollination.

____ 17. Identify the correct order of life cycle stages in a moss.
  a. gametophyte—sporophyte—spores—zygote—protonema
  b. sporophyte—spores—protonema—gametes—zygote
  c. gametophyte—spores—sporophyte—gametes—zygote
  d. zygote—sporophyte—gametophyte—spores—protonema

____ 18. The sporangia of a fern are generally
  a. at the tips of the rhizomes.
  b. on the bottom surface of the prothallus.
  c. inside the fiddleheads.
  d. at the point where leaves join the stem.
  e. on the undersides of fronds.

____ 19. In which of these groups is the gametophyte nutritionally dependent upon the sporophyte?
  a. ferns
  b. angiosperms
  c. gymnosperms
  d. Both b and c are correct.
  e. All of these are correct.

____ 20. In pine trees, _____ develop in separate types of cones.
  a. microgametophytes and megagametophytes
  b. pollen and ovules
  c. microspores and megaspores
  d. Both a and b are correct.
  e. All of these are correct.

## THOUGHT QUESTIONS

Answer in complete sentences.

21. What do you think will be the impact on human life if there is a major extinction of plants during the next century?

22. Should algae be considered plants? Offer reasons why they should and should not be classified this way.

**Test Results:** _____ number correct ÷ 22 = _____ × 100 = _____ %

1. **a.** T **b.** F **c.** T **d.** T **2. a.** 2 **b.** 3 **c.** 1 **d.** 4 **e.** 2 **f.** 3 **3. a.** antheridium **b.** egg **c.** archegonium **d.** sperm **4. a.** gametophyte **b.** yes **c.** yes **5. a.** yes **b.** no **c.** meiosis **d.** spores **e.** protonema **f.** gametophyte **6. a.** rhizoids **b.** water **c.** antheridia **d.** archegonia **7. a.** capsule **b.** sporophyte generation **c.** gametophyte generation **d.** rhizoids **8. a.** sporophyte **b.** diploid **c.** Xylem **d.** Phloem **e.** mask **9. a.** F, Only gymnosperms and angiosperms produce pollen grains and seeds. **b.** T **c.** F, In most seed vascular plants, the gametophyte generation is dependent on the sporophyte. **10. a.** F **b.** F **c.** T **11. a.** 2 **b.** 3 **c.** 1 **d.** 4 **12. a.** archegonium **b.** antheridium **13. a.** prothallus **b.** heart **c.** gametophyte **d.** does not **e.** does **f.** is **g.** Spores **14. a.** frond **b.** sori **c.** fiddlehead **d.** rhizome **15. a.** sporophyte **b.** does **c.** diploid **16. a.** B **b.** F **c.** F **17. a.** 2n **b.** 2n **c.** 2n **d.** n **e.** n **f.** n **g.** n **h.** n **i.** n **j.** 2n **18. a.** yes **b.** yes **c.** pollen grain **d.** seed **19. a.** cycad **b.** ginkgo **c.** conifer **20.** (1) Sporophyte is dominant. (2) pollen cones and seed cones (3) microsporangia on lower surface of pollen cone; megasporangia on upper surface of seed cone inside ovule (4) Meiosis produces microspores/megaspores. (5) Each **27.**

microspore develops in a microgametophyte (pollen grain); one megaspore develops into a megagametophyte. (6) Pollen tube delivers sperm to egg. (7) Ovule becomes seed. **21. a.** tree **b.** microsporangium on scale of pollen cone **c.** megasporangium inside ovule on scale of seed cone **d.** pollen grain **e.** ovule **f.** mature microgametophyte (pollen grain) **g.** megagametophyte inside ovule **h.** seed **22. a.** M **b.** D **c.** D **d.** M **e.** D **f.** M **g.** D **h.** M **23. a.** anther **b.** ovary **c.** ovule **24.** (1) Flower contains stamens and pistil. (2) Megasporocyte produces four haploid megaspores; microsporocyte produces four haploid microspores. (3) One functional megaspore survives and divides mitotically; microgametophyte contains generative cell nucleus and tube cell nucleus. (4) Megagametophyte results and consists of eight haploid nuclei; pollen grain germinates and produces a pollen tube. (5) Double fertilization occurs; one sperm fertilizes the egg, and the other joins with the polar nuclei. (6) Seed contains endosperm, embryo, and seed coat. **25. a.** flowering plant **b.** pollen sac **c.** megasporangium inside ovule **d.** pollen grain **e.** ovule **f.** pollen tube **g.** megagametophyte (embryo sac) **h.** seed **i.** ovary **26. a.** attract pollinators **b.** micro- and megasporangia where microspores and megaspores are produced **c.** fruit **d.** disperse the species

| Vascular Tissue (yes or no) | Dominant Generation | Spores or Seeds Disperse Species | Fruit (yes or no) |
|---|---|---|---|
| no | gametophyte | spores | no |
| yes | sporophyte | spores | no |
| yes | sporophyte | seeds | no |
| yes | sporophyte | seeds | fruit |

1. **a.** Bacteria **b.** plants, fungi, protists **c.** nonvascular plants, vascular plants, seed plants **d.** all the rest **2. a.** plants, mosses, flowers, fern, pine, nonvascular plants, vascular plants **b.** protozoans, algae **c.** bacteria, cyanobacterium **d.** fungi, slime molds, mushrooms **3. a.** alga **b.** protozoan **c.** plants **d.** vascular plants **e.** bacteria **f.** plants **g.** fungi **h.** plants **i.** plants **j.** protists **k.** protozoan **l.** protozoan **m.** fungi **n.** seed plants **o.** flower **p.** fungi **q.** fungi **r.** mosses **s.** protists **t.** pine **u.** flower **v.** bacteria **w.** flowering plants **x.** fern **y.** slime mold **z.** mushroom

**a.** megaspore **b.** gymnosperm **c.** sporophyte **d.** seed **e.** ovule **f.** phloem **g.** pollination

1. a **2.** a **3.** c **4.** d **5.** d **6.** d **7.** d **8.** d **9.** a **10.** d **11.** e **12.** b **13.** a **14.** c **15.** b **16.** b **17.** b **18.** e **19.** d **20.** e **21.** The ecological and economical contributions of plants will be lost. Examples include loss of food production and inability to maintain a balance of gases in the atmosphere. **22.** They are photosynthetic producers, so in this way they should be considered plants. However, they lack the adaptations for terrestrial life, so in this way they should not be considered plants.

# 30

# ANIMALS: PART 1

## CHAPTER REVIEW

Animals (domain Eukarya, kingdom Animalia) are multicellular, locomote by contracting fibers, and ingest their food. They all have the diplontic life cycle, but differ in a number of ways by which they are classified. Many animals, such as those discussed in this chapter, are **invertebrates.**

Sponges have the cellular level of organization and lack tissues and symmetry. Sponges are **sessile filter feeders** that depend on a flow of water through the body to acquire food, which is digested in vacuoles within collar cells that line a central cavity.

**Cnidarians** have a sac body plan, and the original life cycle alternated between a polyp and medusa phase. Today, many cnidarians exist as either polyps (e.g., *Hydra*) or medusae (e.g., jellyfishes).

Flatworms have **bilateral symmetry** and the organ level of organization. Flatworms may be free living or parasitic. Freshwater planaria have muscles, a ladderlike nervous organization, and **cephalization** consistent with a predatory way of life. Parasitic flatworms, namely, tapeworms and flukes, lack cephalization and are otherwise modified for a parasitic lifestyle.

Roundworms are **pseudocoelomates** and have a tube-within-a-tube body plan. Many roundworms, ranging from pinworms to *Ascaris* and *Trichinella,* are parasitic.

The **protostomes** (and **deuterostomes**) have a **coelom** completely lined with mesoderm. In protostomes, the mouth appears at or near the blastopore, the first embryonic opening. The body of a **mollusc** is composed of a foot, a visceral mass, and a mantle cavity. Molluscs are adapted to various ways of life; for example, clams are sessile filter feeders, squids are active predators of the open ocean, and snails are terrestrial. **Annelids** are the segmented worms. Polychaetes (e.g., clam worms, tube worms) are marine animals with parapodia for gas exchange; earthworms live on land; and the **leech** is parasitic. In the earthworm, the coelom and the nervous, excretory, and circulatory systems all provide evidence of **segmentation.**

**Arthropods** have a ventral nerve cord and an external exoskeleton made of **chitin,** which is periodically shed by **molting;** this is different from the internal skeleton of **vertebrates.** Like vertebrates, however, arthropods are segmented and have jointed appendages. This combination has led to specialization of parts, and in some arthropods, the body is divided into special regions, each with its own particular type of appendages. Whereas the crayfish, a **crustacean,** is adapted to a marine existence, the grasshopper, an **insect,** is adapted to a terrestrial existence. Insects have wings and breathe by means of air tubes called **tracheae.** Their circulatory system does not contain a respiratory pigment.

## STUDY QUESTIONS

Study the text section by section. Answer the study questions so that you can fulfill the learning objectives for each section.

## 30.1 EVOLUTION AND CLASSIFICATION OF ANIMALS (PP. 610–12)

The learning objectives for this section are:
- Describe characteristics that animals have in common.
- List the criteria for classifying animals.
- Diagram the evolutionary tree of animals.

1. Place a check in front of those phrases that correctly describe animal characteristics.
   a. _____ autotrophic nutrition
   b. _____ haplontic life cycle
   c. _____ ingest food
   d. _____ meiosis produces gametes
   e. _____ multicellular organisms

2. Match the definitions to these types of symmetry:

   asymmetry     bilateral symmetry     radial symmetry

   a. _____ A definite right and left half; one longitudinal cut down the center of the animal produces two equal halves.
   b. _____ Animal has no particular symmetry.
   c. _____ Animal is organized circularly; two identical halves are obtained no matter how the animal is sliced longitudinally.

3. The three germ layers are called a. _____, b. _____, and c. _____.

4. Animals with the tissue level of organization have a. _____ germ layers. Animals with the b. _____ level of organization have three germ layers.

5. Match the definitions to these terms:

   acoelomate     pseudocoelomate     protostome     deuterostome

   a. _____ A cavity is incompletely lined with mesoderm.
   b. _____ The blastopore becomes the mouth.
   c. _____ A coelom is lacking.
   d. _____ The blastopore becomes the anus, and a second opening becomes the mouth.

6. The evolutionary tree of animals shows that all animals are believed to have evolved from a. _____ ancestors, most likely b. _____.

7. Write the names of the animal phyla depicted in the arrows. Then fill in the ovals, using the following alphabetized list of terms.

acoelomates
bilateral symmetry
coelomates
deuterostomes
multicellularity
protostomes
pseudocoelomates
radial symmetry

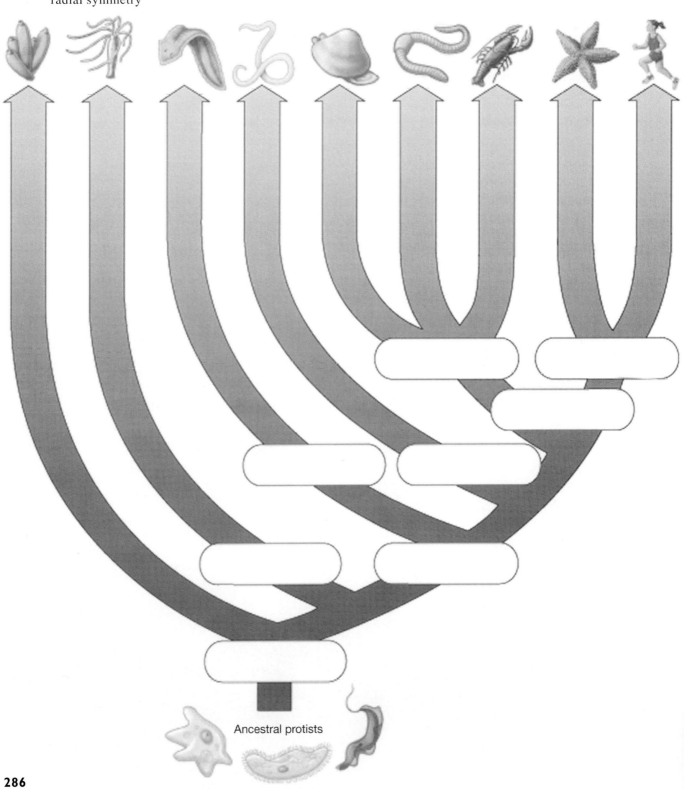

Ancestral protists

The learning objective for this section is:
• Describe the characteristics of sponges, cnidarians, planarians, and roundworms.

8. Label the cells in this diagram of the body wall of the sponge, using the following alphabetized list of terms. Write the function of each cell next to its label.

amoeboid cell
collar cell
epidermal cell
pore
spicule

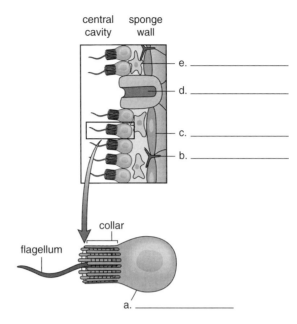

central sponge
cavity wall

e. _____

d. _____

c. _____

b. _____

collar

flagellum

a. _____

9. Sponges are the only animals to have the a._____ level of organization. Their bodies are perforated by many b._____; the phylum name Porifera means c._____. The beating of the flagella of d._____ creates water currents, which flow through the pores into the central cavity and out through the e._____. Food particles brought in by water are digested in food vacuoles and f._____ cells. The result of sexual reproduction is a ciliated larva that moves (disperses) by g._____. The two methods of asexual reproduction of sponges are called h._____ and fragmentation. If a sponge fragments, it can grow whole again by a process called i._____. The classification of sponges is based on the type of j._____ they have.

10. Embryonic comb jellies and cnidarians have a._____ germ layers, the b._____ and the c._____. Because of this, they are said to be d._____. They exhibit e._____ symmetry.

11. What is the function of a nematocyst? _____

12. Label this diagram with these terms:

hydra    jellyfish    medusa    polyp

oral

a._____

b._____

_____ c.

_____ d.

oral

13. In cnidarians, which demonstrate the alternation of generations life cycle, the polypoid stage is sessile and produces a._____. The medusan stage is motile and produces b._____.

14. The hydra is a solitary a._____ that lives in b._____. Because they contain muscle fibers, the cells of the epidermis of the hydra are called c._____. Cells capable of becoming other types of cells, allowing the animal to regenerate, are called d._____. e._____ secrete digestive juices. The main type of gastrodermal cell is the f._____ cell. Hydras can reproduce either asexually by g._____ or sexually.

15. The cnidarians whose calcium-carbonate skeletons provide a major ocean habitat for thousands of other animals are the _____.

16. Animals with three germ layers are said to be a._____, and they exhibit b._____ symmetry.

17. Flatworms have the a._____ body plan; ribbon worms have a b._____ body plan. Of the two types of body plans, which is the more advanced? c._____ Why? d._____
_____

18. Complete the following table to compare a hydra and a planarian.

| | Hydra | Planarian |
|---|---|---|
| Body plan | | |
| Cephalization | | |
| Number of germ layers | | |
| Organs | | |
| Symmetry | | |

19. The condition schistosomiasis is caused by a(n) a._____. The Chinese liver fluke requires two hosts, the snail and the fish. Humans contract liver flukes when they eat b._____.

20. The body of the tapeworm is made up of a head region called a(n) a._____ and numerous segments called b._____. Each segment is a(n) c._____ factory and produces many fertilized d._____.

21. Label this diagram of the life cycle of a human tapeworm with these terms:
humans    meat    pigs/cattle

288

22. Complete the following table to compare the structure of the parasitic tapeworm with that of the planarian.

| Body Structure | Tapeworm | Planarian |
|---|---|---|
| Eyes | | |
| Nervous system | | |
| Digestive system | | |
| Reproductive system | | |

23. Two animal phyla that demonstrate bilateral symmetry are a._____
    and b._____.

24. Complete the following table to compare the structure of flatworms and roundworms.

| | Flatworms | Roundworms |
|---|---|---|
| Number of germ layers | | |
| Organs | | |
| Sexes separate | | |
| Pseudocoelom | | |
| Body plan | | |

25. Trichinosis is a condition humans get by eating rare a._____
    containing b._____.

26. Complete the following table to compare the four animal phyla discussed in this chapter.

| | Sponges | Cnidarians | Flatworms | Roundworms |
|---|---|---|---|---|
| Level of organization | | | | |
| Germ layers | | | | |
| Type of body plan | | | | |
| Type of symmetry | | | | |
| Type of coelom | | | | |
| Segmentation | | | | |

## 30.3 MOLLUSCS (PP. 620–22)

The learning objective for this section is:
• Describe the features of molluscs.

27. All molluscs have a(n) a._____, a(n) b._____, and a(n)
    c._____. Another feature often present is a(n) d._____, an organ that bears
    rows of teeth. The nervous system is one of e._____ connected by nerve cords. Slow-moving
    molluscs tend to lack a(n) f._____, but the active ones do undergo g._____.

28. Give an example of each of the following animals:

   a. bivalve _____

   b. cephalopod _____

   c. gastropod _____

29. State the manner in which a clam, a squid, and a snail are adapted to their way of life.

   a. clam _____

   _____

   b. squid _____

   _____

   c. snail _____

   _____

   _____

   _____

30. Complete the following table to compare the clam and the squid.

| Characteristic | Clam | Squid |
|---|---|---|
| Skeleton | | |
| Food procurement | | |
| Locomotion | | |
| Cephalization | | |
| Reproduction | | |

## 30.4 ANNELIDS (PP. 623–25)

The learning objective for this section is:
• Describe the features of annelids.

31. Annelids are the a._____ worms. The tube-within-a-tube body plan has resulted in a digestive system with b._____ parts. Annelids have an extensive c._____ circulatory system. The nervous system is a brain and d._____ nerve cord.
   Paired e._____ in each segment collect and excrete waste.

32. Complete the following table to compare annelids.

| Type | Representative Organism | Setae | Parapodia |
|---|---|---|---|
| Marine worms | | | |
| Earthworms | | | |
| Leeches | | | |

33. Describe the following systems to show that an earthworm is segmented:

   a. nervous system _____

   b. excretory system _____

   c. circulatory system _____

34. Place a check in front of the item(s) that correctly describe(s) earthworm reproduction.
    a. _____ Worms are hermaphroditic.
    b. _____ Worms have separate sexes.
    c. _____ Glands in every segment provide mucus.
    d. _____ The clitellum provides mucus.
    e. _____ Worms exchange sperm and eggs.
    f. _____ Worms exchange sperm.
    g. _____ The embryo develops externally.

## 30.5 ARTHROPODS (PP. 626–31)

The learning objective for this section is:
• Describe the features of arthropods.

35. Like the annelids, arthropods are a._____, but there is specialization of parts. The segments
    are fused into three regions: b._____, c._____, and d._____.
    The arthropods have an exoskeleton that contains e._____, and there
    are f._____ appendages. Because they have an external skeleton, arthropods have
    to g._____ to grow larger. Arthropods have a(n) h._____ nerve
    cord; i._____ is apparent, and the head bears sense organs, including j._____
    eyes in most species.

36. Label these diagrams of the grasshopper, using the following alphabetized list of terms. (Some terms are used
    more than once.)
    antenna
    compound eye
    crop
    Malpighian tubules
    spiracle(s)
    tracheae
    tympanum

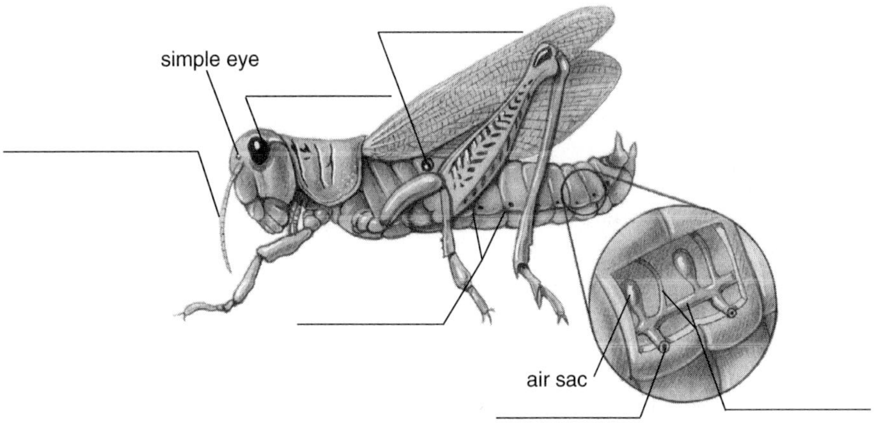

simple eye

air sac

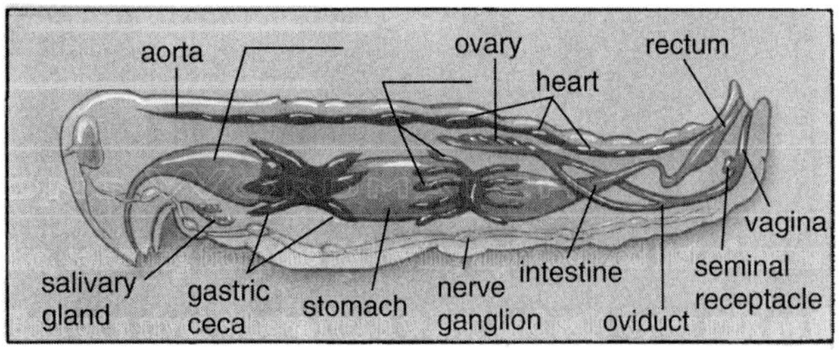

aorta    ovary    rectum
         heart
                                      vagina
                                      seminal
salivary  gastric  stomach  nerve  intestine  receptacle
gland     ceca             ganglion      oviduct

37. Give the function of each of the following structures:
    a. crop and gizzard _____
    b. Malpighian tubules _____
    c. tracheae _____
    d. hemocoel _____
    e. ovipositor _____

38. Complete the following table to compare the crayfish to the grasshopper, indicating how each is adapted to its way of life.

| System | Crayfish | Grasshopper |
|---|---|---|
| Locomotion | | |
| Excretion | | |
| Digestion | | |
| Reproduction | | |
| Respiration | | |

39. To demonstrate a knowledge of arachnids,
   a. circle the arachnids: snails, chiggers, spiders, millipedes
   b. circle the appendages of arachnids: walking legs, chelicerae, pedipalps, swimmerets
   c. circle the internal organs of arachnids: Malpighian tubules, green glands, salt glands, book lungs
   d. circle where you expect to find an arachnid: marine waters only, on land, in the frozen tundra, in the tropics

## DEFINITIONS WORDMATCH

Review key terms by completing this matching exercise, selecting from the following alphabetized list of terms:

*arthropod*
*bilateral symmetry*
*cephalization*
*coelom*
*invertebrate*
*metamorphosis*
*molting*
*nephridium*
*pseudocoelom*
*radial symmetry*
*segmentation*
*trachea*

a. _____ Air tube in insects located between the spiracles and the tracheoles.
b. _____ Body cavity lying between the digestive tract and body wall that is completely lined by mesoderm.
c. _____ Body plan having two corresponding or complementary halves.
d. _____ Periodic shedding of the exoskeleton in arthropods.
e. _____ Segmentally arranged, paired excretory tubules of many invertebrates, as in the earthworm.
f. _____ Having a well-recognized anterior head with a brain and sensory receptors.
g. _____ Change in shape and form that some animals, such as amphibians and insects, undergo during development.

## CHAPTER TEST

### OBJECTIVE QUESTIONS

Do not refer to the text when taking this test.
In questions 1–5, match each description with these animals:
   a. sponges
   b. cnidarians
   c. flatworms
   d. roundworms
____ 1. tube-within-a-tube body plan
____ 2. two germ layers present
____ 3. pseudocoelom present
____ 4. include planarians
____ 5. feed by filter feeding

____ 6. The gastrovascular cavity functions for
   a. digestion only.
   b. transport only.
   c. digestion and transport.
   d. neither digestion nor transport.
____ 7. In cnidarians, the epidermis is separated from the cell layer lining the internal cavity by
   a. mesoderm.
   b. mesoglea.
   c. a coelom.
   d. a pseudocoelom.
   e. endoderm.

8. In contrast to cnidarians, flatworms have
   a. a complete digestive tract.
   b. sexual reproduction.
   c. a mesoderm layer that gives rise to organs.
   d. a nervous system.
   e. specialized tissues for gas exchange.

9. A fluke is responsible for the condition of
   a. pinworms.
   b. schistosomiasis.
   c. trichinosis.
   d. elephantiasis.

10. A true coelom differs from a pseudocoelom in that it
    a. has a body cavity for internal organs.
    b. is lined completely by mesoderm.
    c. is incompletely lined by mesoderm.
    d. is found only in segmented worms.

11. Humans may become infected with *Ascaris* by
    a. contact with soil that contains eggs.
    b. consuming muscle tissue that contains a cyst.
    c. drinking water contaminated with eggs.
    d. Both *a* and *c* are correct.

12. The digestive and reproductive systems of a mollusc are located within the
    a. mantle cavity.
    b. visceral mass.
    c. gastrovascular cavity.
    d. highly branched coelom.
    e. water vascular cavity.

13. Earthworms cannot live above ground because
    a. their reproduction takes place in the water.
    b. they have to exchange gases with water.
    c. they tend to dry out if exposed to dry air.
    d. their setae have to be kept wet.

14. The _____ in the earthworms secretes mucus for deposition of eggs and sperm.
    a. typhlosole
    b. nephridia
    c. clitellum
    d. setae
    e. gizzard

15. What do a scallop, a nautilus, and a chiton have in common?
    a. cephalization
    b. vertebrate-type eyes
    c. a mantle cavity
    d. a jointed external skeleton

16. Molting by arthropods means that they
    a. circulate blood through a closed system.
    b. move with jointed appendages.
    c. reproduce sexually.
    d. shed their exoskeletons.

17. What do a crab, an insect, and a spider have in common?
    a. cephalization
    b. external skeleton
    c. molting
    d. All of these are correct.

18. Which of these pairs is mismatched?
    a. insect—scorpion
    b. arachnid—spider
    c. crustacean—crab
    d. insect—grasshopper

19. Which is NOT generally true of arthropods?
    a. exoskeleton contains chitin
    b. breathe with tracheae
    c. have jointed appendages
    d. have compound eyes

20. Which of these pairs is mismatched?
    a. annelid—earthworm
    b. mollusc—spider
    c. arthropod—grasshopper
    d. annelid—clam worm

21. Which is NOT a general feature of insects?
    a. body divided into head, thorax, and abdomen
    b. three pairs of legs
    c. respiration typically by book lungs
    d. one or two pairs of wings

22. Which of these is NOT a characteristic of molluscs?
    a. body in three parts—head, thorax, and abdomen
    b. a foot modified in various ways
    c. a visceral mass that contains the internal organs
    d. usually an open circulatory system

23. What do an earthworm, a clam worm, and a leech have in common?
    a. a clitellum
    b. parapodia
    c. a closed circulatory system
    d. All of these are correct.

Answer in complete sentences.

24. A loss of body complexity accompanies parasitism among flatworms. What evidence supports this?

25. The arthropods are considered the most successful animal phylum inhabiting the earth. What justifies this claim?

Test Results: _____ number correct ÷ 25 = _____ × 100 = _____ %

# ANSWER KEY

## STUDY QUESTIONS

1. c, d, e  2. a. bilateral symmetry  b. asymmetry  c. radial symmetry  3. a. ectoderm  b. mesoderm  c. endoderm  4. a. two  b. organ  5. a. pseudocoelomate  b. protostome  c. acoelomate  d. deuterostome  6. a. protistan  b. protozoan  7. See Figure 30.2, p, 611 in text.  8. a. collar—produces water currents and captures food  b. spicule—internal skeleton  c. epidermal—protection  d. pore—entrance of water  e. amoeboid—distributes nutrients and produces gametes  9. a. cellular  b. pores  c. pore bearing  d. collar cells  e. osculum  f. amoeboid  g. swimming  h. budding  i. regeneration  j. skeleton  10. a. two  b. ectoderm  c. endoderm  d. diploblasts  e. radial  11. to trap or paralyze prey  12. a., b. polyp, hydra  c., d. medusa, jellyfish  13. a. medusae  b. eggs and sperm  14. a. polyp  b. freshwater  c. epitheliomuscular cells  d. interstitial (embryonic)  e. Gland cells  f. nutritive-muscular  g. budding  15. corals  16. a. triploblasts  b. bilateral  17. a. sac  b. tube-within-a-tube  c. tube-within-a-tube  d. With a one-way flow of contents, each part can take on a particular function.
18.

| Hydra | Planarian |
|-------|-----------|
| sac | sac |
| no | yes |
| two | three |
| no | yes |
| radial | bilateral |

19. a. blood fluke  b. raw fish  20. a. scolex  b. proglottids  c. reproductive  d. eggs  21. a. meat  b. humans  c. pigs/cattle
22.

| Tapeworm | Planarian |
|----------|-----------|
| no | yes |
| much reduced | extensive |
| not present | branches |
| well developed | present |

23. a. Nematoda (roundworms)  b. Platyhelminthes (flatworms)
24.

| Flatworms | Roundworms |
|-----------|------------|
| three | three |
| yes | yes |
| no | yes |
| no | yes |
| sac | tube-within-a-tube |

25. a. pork  b. encysted roundworm larvae  26. See Table 30.2, page 632, in text.  27. a. mantle  b. visceral mass  c. foot  d. radula  e. ganglia  f. head  g. cephalization  28. a. clam, oyster, mussel, scallop  b. squid, cuttlefish, octopus, nautilus  c. snail, whelk, conch, periwinkle, sea slug  29. a. A clam has a hatchet foot for burrowing in

the sand and is a sessile filter feeder that lacks cephalization. **b.** A squid has a head-foot—that is, tentacles about the head with vertebrate-type eyes that help the squid actively capture food. **c.** A snail has a broad, flat foot for moving over flat surfaces and a head with a radula for scraping up food from a surface.

**30.**

| Clam | Squid |
|------|-------|
| external shell | no external skeleton |
| filter feeder | active predator |
| hatchet foot | jet propulsion and fins |
| no | yes |
| separate sexes | separate sexes |

**31. a.** segmented **b.** specialized **c.** closed **d.** ventral **e.** nephridia

**32.**

| Representative Organism | Setae | Parapodia |
|-------------------------|-------|-----------|
| clam worm | many | yes |
| earthworm | few | no |
| leech | no | no |

**33. a.** ganglia and lateral nerves in every segment **b.** paired nephridia in every segment **c.** branched blood vessels in every segment **34.** a, d, f, g **35. a.** segmented **b.** head **c.** thorax **d.** abdomen **e.** chitin **f.** jointed **g.** molt **h.** ventral **i.** cephalization **j.** compound **36.** See Figure 30.18, page 629, in text. **37. a.** crop stores food, and gizzard grinds it **b.** excretory tubules that concentrate nitrogenous waste **c.** air tubes that deliver oxygen to muscles **d.** cavity where blood is found **e.** special female appendage for depositing eggs in soil

**38.**

| Crayfish | Grasshopper |
|----------|-------------|
| swimmerets | hopping legs, wings |
| liquid | solid |
| gastric mill | crop |
| male uses swimmeret to pass sperm | male uses penis to pass sperm |
| gills | tracheae |

**39. a.** chiggers, spiders **b.** walking legs, chelicerae, pedipalps **c.** Malpighian tubules, book lungs **d.** on land, in the tropics

## DEFINITIONS WORDMATCH

**a.** trachea **b.** coelom **c.** bilateral symmetry **d.** molting **e.** nephridia **f.** cephalization **g.** metamorphosis

## CHAPTER TEST

**1.** d **2.** b **3.** d **4.** c **5.** a **6.** c **7.** b **8.** c **9.** b **10.** b **11.** d **12.** b **13.** c **14.** c **15.** c **16.** d **17.** d **18.** a **19.** b **20.** b **21.** c **22.** a **23.** c **24.** Organ systems are lost partially or completely in flatworm parasites. Only the reproductive system remains well developed. The head region bears hooks and/or suckers instead of sense organs. **25.** They are the most diversified phylum, with a wide variety of species filling numerous ecological niches. Insects far outnumber any other type of animal.

# 31

# ANIMALS: PART II

In **deuterostomes,** the blastopore develops into an anus. Cleavage is radial, and an enterocoelom develops.

The **echinoderms** have evolved radial symmetry, an internal skeleton, **gills,** a nerve ring, and a **water vascular system.** The sea star is a major echinoderm showing these specializations.

**Chordates** have evolved a **notochord,** a dorsal tubular nerve cord, and pharyngeal pouches at some point in their life history. Only **lancelets** show all of these characteristics as adults. The **vertebrates** develop a vertebral column in place of a notochord.

There are several groups of **fishes.** The hagfishes and lampreys are descendants of the original jawless fishes. Sharks are modern representatives of the cartilaginous fishes. The original bony fishes diverged into two groups: the ray-finned fishes and the lobe-finned fishes.

**Amphibians,** which evolved from lobe-finned fishes during the Devonian period, reached their greatest size and diversity in the swamp forests of the Carboniferous period. Most amphibians return to the water to reproduce.

**Reptiles,** which are believed to have evolved from the amphibians, have a shelled amniote egg. A shelled egg, along with extraembryonic membranes, makes reproduction on land possible.

While amphibians and reptiles are ectothermic—their body temperature is the same as that of the environment— both birds and mammals are homeothermic—they metabolically produce a constant body temperature. Adaptations of **birds** include feathers and hollow bones, both adaptations for flight. **Mammals** have evolved hair and mammary glands. **Monotremes** are egg-laying mammals, and **marsupials** are pouched mammals that give birth to very immature young. While the young of placental mammals are more developed at birth, they still require parental care to survive.

Among mammals, **primates—prosimians,** monkeys, apes, and humans—are adapted for living in trees. Monkeys leap, but apes swing from limb to limb. This may have been an adaptation that led to bipedalism in **hominids.** The first hominid (humans and immediate ancestors) was *Australopithecus afarensis,* which could walk erect but had only a small brain. Later-appearing **australopithecines** may have manufactured stone tools, but *H. habilis* certainly did. *H. erectus*, the first fossil to have a brain size of more than 1,000 cc, migrated from Africa into Europe and Asia. They used fire and may have been big-game hunters.

Two contradicting hypotheses have been suggested about the origination of modern humans. The **multiregional continuity hypothesis** says that modern humans originated separately in Asia, Europe, and Africa. The **out-of-Africa hypothesis** says that modern humans originated in Africa and, after migrating into Europe and Asia, replaced the archaic *Homo* species found there. Cro-Magnon is a name often given to modern humans.

Study the text section by section. Answer the study questions so that you can fulfill the learning objectives for each section.

## 31.1 ECHINODERMS (PP. 636—37)

The learning objectives for this section are:
- Explain why echinoderms and chordates are called deuterostomes.
- Describe characteristics of echinoderms.

1. Complete the following table to describe the characteristics of echinoderms.

| Characteristic | Description |
| --- | --- |
| Type of symmetry | |
| Skeletal system | |
| Respiration | |
| Nervous system | |
| Water vascular system | |

2. Label this diagram of a sea star with the following terms. (Some terms are used more than once.)

ampulla
anus
arm
cardiac stomach
central disk
digestive gland
eyespot
gonads
pyloric stomach
sieve plate
skin gill
spine
tube feet

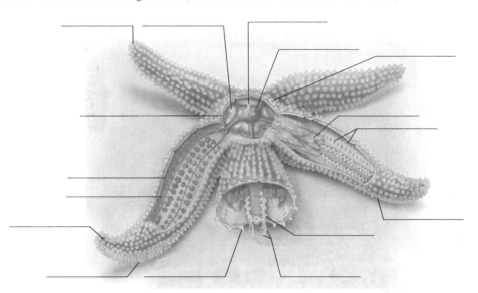

3. Trace the path of water in the water vascular system: sieve plate to a._____canal
   to b._____canal to radial canal to c._____feet. Each of these feet has
   a(n) d._____. The function of the water vascular system is e._____.

4. A sea star has a two-part stomach. Describe how the sea star feeds on a clam, mentioning both parts of the
   stomach. _____

## 31.2 CHORDATES (PP. 638–40)

The learning objective for this section is:
• Describe the four basic characteristics an animal must have to be considered a chordate.

5. Label this diagram with the three primary chordate characteristics, plus another that also distinguishes chordates.

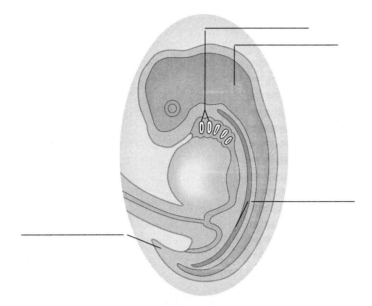

6. What type of evidence suggests that chordates and echinoderms are related? _____

7. Complete the following table to describe the invertebrate chordates.

| Name | Chordate Characteristics | Appearance |
|------|--------------------------|------------|
| Tunicates (subphylum Urochordata) | | |
| Lancelets (subphylum Cephalochordata) | | |

8. Complete this evolutionary tree of the chordates by filling in the name of the subphylum/class in the arrowheads. Then place these terms in the ovals:

limbs    mammary glands and hair    jaws    lungs    amnion    feathers    vertebrae

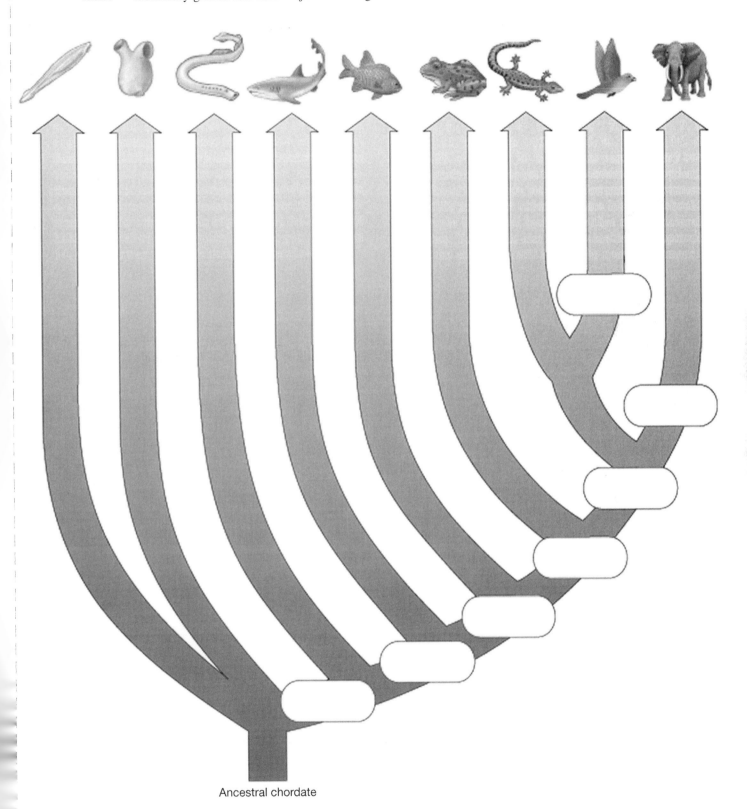

Ancestral chordate

The learning objectives for this section are:
- Describe the three groups of fishes.
- Give characteristics of amphibians.
- List features that distinguish reptiles.
- Give characteristics of birds.

9. Place a check in front of the characteristics that distinguish the vertebrates.
   a. _____ bilaterial symmetry in all
   b. _____ radial symmetry in some
   c. _____ tube-within-a-tube plus coelom
   d. _____ segmented
   e. _____ vertebral column replaces embryonic notochord
   f. _____ open/closed circulatory system
   g. _____ cephalization with compound eyes
   h. _____ living endoskeleton

10. Place the appropriate letters next to each statement.

    JF—jawless fishes     CF—cartilaginous fishes     BF—bony fishes

    a. _____ have jaws (choose two)
    b. _____ are parasitic
    c. _____ sharks, rays, and skates
    d. _____ hagfishes and lampreys
    e. _____ ray-finned and lobe-finned

11. In bony fishes, state the purpose of the following:

    a. lateral line system _____

    b. swim bladder _____

    c. gills _____

    d. single circulatory loop _____

    e. paired fins _____

12. Describe each of these features of amphibians:

    a. skin _____

    b. lungs _____

    c. body temperature _____

    d. life cycle _____

    e. heart _____

13. Describe each of these features of reptiles:

    a. skin _____

    b. lungs _____

    c. body temperature _____

    d. type of egg _____

    e. heart _____

14. How do the following characteristics of birds contribute to their ability to fly?

    a. feathers  _____

    b. horny beak  _____

    c. keel  _____

    d. four-chambered heart  _____

    e. one-way flow of air in lungs  _____

15. Match the types of mammals to these descriptions (some descriptions are used more than once):

    1 All have hair and mammary glands.
    2 All lay eggs.
    3 All have pouches.
    4 All have internal development to term.

    a. _____ monotremes
    b. _____ marsupials
    c. _____ placental mammals

16. Name a type of mammal adapted to each of the following:

    a. flying in air  _____

    b. running on land  _____

    c. swimming in the ocean  _____

    d. preying on other animals  _____

    e. living in trees  _____

## 31.4 HUMAN EVOLUTION (PP. 650–56)

The learning objectives for this section are:
- List characteristics that distinguish primates from other mammals.
- Diagram the primate evolutionary tree.
- Name the first hominid.
- Explain the two opposing hypotheses for the evolution of modern humans.

17. List an adaptation for arboreal life in relation to the following:

    a. vision _____

    b. digits _____

    c. brain size _____

    d. birth number _____

18. Australopithecines are well known [a.]_____. The oldest ones, MYA represents a transitional stage

    between [b.]_____ and [c.]_____. Australopithecines walked [d.]_____, but their brain

    was [e.]_____. They exhibit mosaic evolution, meaning that their [f.]_____

    _____.

    The next australopithecine named [g.]_____, may have been a common ancestor for the rest of the

    australopithecines and [h.]_____. A. afarensis and A. africanus are termed gracile because they are

    slight in appearance. [i.]_____ and [j.]_____ are termed [k.]_____ because they had

    larger facial bones, most likely due to larger teeth and chewing muscles.

19. Match the *Homo* species with the following phrases that describe their way of life. (Numbers can be used more than once.)

 1  brain size less than 1,000 cc
 2  brain size 1,000 cc or larger
 3  more likely scavenged meat
 4  more likely hunted animals
 5  certainly had speech and culture
 6  most likely had speech and culture
 7  perhaps had speech and culture
 8  made tools
 9  had upright posture

a. *Homo habilis* _____

b. *Homo erectus* _____

c. Neanderthals _____

d. Cro-Magnons _____

20. Which hypothesis—the out-of-Africa hypothesis or the multiregional continuity hypothesis—states that *H. erectus* and then, later on, humans left Africa? a._____ Which hypothesis states that *H. erectus* left Africa and that modern humans then simultaneously arose in Europe, Asia, and Africa? b._____ With which hypothesis would you expect more similarity between fossils dated 100,000 BP? c._____ The fossil record shows several varieties of humans in Asia and Europe dated prior to 100,000 BP. These are called d. "_____ *Homo*." One example of an "archaic *Homo*" is e._____.

# Constructing Office Buildings

The object of this game is to construct an office building by matching the numbered terms with the organisms in the key. (Some numbers should be matched to more than one letter.) Five correct answers in a row gives you one story. Any wrong answer is a natural disaster that forces you to start from the ground again.

## OFFICE BUILDING ONE

A fourteen-story office building is possible.

**Key One**

 a. Protozoa
 b. Porifera
 c. Cnidaria
 d. Platyhelminthes
 e. Nematoda
 f. Annelida
 g. Mollusca
 h. Arthropoda
 i. Echinodermata

_____ 1. sea star (starfish)
_____ 2. clitellum
_____ 3. flame cells
_____ 4. egg—nymph—adult
_____ 5. organ system level of organization
_____ 6. ampulla
_____ 7. fluke
_____ 8. jellyfish
_____ 9. leech
_____ 10. octopus
_____ 11. jointed appendages
_____ 12. tube feet
_____ 13. elephantiasis
_____ 14. soft body
_____ 15. five hearts
_____ 16. bilateral symmetry
_____ 17. muscles
_____ 18. water vascular system
_____ 19. medusa
_____ 20. cellular level of organization
_____ 21. muscular foot
_____ 22. five arms
_____ 23. earthworm
_____ 24. nerve net
_____ 25. clam
_____ 26. setae
_____ 27. *Trichinella*
_____ 28. ladderlike nervous organization
_____ 29. horseshoe crab
_____ 30. stone canal
_____ 31. pseudocoelom
_____ 32. coelom
_____ 33. mesoglea
_____ 34. pore bearers
_____ 35. polychaete

_____ 36. squid
_____ 37. segmentation
_____ 38. closed circulatory system
_____ 39. open circulatory system
_____ 40. mesoderm
_____ 41. collar cells
_____ 42. trachea
_____ 43. trochophore larva
_____ 44. hydra
_____ 45. *Hirudo*
_____ 46. Malpighian tubules
_____ 47. mantle
_____ 48. *Dirofilaria*—filarial worm
_____ 49. nematocysts
_____ 50. metamorphosis
_____ 51. gills
_____ 52. typhlosole
_____ 53. *Ascaris*
_____ 54. sieve plate (madreporite)
_____ 55. hermaphroditic
_____ 56. acoelomate
_____ 57. pseudocoelomate
_____ 58. pyloric stomach
_____ 59. worms
_____ 60. molting
_____ 61. nephridia
_____ 62. visceral mass
_____ 63. cephalization
_____ 64. spicules
_____ 65. green gland
_____ 66. exoskeleton
_____ 67. wings
_____ 68. sessile
_____ 69. sac body plan
_____ 70. tube-within-a-tube body plan
How many stories is your building? _____

_____ 6. tube-within-a-tube body plan
_____ 7. bilateral symmetry
_____ 8. radial symmetry
_____ 9. organs
_____ 10. closed circulatory system
_____ 11. insect
_____ 12. belong to the same phylum
_____ 13. flame cells
_____ 14. Malpighian tubules
_____ 15. green gland
_____ 16. nephridia
_____ 17. tracheal tubes
_____ 18. gills
_____ 19. body wall for respiration
_____ 20. skin gills
_____ 21. spicules
_____ 22. shell
_____ 23. carapace
_____ 24. "spiny skin"
_____ 25. planula larva
_____ 26. metamorphosis
_____ 27. nymph
_____ 28. molt
_____ 29. ovipositor
_____ 30. asexual reproduction
_____ 31. alternation of generations
_____ 32. clitellum
_____ 33. bilateral larva but radial adult
_____ 34. nerve ring
_____ 35. ganglia in foot and visceral mass
_____ 36. dorsal tubular nerve cord
_____ 37. ventral solid nerve cord
_____ 38. no nervous system
_____ 39. nerve net
_____ 40. ladderlike nervous organization
How many stories is your building? _____

## OFFICE BUILDING TWO

An eight-story office building is possible.
**Key Two**
    a. sponge
    b. hydra
    c. planarian
    d. *Ascaris*
    e. clam
    f. earthworm
    g. lobster
    h. sea star
    i. grasshopper
    j. *Obelia*
_____ 1. diploblastic
_____ 2. coelomate
_____ 3. segmented
_____ 4. nematocyst
_____ 5. sac body plan

## OFFICE BUILDING THREE

A ten-story office building is possible.
**Key Three**
    a. jawless fishes
    b. cartilaginous fishes
    c. bony fishes
    d. amphibians
    e. reptiles
    f. birds
    g. mammals
_____ 1. four-chambered heart
_____ 2. frogs and salamanders
_____ 3. air sacs
_____ 4. lampreys and hagfish
_____ 5. infant dependency
_____ 6. Chondrichthyes
_____ 7. hair
_____ 8. smooth, nonscaly skin

_____ 9. ectothermic
_____ 10. two-chambered heart
_____ 11. differentiated teeth
_____ 12. some are parasitic
_____ 13. highly developed brain
_____ 14. feathers
_____ 15. fish, but no operculum
_____ 16. evolved from amphibians
_____ 17. homeothermic
_____ 18. snakes, lizards
_____ 19. primates
_____ 20. epidermal placoid (toothlike) scales
_____ 21. shelled egg
_____ 22. whales and dolphins
_____ 23. class Aves
_____ 24. lateral line system
_____ 25. mammary glands
_____ 26. some are filter feeders
_____ 27. metamorphosis
_____ 28. sharks, rays, and skates
_____ 29. marsupials
_____ 30. Osteichthyes

_____ 31. ray-finned fishes
_____ 32. monotremes
_____ 33. scales of bone
_____ 34. dinosaurs
_____ 35. operculum
_____ 36. evolved from reptiles
_____ 37. lobe-finned fishes
_____ 38. double circulatory loop
_____ 39. paired pelvic and pectoral fins
_____ 40. one-way path through lungs
_____ 41. three-chambered heart
_____ 42. lungs
_____ 43. expandable rib cage
_____ 44. gills as an adult
_____ 45. wings
_____ 46. single circulatory loop
_____ 47. molt
_____ 48. usually four limbs
_____ 49. amniote egg
_____ 50. cartilaginous skeleton

How many stories is your building? _____

## DEFINITIONS WORDMATCH

Review key terms by completing this matching exercise, selecting from the following alphabetized list of terms:

*australopithecines*
*chordate*
*deuterostome*
*echinoderm*
*lancelet*
*mammal*
*marsupial*
*Neanderthal*
*placental mammal*
*primate*
*water vascular system*

a. _____ Series of canals that takes water to the tube feet of an echinoderm, allowing them to expand.

b. _____ Group of coelomate animals in which the first embryonic opening is associated with the anus, and the second embryonic opening is associated with the mouth.

c. _____ Invertebrate chordate that has a body resembling a lancet and retains the four chordate characteristics as an adult.

d. _____ Homeothermic vertebrate characterized especially by the presence of hair and mammary glands.

e. _____ Member of a group of mammals bearing immature young nursed in a marsupium, or pouch—for example, kangaroo and opossum.

f. _____ Member of the order that includes prosimians, monkeys, apes, and humans, all of whom have adaptations for living in trees.

## OBJECTIVE QUESTIONS

Do not refer to the text when taking this test.

_____ 1. Each is a vertebrate characteristic EXCEPT
   a. bilateral symmetry.
   b. coelom development.
   c. open circulatory system.
   d. segmentation.

_____ 2. The earliest vertebrate fossils came from the
   a. amphibians.
   b. bony fishes.
   c. cartilaginous fishes.
   d. jawless fishes.

_____ 3. Modern humans are more closely related to
   a. monkeys than apes.
   b. African apes than Asian apes.
   c. Neanderthals than Cro-Magnon.
   d. whales than prosimians.

_____ 4. The skin of amphibians functions mainly for
   a. circulation.
   b. excretion.
   c. reproduction.
   d. respiration.

_____ 5. The extraembryonic membrane in the reptile egg promotes
   a. additional reinforcement from drying out.
   b. complete independence from the water for re-production.
   c. enhanced elimination of wastes from the embryo.
   d. increased hardness to prevent breakage.

_____ 6. Bird feathers are modified
   a. fish fins.
   b. mammalian hair.
   c. reptilian scales.
   d. vertebrate teeth.

_____ 7. The hair of mammals is an adaptation for
   a. camouflage in all species.
   b. control of body temperature.
   c. faster locomotion.
   d. regulation of waste elimination.

_____ 8. The most successful mammals are the
   a. marsupials.
   b. monotremes.
   c. lancelets.
   d. placentals.

_____ 9. Which is true of echinoderms?
   a. contain a dorsal tubular nerve cord
   b. have internal organs in a visceral mass
   c. move by a water vascular system
   d. All of these are true.

_____ 10. Which is found only among echinoderms?
   a. deuterostome developmental pattern
   b. radial symmetry
   c. exoskeleton
   d. tube feet
   e. ventral mouth

_____ 11. Which feature is NOT found among fishes?
   a. endoskeleton
   b. closed circulatory system
   c. warm blood
   d. dorsal tubular nerve cord

_____ 12. The most important reason amphibians are in-completely adapted to life on land is that
   a. they depend on water for external fertilization.
   b. they must reproduce in water.
   c. the skin is more important than the lungs for gas exchange.
   d. their means of locomotion is poorly developed.

_____ 13. A four-chambered heart is seen among
   a. fishes.
   b. amphibians.
   c. birds.
   d. mammals.
   e. Both _c_ and _d_ are correct.

_____ 14. Which pair of statements correctly contrasts birds and mammals?
   a. Birds are cold-blooded. Mammals are warm-blooded.
   b. Birds are egg-laying. No mammals are egg-laying.
   c. Birds have air sacs in addition to lungs. Mam-mals have no such sacs.
   d. Birds lack a septum between the ventricles. Mammals have such a septum.

_____ 15. What do echinoderms and chordates have in common?
   a. radial symmetry
   b. pharyngeal pouches
   c. second embryonic opening is the mouth
   d. All of these are correct.

_____ 16. Extraembryonic membranes
   a. are found during the development of all vertebrates.
   b. are found during the development of reptiles, birds, and mammals.
   c. have exactly the same function in all vertebrates.
   d. Both _a_ and _c_ are correct.
   e. Both _a_ and _b_ are correct.

____17. Which is NOT a distinguishing feature of vertebrates?
   a. dorsal notochord
   b. jointed internal skeleton
   c. extreme cephalization
   d. open circulatory system
   e. efficient respiration

____18. The type of mammal that lays eggs while nourishing its young with milk is called
   a. a monotreme.
   b. a marsupial.
   c. placental.
   d. hermaphroditic.

____19. Which is NOT characteristic of echinoderms?
   a. external skeleton
   b. tube feet
   c. skin gills
   d. gonads in arms

____20. Which is NOT an echinoderm?
   a. sea lily
   b. sea urchin
   c. sea cucumber
   d. sea horse

____21. Australopithecines
   a. were apelike below the waist and humanlike above the waist.
   b. were humanlike below the waist and apelike above the waist.
   c. were generally apelike.
   d. were generally humanlike.

____22. The out-of-Africa hypothesis says that
   a. *Homo erectus* migrated out of Africa and replaced archaic *Homo* species in Europe and Asia.
   b. no interbreeding took place between different types of humans.
   c. modern humans arose in Africa in several different places.
   d. Neanderthals, and also *H. erectus,* migrated out of Africa.
   e. All of these are correct.

## THOUGHT QUESTIONS

Answer in complete sentences.

23. Compare the success of chordate evolution to arthropod evolution. What are the similarities and differences?

24. Compare the adaptations of amphibians and reptiles to a land existence.

**Test Results:** _____ number correct ÷ 24 = _____ × 100 = _____ %

## ANSWER KEY

## STUDY QUESTIONS

**1.**

| Description |
| --- |
| radial |
| spine-bearing, calcium-rich plates in an endoskeleton |
| gas exchange across skin gills and tube feet |
| central nerve ring plus radial nerves |
| a series of canals that ends at tube feet; a means of locomotion |

**2.** See Figure 31.2b, page 637, in text.   **3. a.** stone **b.** ring **c.** tube **d.** ampulla **e.** locomotion   **4.** The sea star everts its cardiac stomach, puts it in the shell, and secretes enzymes; partly digested food is taken up, and digestion is completed in the pyloric stomach.   **5.** See page 638 in text.   **6.** embryological evidence

**7.**

| Chordate Characteristics | Appearance |
| --- | --- |
| all three (larva); gill slits (adult) | thick-walled, squat sac |
| all three | lancet-shaped |

**8. 2.** See Figure 31.3, page 638, in text. **9.** a, c, d, e, h **10. a.** CF, BF **b.** JF **c.** CF **d.** JF **e.** BF **11. a.** senses presence of other organisms **b.** buoyancy **c.** respiration **d.** movement of blood to gills first **e.** balancing and propelling body in water **12. a.** smooth, nonscaly, and used for respiration **b.** small and poorly developed **c.** ectothermic **d.** undergo metamorphosis from tadpole to adult **e.** three chambers **13. a.** thick, dry, scaly **b.** more developed than in amphibians **c.** ectothermic **d.** amniote **e.** nearly or completely four-chambered **14. a.** provide broad, flat surfaces **b.** Replaces jaws and teeth, making bird lighter. **c.** attaches flight muscles **d.** provides good delivery of $O_2$-rich blood to muscles **e.** provides good oxygenation of blood **15. a.** 1, 2 **b.** 1, 3 **c.** 1, 4 **16. a.** bat **b.** horse **c.** whale **d.** lion **e.** monkey **17. a.** eyes forward with stereoscopic vision **b.** nails, not claws, and opposable thumb **c.** large, well developed **d.** single offspring at a time **18. a.** hominids **b.** apes **c.** humans **d.** erect **e.** small **f.** body parts evolved at different rates **g.** *A. afarensis* **h.** humans **i.** *A. robustus* **j.** *A. boisei* **k.** robust **19. a.** 1, 3, 7, 8, 9 **b.** 2, 4, 6, 8, 9 **c.** 2, 4, 6, 8, 9 **d.** 2, 4, 5, 8, 9 **20. a.** out-of-Africa **b.** multiregional continuity **c.** multiregional continuity **d.** archaic **e.** Neanderthal

## GAME: OFFICE BUILDING ONE

**1.** i **2.** f **3.** d **4.** h **5.** d, e, f, g, h **6.** i **7.** d **8.** c **9.** f **10.** g **11.** h **12.** i **13.** e **14.** c, d, f, g **15.** f **16.** d, e, f, g, h **17.** d, e, f, g, h, i **18.** i **19.** c **20.** a, b **21.** g **22.** i **23.** f **24.** c **25.** g **26.** f **27.** e **28.** d **29.** h **30.** i **31.** e **32.** e, f, g, h, i **33.** c **34.** b **35.** f **36.** g **37.** f, g, h **38.** f **39.** g, h, i **40.** d, e, f, g, h, i **41.** b **42.** h **43.** f, g, h **44.** c **45.** f **46.** h **47.** g **48.** e **49.** c **50.** h **51.** g, h, i **52.** f **53.** e **54.** i **55.** d, f **56.** b, c **57.** e **58.** i **59.** d, e, f **60.** h **61.** f **62.** g **63.** d, h **64.** b **65.** h **66.** g, h **67.** h **68.** b, c **69.** c, d **70.** e, f, g, h, i

## GAME: OFFICE BUILDING TWO

**1.** b, j **2.** d, e, f, g, h, i **3.** f, g, i **4.** b **5.** b, c **6.** d, e, f, g, h, i **7.** c, d, e, f, g, i **8.** b, h, j **9.** c, d, e, f, g, h, i **10.** f **11.** i **12.** b and j; g and i **13.** c **14.** i **15.** g **16.** f **17.** i **18.** e, g, h **19.** f **20.** h **21.** a **22.** e **23.** g **24.** h **25.** b **26.** i **27.** i **28.** g, i **29.** i **30.** b **31.** j **32.** f **33.** h **34.** h **35.** e **36.** none **37.** f, g **38.** a **39.** b, j **40.** c

## GAME: OFFICE BUILDING THREE

**1.** e, f, g **2.** d **3.** f **4.** a **5.** g **6.** b **7.** g **8.** a, d **9.** a, b, c, d, e **10.** a, b, c **11.** g **12.** a **13.** g **14.** f **15.** a, b **16.** e **17.** f, g **18.** e **19.** g **20.** b **21.** e, f **22.** g **23.** f, g **24.** b, c **25.** g **26.** b, g **27.** d **28.** b **29.** g **30.** c **31.** c **32.** g **33.** c **34.** e **35.** c **36.** f, g **37.** c **38.** c, d, e, f, g **39.** c **40.** f **41.** d **42.** d, e, f, g **43.** e, f, g **44.** a, b, c **45.** f **46.** a, b, c **47.** e **48.** d, e, f, g **49.** e, f **50.** a, b

## DEFINITIONS WORDMATCH

**a.** water vascular system **b.** deuterostome **c.** lancelet **d.** mammal **e.** marsupial **f.** primate

## CHAPTER TEST

**1.** c **2.** d **3.** b **4.** d **5.** b **6.** c **7.** b **8.** d **9.** c **10.** d **11.** c **12.** a **13.** e **14.** c **15.** c **16.** b **17.** d **18.** a **19.** a **20.** d **21.** b **22.** a **23.** Each phylum contains numerous diversified species adapted to a variety of environments. Arthropods have more species. **24.** Amphibians reproduce in the water and have a larval stage that develops in the water. The skin must be kept moist because it supplements the lungs for gas exchange. Reptiles reproduce on land because they lay a shelled egg with extraembryonic membranes. The skin can prevent desiccation because it is dry and scaly. The lungs are moderately developed, and a rib cage helps ventilate the lungs.

# 32

# ANIMAL BEHAVIOR

## CHAPTER REVIEW

An animal is organized to carry out **behaviors** that help it survive and reproduce. Hybrid studies with lovebirds and blackcap warblers show that behavior has a genetic basis. The nervous and endocrine systems control behavior, as shown by garter snake experiments and snail (*Aplysia*) DNA studies.

A behavior sometimes undergoes development after birth, as exemplified by improvement in laughing gull chick begging behavior. **Learning** occurs when a behavior changes with practice. Experiments teaching male birds to sing in their species dialect show that various factors—such as social experience—influence whether or not learning takes place.

Since genes influence behavior, it is reasonable to assume that adaptive behavioral traits will evolve. Both sexes are expected to behave in a manner that will raise their reproductive success. Females who produce only one egg a month are expected to choose the best mate, and males who produce many sperm are expected to inseminate as many females as possible. **Sexual selection** is natural selection due to mate choice by females and competition among males. Do females choose mates who have the best traits for survival or simply the ones to whom they are attracted? Or are these hypotheses one and the same? Experiments with satin bowerbirds and birds of paradise support these bases for sexual selection.

A cost-benefit analysis is particularly applicable to male competition. A **dominance hierarchy**, as seen in baboons, and establishment of a **territory**, as seen in red deer, are two ways in which strong males get to reproduce more than weaker males. Do the nonmating males receive a benefit? And are the costs within reason for the mating males? Experiments are still being conducted.

**Communication** between animals consists of chemical, auditory, visual, and tactile signals. Social living can help an animal avoid predators, rear offspring, and find food. Disadvantages include fighting among members, spread of contagious diseases, and the possibility of subordination to others.

**Altruistic** behavior seems self-sacrificing until we consider the concept of **inclusive fitness,** which includes personal reproductive success and the reproductive success of relatives. Among social insects, sisters share 75% of their genes rather than 50%. This makes it more likely that they will help raise siblings.

## STUDY QUESTIONS

Study the text section by section. Answer the study questions so that you can fulfill the learning objectives for each section.

### 32.1 GENETIC BASIS OF BEHAVIOR (PP. 662–63)

The learning objectives for this section are:
- Describe examples supporting the hypothesis that behavior has a genetic basis.
- Describe the role of neural and hormonal mechanisms in the control of behavior.

1. Describe the experiment and results concerning Cape Verde blackcap warblers, which do not migrate, and German blackcap warblers, which migrate to Africa.

   a. experiment: _____

   b. results: _____

   c. conclusion: _____

2. Inland garter snakes eat frogs, and coastal garter snakes eat slugs. Investigators discovered that inland snakes do not respond to the smell of slugs, and hybrids generally have only an intermediate ability to respond to the smell of slugs. What conclusion was reached? _____

3. Egg-laying behavior of *Aplysia* involves a set sequence of movements. Investigators found that the gene that controls behavior codes for hormones. What conclusion was reached?_____

## 32.2 DEVELOPMENT OF BEHAVIOR (PP. 664–65)

The learning objective for this section is:
• Give examples of behaviors that undergo development after birth, as when learning affects behavior.

4. Indicate whether these statements about the experiment with laughing gull chicks are true (T) or false (F).
   a. _____ Laughing gull chicks seek their own food.
   b. _____ Motor development helps explain why the pecking behavior of chicks improves.
   c. _____ Operant conditioning—a form of learning—helps explain why older chicks choose a model that looks more like the parent.

5. Explain statement 4c here: _____
   _____

6. Due to imprinting, chicks follow the first moving object they see. In relation to this observation, explain the following:
   a. sensitive period _____
   _____
   b. need for social interaction _____
   _____

7. The following diagram illustrates experiments studying how birds learn to sing:

Explain each of the frames.
   a. first _____
   b. second _____
   c. third _____
   d. What conclusion is appropriate? _____

## 32.3 ADAPTIVENESS OF BEHAVIOR (PP. 667–69)

The learning objective for this section is:
• Describe how natural selection influences behaviors such as methods of feeding, selecting a home, and reproducing.

8. Indicate whether these statements about the adaptive nature of behavior are true (T) or false (F).
    a. _____ Behavior has a genetic basis.
    b. _____ Certain behaviors can improve reproductive success.
    c. _____ The nervous and endocrine systems control behavior.

9. With reference to the reproductive behavior of satin bowerbirds, females chose males with well-kept bowers that contained blue objects. Why might this support the good genes hypothesis? _____
_____

10. With reference to the reproductive behavior of birds of paradise, perhaps females choose the males with spectacular plumes because it signifies a._____ or because their sons will be b._____ to females also. It was also found that raggiana offspring are fed a more c._____ food than those of the related species, the trumpet manucode. This seems to correlate with the fact that the male raggiana birds of paradise are d._____, while the trumpet manucode birds are e._____. Birds are monogamous when it takes two parents to f._____ the offspring.

11. a. In terms of a cost-benefit analysis, what is the benefit to dominant males in a baboon troop? _____
_____

    b. What are the costs? _____
_____

    c. What is the benefit to subordinate males in a baboon troop? _____

    d. What are the costs? _____

    e. Why do you predict that the benefits for each must outweigh the costs? _____

## 32.4 ANIMAL SOCIETIES (PP. 671–72)

The learning objective for this section is:
• Describe various means of communication in animal societies.

12. Match the descriptions to these terms:
    1 chemical communication
    2 auditory communication
    3 visual communication
    4 tactile communication
    a. _____ Honeybees do a waggle dance in a dark hive.
    b. _____ Male raggiana birds of paradise do spectacular courtship dances.
    c. _____ Birds sing songs.
    d. _____ Cheetahs spray a pheromone onto a tree.

## 32.5 SOCIOBIOLOGY AND ANIMAL BEHAVIOR (PP. 673–74)

The learning objective for this section is:
• Define the concept of inclusive fitness with regard to altruistic behavior.

13. Indicate whether these statements are true (T) or false (F). Change all false statements to true statements.
    a. _____ Subordinate males have less chance to mate, but group living may help them survive. Rewrite: _____
_____

    b. _____ Animals that live alone may have to spend less time grooming. Rewrite: _____
_____

    c. _____ Animals that capture large prey tend to live alone. Rewrite: _____
_____

    d. _____ The cost of social living outweighs the benefits, but animals like being with others. Rewrite: _____
_____

14. Match the statements to these terms (multiple answers are possible):
    1 altruism
    2 inclusive fitness
    3 helpers at the nest
    4 reciprocity
    a. _____ Males do not prevent receptive female chimpanzees from copulating with several members of a group.
    b. _____ A behavior seems to be self-sacrificing.
    c. _____ Older siblings take care of younger siblings.
    d. _____ Worker bees do not reproduce and instead help raise siblings.
    e. _____ A younger bird helps an older bird raise its young but takes over the territory when the older bird dies.

## DEFINITIONS WORDMATCH

Review key terms by completing this matching exercise, selecting from the following alphabetized list of terms:

*altruism*
*behavior*
*communication*
*imprinting*
*inclusive fitness*
*operant conditioning*
*pheromone*
*sociobiology*
*territoriality*

a. _____ Behavior related to the act of marking or defending a particular area against invasion by another species member; area often used for the purpose of feeding, mating, and caring for young.

b. _____ Social interaction that has the potential to decrease the lifetime reproductive success of the member exhibiting the behavior.

c. _____ Signal by a sender that influences the behavior of a receiver.

d. _____ Chemical substance secreted by one organism that influences the behavior of another.

e. _____ Increase in reproduction that results from direct selection and indirect selection.

f. _____ Observable, coordinated responses to environmental stimuli.

g. _____ Application of evolutionary principles to the study of social behavior of animals, including humans.

## CHAPTER TEST

### OBJECTIVE QUESTIONS

Do not refer to the text when taking this test.

_____ 1. Which of these pertain(s) to behavior?
   a. The heart pumps blood into the arteries.
   b. Ants lay a pheromone trail to guide other ants.
   c. Birds have warning calls.
   d. Both *b* and *c* are correct.

_____ 2. Which of these statements are supported by data?
   a. Males compete for mates.
   b. Operant conditioning brings about learning.
   c. Social interactions control behavior.
   d. Females carry out sexual selection.
   e. All of these statements are supported by data.

For questions 3–7, match the statements to these descriptions of behavior:
   a. Behavior has a genetic basis.
   b. The nervous and endocrine systems control behavior.
   c. Behavior undergoes development.

_____ 3. Hybrid studies with warblers reveal this.

_____ 4. Garter snakes differ in their ability to smell slugs.

_____ 5. Laughing gull chicks improve in their ability to recognize their parent.

_____ 6. Egg-laying behavior in *Aplysia* reveals this.

_____ 7. Caged birds can learn to sing their species' song if they hear a recording of it during a sensitive period.

8. The pecking improvement of laughing gull chicks
   a. can be explained by operant conditioning.
   b. correlates with improved motor skills.
   c. is a form of learning.
   d. All of these are correct.

9. A sensitive period for learning was observed when
   a. hybrid garter snakes were intermediate in their ability to smell slugs.
   b. captive birds learned to sing a more developed song by hearing a recording.
   c. imprinting occurred.
   d. Both *b* and *c* are correct.

10. The adaptiveness of behavior may be associated with which statement(s)?
    a. Behavior has a genetic basis.
    b. The nervous and endocrine systems control behavior.
    c. Altruism is involved in inclusive fitness.
    d. Both *a* and *c* are correct.

11. Which of these is NOT consistent with reproduction in females?
    a. selecting the best mate possible
    b. having a higher potential to produce many offspring
    c. nurturing offspring until they can care for themselves
    d. producing few eggs over a lifetime

12. Male competition leads to
    a. dominance hierarchies and reduction in fighting.
    b. defense of a territory.
    c. neglect of the young.
    d. Both *a* and *b* are correct.

13. A cost-benefit analysis can explain
    a. why subordinate males remain in a group.
    b. why red deer males are large despite the chances of it shortening their life span.
    c. why older siblings take care of younger siblings.
    d. All of these are correct.

14. Inclusive fitness explains
    a. seemingly altruistic behavior.
    b. why older siblings help raise younger siblings.
    c. the benefit of being a worker bee.
    d. All of these are correct.

## THOUGHT QUESTIONS

Answer in complete sentences.

15. What evidence shows that behavior is inherited?

16. According to the tenets of sociobiology, is the behavior of animals altruistic?

**Test Results:** _____ number correct ÷ 16 = _____ × 100 = _____ %

## ANSWER KEY

### STUDY QUESTIONS

**1. a.** Mate the two types of warblers. **b.** Hybrids show migratory restlessness. **c.** Hybrids inherit genes from both parents, and therefore show behavior intermediate between the two. **2.** The nervous system controls the eating behavior of garter snakes. **3.** Hormones also control behavior. **4. a.** F **b.** T **c.** T **5.** Due to operant conditioning, chicks learn to peck correctly (i.e., only at models that closely resemble the parent) because in that way they are rewarded with food. **6. a.** Behavior is best learned during a sensitive period immediately after birth. **b.** Clucking by a hen that has recently had chicks can bring about the behavior even outside the sensitive period. **7. a.** Isolated bird sings but does not learn to sing the species' song. **b.** Bird learns to sing the song if a recording is played during a sensitive period. **c.** Bird learns to sing the song of a social tutor of another species outside a sensitive period. **d.** Social interactions help learning take place. **8. a.** T **b.** T **c.** T **9.** Aggressive males are able to have well-kept bowers, and this behavior, which may be inherited, may lead to reproductive success. **10. a.** health **b.** attractive **c.** nutritious **d.** polygynous **e.** monogamous **f.** feed **11. a.** first chance to mate **b.** might be injured protecting the troop **c.** protection **d.** less frequent chance to mate **e.** because the behavior evolved through natural selection **12. a.** 4 **b.** 3 **c.** 2 **d.** 1 **13. a.** T **b.** T **c.** F, . . . tend to live in a group **d.** F, The benefits of social living outweigh the costs, or else animals would not live in a group. **14. a.** 2 **b.** 1, 3 **c.** 1, 2, 3 **d.** 1, 2, 3 **e.** 4

## DEFINITIONS WORDMATCH

**a.** territoriality **b.** altruism **c.** communication **d.** pheromone **e.** inclusive fitness **f.** behavior **g.** sociobiology

## CHAPTER TEST

**1.** d **2.** e **3.** a **4.** b **5.** c **6.** b **7.** c **8.** d **9.** d **10.** d **11.** b **12.** d **13.** d **14.** d **15.** Experimentation has shown that behavior has a genetic basis. Hybrid warblers show migratory restlessness, a trait intermediate to both parents, indicating that behavior is inherited. Hybrid garter snakes generally have an intermediate ability to smell slugs. Since behavior has a genetic basis, it has to be inherited. **16.** It may appear to be altruistic but may be explainable by inclusive fitness—which depends not only on the number of direct descendants due to personal reproduction but also on the number of offspring produced by relatives that the individual has helped nurture.

# 33

# POPULATION GROWTH AND REGULATION

**Ecology** is the study of the interactions of organisms with other organisms and with the physical environment. Ecology encompasses several levels of study: organism, **population, community, ecosystem,** and finally, the **biosphere.** The interactions of organisms with the abiotic and biotic environment affect their distribution and abundance.

Populations have a certain size that depends, in part, on their net reproductive rate ($r$). There are two patterns of population growth: **exponential growth** results in a J-shaped growth curve, and **logistic growth** results in an S-shaped growth curve. Exponential growth can only occur when resources are abundant; otherwise, logistic growth occurs. When population size reaches the **carrying capacity** of the environment, environmental resistance opposes **biotic potential.**

A survivorship curve describes the mortality (deaths per capita) of a population. There are three idealized survivorship curves: With type I, most individuals survive well past the midpoint of the life span; with type II, survivorship decreases at a constant rate throughout the life span; and with type III, most individuals die young. **Age structure diagrams** tell what proportion of the population is prereproductive, reproductive, and postreproductive.

The human population is currently in the exponential part of its growth curve. **MDCs (more-developed countries)** experienced **demographic transition** some time ago, and the populations of most are either not growing or are decreasing. **LDCs (less-developed countries)** are only now undergoing demographic transition but will still experience much growth because of their pyramid-shaped age structure diagram.

Two types of life history patterns have been observed. The opportunistic pattern occurs in unpredictable environments (density-independent factors such as the weather and natural disasters regulate population size) and favors small adults that reproduce early and do not invest in parental care. The equilibrium pattern occurs in stable environments (density-dependent factors such as competition and predation regulate population size) and favors large adults that reproduce repeatedly during a long life span and invest much energy in parental care.

The **competitive exclusion principle** states that no two species can occupy the same niche at the same time. **Resource partitioning** is believed to be present whenever similar species feed on slightly different foods or occupy slightly different habitats. Predation reduces the size of the **prey** population but can have a feedback effect that limits the **predator** population. Prey species have evolved various means to escape predators—for example, chemical defenses in plants and **mimicry** in animals.

Species in a community may exhibit several types of symbiotic relationships. In **parasitism,** the fitness of the parasite increases, and that of the host decreases. **Commensalism** has a neutral effect. In **mutualism,** the fitness of both species increases.

Study the text section by section. Answer the study questions so that you can fulfill the learning objectives for each section.

## 33.1 SCOPE OF ECOLOGY (PP. 678–80)

The learning objectives for this section are:
- State and define the levels of organization above the organism.
- Compare communities on the basis of composition and diversity.
- Explain two models of community structure.
- Describe primary and secondary succession, and explain three models of succession.

1. Match the statement to these levels of ecological study:

    community
    population
    ecosystem
    biosphere

    a. _____ a group of populations interacting in an area
    b. _____ a community interacting with its physical environment
    c. _____ all the individuals of the same species in an area
    d. _____ portion of earth's surface where living things exist
    e. _____ focus of ecological study is growth and regulation of size
    f. _____ focus of ecological study is interactions such as predation and competition between populations

2. Match the description to these terms:

    composition (of community)     interactive model (of community structure)
    diversity (of community)       individualistic model (of community structure)

    a. _____ The community has 13 different species.
    b. _____ All these species are tied together through interactions.
    c. _____ Among the 13 species, one has 30 species and another has only one species.
    d. _____ All these species are temperature tolerant, or else they would not be in this community.

3. Put these stages in the proper order to describe secondary succession. Indicate by letters. _____
    a. high trees
    b. high shrub
    c. shrub tree
    d. grass
    e. low shrub

4. Explain the various models of succession by telling why they are called by the following names:

    a. climax model _____

    b. facilitation model _____

    c. inhibition model _____

    d. tolerance model _____

## 33.2 POPULATION CHARACTERISTICS AND GROWTH (PP. 681–87)

The learning objectives for this section are:
- Explain the difference between population density and population distribution.
- Describe factors that influence population distribution.
- Describe two patterns of population growth—exponential and logistic.
- Describe different types of survivorship curves.
- Describe human population growth at the present time.

5. Populations vary in density (number of individuals per area) and distribution. Use these labels to describe distribution:

uniform    random    clumped

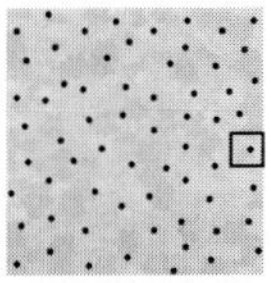

a. _____

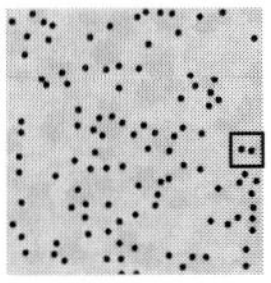

b. _____

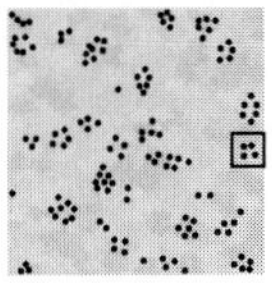

c. _____

6. Give an example for each of these types of factors that can affect distribution:

a. abiotic factor _____

b. biotic factor _____

7. Calculate the per capita rate of increase for a population of 1,000 individuals in which the birth rate is 20 per year and the death rate is 10 per year.

8. What is *biotic potential*?_____

_____

9. Place a check beside all those factors that could contribute to an increase in a population's biotic potential.

a. _____ a reduction in the usual number of offspring per reproduction

b. _____ an increase in the chances of survival until age of reproduction

c. _____ a reduction in how often each individual reproduces

d. _____ an increase in the age at which reproduction begins

Questions 10–15 pertain to the following two population growth curves:

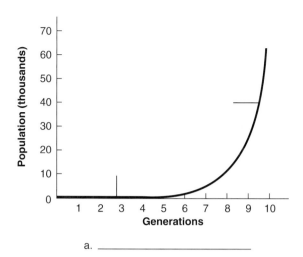

a. _____

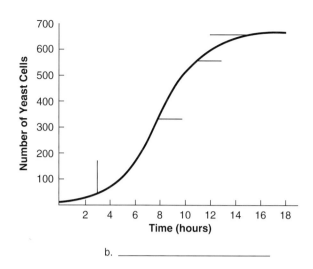

b. _____

10. Place the label *ep* for exponential growth pattern or *lg* for logistic growth pattern beneath the appropriate figure.

11. Which of these two growth curves has a *lag phase?* _____ Place this label where appropriate on the curves.

12. Which of these two growth curves has an *exponential growth phase?* _____ Place this label where appropriate on the curves.

13. During the lag phase, growth is ᵃ·_____; during the exponential growth phase, growth is
    ᵇ·_____.

14. Which of these two growth curves has a *deceleration phase?* _____ Place this label where appropriate on the curves.

15. Which of these two growth curves has a *stable equilibrium phase?* _____ Place this label where appropriate on the curves.

16. Which of the phases of a growth curve (lag, exponential, deceleration, or stable equilibrium) best represents the biotic potential of a population?

17. a. Which of these phases (lag, exponential, deceleration, or stable equilibrium) best represents environmental resistance? _____

    b. What is environmental resistance? _____
    _____

    c. At what point will the stable equilibrium phase occur? _____
    _____

18. Study the following diagram of survivorship curves and then answer the questions:

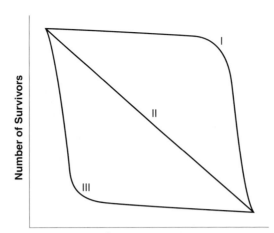

a. Which curve shows that the members of a cohort die at a constant rate? _____

b. Which curve shows that the members of a cohort tend to die early in life? _____

c. Which curve shows that the members of a cohort usually live through their entire allotted life span? _____

19. Which pattern of population growth best describes human population growth? _____

20. The equivalent of a medium-sized city (225,000) is added to the world's population _____ (every year, every six months, every month, every day).

In questions 21–25, indicate whether the statement is true (T) or false (F). Rewrite the false statements to make true statements.

21. _____ a. The demographic transition is accompanied by an increase in growth rate. Rewrite: _____

_____

_____ b. The more-developed countries have not undergone the demographic transition. Rewrite:_____

_____

_____ c. Great Britain is an example of a less-developed country. Rewrite:_____

_____

22. _____ The birthrate has declined in less-developed countries, but the death rate has not. Rewrite: _____

_____

23. _____ The age structure diagram for less-developed countries has the appearance of a pyramid because postreproductive citizens outnumber the prereproductive citizens. Rewrite: _____

_____

24. Even though the growth rate is declining, the populations of countries in the areas of
a._____, b._____, and c._____
are still expanding because of d._____ growth.

25. What three ways have been suggested to help reduce the growth rate in the less-developed countries?

a. _____

b. _____

c. _____

The learning objectives for this section are:
- Relate two types of life history patterns to regulation of growth by density-independent and density-dependent factors.
- Explain the effects of competition, predation, and symbiosis on population growth.

Match questions 26–29 to these descriptions:

    a. unpredictable environment and density-independent regulation of population size
    b. stable environment and density-dependent regulation of population size

26. _____ large adults, long life span, slow to mature, repeated reproduction, and much care of offspring

27. _____ opportunistic pattern

28. _____ equilibrium pattern

29. _____ small adults, short life span, fast to mature, many offspring during a burst of reproduction, and little or no care of offspring

30. You would expect competition to be a lose/lose situation because both species are <sup>a.</sup>_____.

    According to the <sup>b.</sup> _____,

    no two species can occupy the same niche. <sup>c.</sup> _____ is a way for two species to

    ensure different niches. For example, swallows, swifts, and martins can coexist because each type of bird
    <sup>d.</sup> _____.

31. The barnacle *Chthamalus* is able to exist on rocks in the entire intertidal zone. When *Balanus* is present, *Chthamalus* only exists in the upper intertidal zone.

    a. Why isn't *Chthamalus* found in the lower intertidal zone? _____

    _____

    b. Why is *Chthamalus* found in the upper intertidal zone? _____

    _____

32. The Canadian lynx and the snowshoe hare population sizes cycle.

    a. Why would they cycle if the predator-prey relationship is causing the cycling? _____

    _____

    b. Why would they cycle if a hare-food relationship is involved? _____

    _____

    _____

33. Identify the antipredator device in the following:

    a. Inchworms resemble twigs. _____

    b. Dart-poison frogs are brightly colored. _____

    c. Frilled lizards open up folds of skin around the neck. _____

34. If appropriate, label the following as describing Batesian (B) or Müllerian (M) mimicry; if neither is appropriate, leave blank.

    a. _____ A predator mimics another species that has a successful predatory style.
    b. _____ A prey mimics another species that has a successful defense.
    c. _____ A predator captures food.
    d. _____ Several different species with the same defense mimic one another.

35. Commensalism, mutualism, and parasitism are all different types of _____ relationships.

36. Complete this table to compare types of symbiosis by marking a plus (+) if the species *benefits,* a minus (–) if the species is *harmed,* and a zero (0) if the symbiotic relationship has *no effect* on the species.

| | First Species | Second Species |
|---|---|---|
| Parasitism | | |
| Commensalism | | |
| Mutualism | | |

37. Label each of the following as describing commensalism (C), mutualism (M), or parasitism (P).
    a. _____ The clownfish lives safely within the poisonous tentacles of the sea anemone, which other fish avoid.
    b. _____ Certain bacteria cause pneumonia.
    c. _____ Humans get a tapeworm from eating raw pork.
    d. _____ Epiphytes grow in branches of trees but get no nourishment from the trees.
    e. _____ Flowers provide nourishment to a pollinator, and the pollinator carries pollen to another flower.

38. Label the symbiotic relationships of species A to species B with these terms:

commensalism    competition    mutualism    parasitism    predation

    a. _____ Species A consumed more of the resource than species B.
    b. _____ Species A eats species B.
    c. _____ Species A is cultivated by species B as a source of food.
    d. _____ Species A infects species B.
    e. _____ Species A rides along with species B to get food while species B hunts.

## DEFINITIONS WORDMATCH

Review key terms by completing this matching exercise, selecting from the following alphabetized list of terms:

biosphere
biotic potential
carrying capacity
cohort
demographic transition
doubling time
ecology
environmental resistance
exponential growth
logistic growth
population

a. _____ Group of organisms of the same species occupying a certain area.

b. _____ Maximum population growth rate under ideal conditions.

c. _____ Growth, particularly of a population, in which the increase occurs in the same manner as compound interest.

d. _____ Due to industrialization, a decline in the birthrate following a reduction in the death rate so that the population growth rate is lowered.

e. _____ Largest number of organisms of a particular species that can be maintained indefinitely by a given environment.

f. _____ That portion of the surface of the earth (air, water, and land) where living things exist.

g. _____ Group of individuals having a statistical factor in common, such as year of birth, in a population study.

Do not refer to the text when taking this test.

___ 1. Which of these pairs is mismatched?
   a. population—all the members of a species in same area
   b. community—populations interacting with the physical environment
   c. biosphere—surface of the earth where organisms live
   d. ecosystem—energy flow and chemical cycling occur

___ 2. Distribution of organisms tends to be
   a. clumped.
   b. determined by limiting factors.
   c. the same as the population density.
   d. Both *a* and *b* are correct.

___ 3. If the birthrate is 10 per 1,000 and the death rate is 10 per 1,000, then the net reproductive rate is
   a. 0.
   b. 10.
   c. 20.
   d. 100.

___ 4. If the per capita rate of increase is positive, then
   a. population growth will occur.
   b. the size of the population will increase.
   c. environmental resistance is likely to come into play.
   d. All of these are correct.

___ 5. Which is true of a J-shaped growth curve?
   a. represents the logistic growth pattern
   b. does not usually occur for long in nature
   c. is seen in populations that have an equilibrium life history pattern
   d. Both *b* and *c* are correct.

___ 6. During exponential growth,
   a. growth remains steady.
   b. growth is accelerating.
   c. growth is declining.
   d. growth depends on the environment.

___ 7. Which sentence is most appropriate?
   a. Environmental resistance encourages biotic potential.
   b. If population size is at carrying capacity, growth is unlikely.
   c. Environmental resistance consists only of density-independent factors.
   d. Exponential growth can usually occur indefinitely.

___ 8. Which is true of an S-shaped growth curve?
   a. represents the exponential growth pattern
   b. represents the logistic growth pattern
   c. does not usually occur in nature
   d. Both *b* and *c* are correct.

___ 9. Survivorship in a population is related to
   a. age of death.
   b. biotic potential.
   c. age structure diagram.
   d. All of these are correct.

___ 10. If the survivorship curve is a straight diagonal line, then
   a. the rate of death is constant, regardless of age.
   b. most individuals live out the expected life span.
   c. most individuals die early.
   d. environmental resistance has occurred.

___ 11. Populations with an equilibrium life history pattern tend to have a(n) _____ growth curve.
   a. J-shaped
   b. S-shaped

___ 12. Select the characteristic that is NOT consistent with the opportunistic life history pattern.
   a. large body size
   b. many offspring
   c. short life span
   d. fast to mature

___ 13. The countries in Asia and Africa are
   a. MDCs that are experiencing rapid growth.
   b. LDCs that are experiencing rapid growth.
   c. MDCs that are experiencing slow growth.
   d. LDCs that are experiencing slow growth.

___ 14. The world population increases by the number of people found in a medium-sized city (200,000) every
   a. year.
   b. six months.
   c. month.
   d. day.

___ 15. Competition
   a. always eliminates one or the other species.
   b. widens niche breadth.
   c. narrows niche breadth and increases species diversity.
   d. None of these is correct.

____16. Predator population size is limited in part by available _____, while prey population size is limited by _____.
   a. living space; food
   b. predators; prey
   c. food; predators
   d. food; food
____17. Antipredator defenses may include
   a. camouflage.
   b. fright.
   c. warning.
   d. All of these are correct.
____18. Mimicry can help
   a. a predator capture food.
   b. prey avoid capture.
   c. Both *a* and *b* are correct.
   d. None of these is correct.
____19. Bees, wasps, and hornets are examples of
   a. Batesian mimicry.
   b. Müllerian mimicry.
   c. mimicry for predation.
   d. Both *a* and *c* are correct.

For questions 20–22, match the organisms with these terms:
   a. parasitism
   b. commensalism
   c. mutualism
____20. virus
____21. termites and protozoans
____22. barnacle
____23. In which relationship do both species benefit?
   a. mutualism
   b. commensalism
   c. symbiosis
   d. Both *b* and *c* are correct.
____24. Which can be a parasite?
   a. bacteria
   b. plants
   c. animals
   d. All of these are correct.

## THOUGHT QUESTIONS

Answer in complete sentences.
25. Under what conditions could the growth of a population be infinite?

26. What is the relationship between competition and diversity?

**Test Results:** _____ number correct ÷ 26 = _____ × 100 = _____ %

## ANSWER KEY

### STUDY QUESTIONS

**1. a.** community **b.** ecosystem **c.** population **d.** biosphere **e.** population **f.** ecosystem **2. a.** composition **b.** interactive model **c.** diversity **d.** individualistic model **3.** d, e, b, c, a **4. a.** End stage is a climax community **b.** Each stage facilitates the onset of the next stage **c.** Each stage inhibits the onset of the next stage. **d.** The various species tolerate the entrance of other species. **5. a.** uniform **b.** random **c.** clumped **6. a.** temperature **b.** presence of prey **7.** 1% **8.** the highest possible per capita rate of increase **9.** b **10. a.** ep **b.** lg **11.** both; see Figures 33.6*a,b*, p. 682, in text. **12.** both; see Figure 33.6*a,b*, p. 682, in text. **13. a.** slow **b.** ac-

celerating **14.** b; see Figure 33.6*b*, p. 682, in text. **15.** b; see Figure 33.6*b*, p. 682, in text. **16.** exponential **17. a.** deceleration **b.** encompasses all environmental factors that oppose biotic potential **c.** when biotic potential and environmental resistance are equal **18. a.** II **b.** III **c.** I **19.** exponential growth pattern **20.** every day **21. a.** F, The demographic transition is accompanied by a decrease in growth rate. **b.** F, The more-developed countries have undergone the demographic transition. **c.** F, Great Britain is an example of a more-developed country. **22.** F, The birthrate has declined in less-developed countries, and the death rate has also declined. **23.** F, The age structure diagram for less-developed countries has the appearance of a pyramid because prereproductive citizens outnumber postreproductive citizens. **24. a.** Africa **b.** Asia **c.** Latin America **d.** exponential **25. a.** Begin and/or strengthen family planning programs. **b.** Help reduce the desire for large families through social progress **c.** Help women delay the onset of childbearing by several to many years. **26.** b **27.** a **28.** b **29.** a **30. a.** competing for the same resource **b.** competitive exclusion principle **c.** Resource partitioning **d.** occupies a different spruce tree zone **31. a.** *Balanus* competes successfully and takes over in the lower intertidal zone. **b.** *Chthamalus* is better able to withstand drying out and takes over in the upper intertidal zone. **32. a.** The predator overkills the prey, and as the prey population declines, so does the predator population. **b.** As the hares die off due to lack of food, the predator population would decline, and as the hare population recovers, so would the predator population recover. **33. a.** camouflage **b.** warning coloration **c.** fright **34. b.** B **d.** M (Both *a* and *c* should be left blank.) **35.** symbiotic **36.**

| First Species | Second Species |
| --- | --- |
| + | – |
| + | 0 |
| + | + |

**37. a.** C **b.** P **c.** P **d.** C **e.** M **38. a.** competition **b.** predation **c.** mutualism **d.** parasitism **e.** commensalism

## DEFINITIONS WORDMATCH

**a.** population **b.** biotic potential **c.** exponential growth **d.** demographic transition **e.** carrying capacity **f.** biosphere **g.** cohort

## CHAPTER TEST

**1.** b **2.** d **3.** a **4.** d **5.** b **6.** b **7.** b **8.** b **9.** a **10.** a **11.** b **12.** a **13.** b **14.** d **15.** c **16.** c **17.** d **18.** c **19.** b **20.** a **21.** c **22.** b **23.** a **24.** d **25.** If environmental resistance were absent, the population size could continue to increase without end. This is unlikely, because the increase in population size without an increase in resources would force environmental resistance to be present. **26.** Competition increases diversity because no two species can occupy the same niche.

# 34

# ECOSYSTEMS AND HUMAN INTERFERENCES

An **ecosystem** is a community of organisms plus the physical and chemical environment. Some populations are **producers** and some are **consumers.** Producers are autotrophs that produce their own organic food. Consumers are heterotrophs that take in organic food. Consumers may be **herbivores, carnivores, omnivores,** or **decomposers.**

Energy flows through an ecosystem. Producers transform solar energy into food for themselves and all consumers. As herbivores feed on plants and carnivores feed on herbivores, energy is converted to heat. Feces, urine, and dead bodies become food for decomposers. Eventually, all the solar energy that enters an ecosystem is converted to heat. Therefore, ecosystems require a continual supply of solar energy.

Inorganic nutrients are not lost from the biosphere as is energy. They recycle within and between ecosystems. Decomposers return some portion of inorganic nutrients to autotrophs, and other portions are imported or exported between ecosystems in global cycles.

The **food webs** of ecosystems contain **grazing food chains** (begin with a producer) and **detrital food chains** (begin with detritus). A **trophic level** includes all the organisms that feed at a particular link in food chains. In general, biomass and energy content decrease from one trophic level to the next, as is depicted in an **ecological pyramid.**

The global cycling of inorganic elements involves the biotic and abiotic parts of an ecosystem. Cycles usually contain (1) a reservoir (a source normally unavailable to organisms), (2) an exchange pool (a source available to organisms), and (3) the biotic community.

In the **water cycle,** evaporation of ocean waters and transpiration from plants contribute to aerial moisture. Rainfall over land results in bodies of fresh water plus groundwater. Eventually, all water returns to the oceans.

In the **carbon cycle,** respiration by organisms adds as much carbon dioxide to the atmosphere as photosynthesis removes. Human activities such as the burning of fossil fuels and trees add carbon dioxide to the atmosphere. Carbon dioxide and other gases trap heat, leading to **global warming.** The effects of global warming could be a rise in sea level and a change in climate patterns, with disastrous effects.

In the **nitrogen cycle,** the biotic community keeps nitrogen recycling back to producers. Human activities convert atmospheric nitrogen to fertilizer, which when broken by soil bacteria adds nitrogen oxides to the atmosphere. Nitrogen oxides and sulfur dioxide react with water vapor to form acids that contribute to **acid deposition.** Acid deposition is killing lakes and forests, and it also corrodes marble, metal, and stonework. Nitrogen oxides and hydrocarbons (HC) react to form smog, which contains ozone and PAN. These oxidants are harmful to animal and plant life.

In the **phosphorus cycle,** the biotic community recycles phosphorus back to the producers, and only limited quantities are made available by the weathering of rocks. Phosphates are mined for fertilizer production, and fertilizers overenrich lakes and ponds.

Global warming, acid deposition, and water pollution reduce biodiversity. **Ozone shield** destruction, which is associated with CFCs, is expected to result in decreased productivity of the oceans. Tropical rain forests are being destroyed in numerous ways, and many organisms that could possibly benefit humans are threatened. **Conservation biology** is a new discipline that pulls together information from a number of biological fields to determine how to manage ecosystems for the benefit of all species.

Study the text section by section. Answer the study questions so that you can fulfill the learning objectives for each section.

The learning objectives for this section are:

- Give examples of both the abiotic and biotic components of ecosystems.
- Distinguish between producers and consumers, and distinguish between four types of consumers.
- State the two fundamental characteristics of ecosystems: energy flow and chemical cycling.
- Explain why ecosystems require a continual input of solar energy.
- Explain why the same chemicals can be used over and over again.

1. Match the descriptions to these biotic components of an ecosystem:

   carnivores    consumers    decomposers    herbivores    omnivores    autotrophs

   a. _____ organisms of decay
   b. _____ feed only on other animals
   c. _____ producers in an ecosystem
   d. _____ heterotrophs
   e. _____ feed directly on green plants
   f. _____ feed on both plants and animals

2. Label this diagram using these terms:

   decomposers    consumers    producers    inorganic nutrient pool

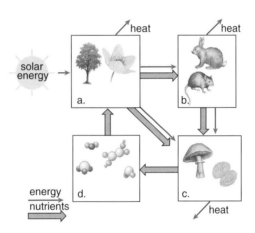

3. Energy doesn't cycle in ecosystems.

   a. Which populations contain the most energy? _____

   b. Which populations contain the least energy? _____

   c. What happened to the energy? Explain on the basis of the second law of thermodynamics. _____
   
   _____

4. Chemicals do cycle in an ecosystem.

   a. What two molecules do plants use to make glucose? _____

   b. Glucose could conceivably pass from a producer poplation to the _____

   populations to the _____ populations.

   c. Eventually, the glucose is broken down, and what is returned to plants? _____
   
   _____

## 34.2 ENERGY FLOW (PP. 701–2)

The learning objecitves for this section are:

- Describe a food web, and use a food web to construct grazing food chains and detrital food chains.
- Explain an ecological pyramid in terms of trophic levels, and explain why each succeeding feeding level is smaller in biomass/energy than the one before.

This is a food web diagram.

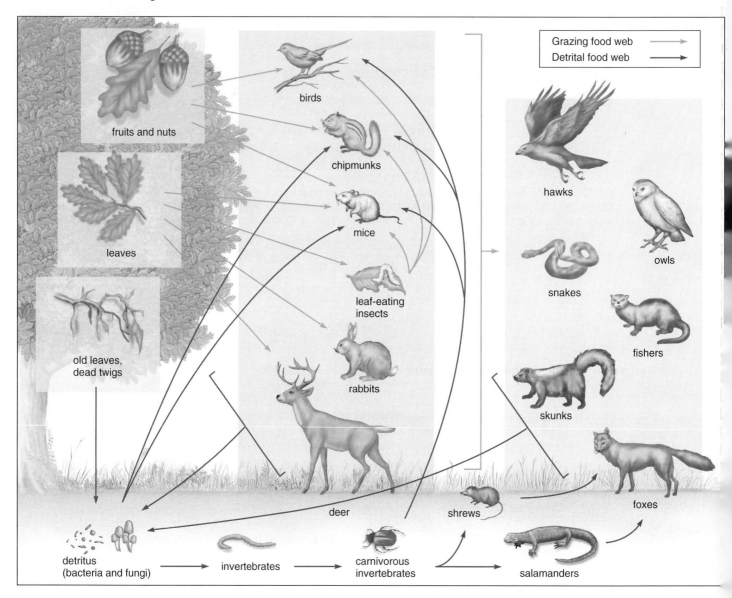

5.  a. How many trophic levels do you see in the grazing portion of this food web diagram?_____

   b. Name a population at the first trophic level. _____

   c. Name two populations at the second trophic level. _____

   d. Name two populations at the third trophic level. _____

   e. From these trophic levels, construct a grazing food chain. _____

6.  Construct a detrital food chain from the food web diagram.

   _____

7.  Explain one way in which the detrital food web and the grazing food web are always connected.

   _____

This is an ecological pyramid diagram.

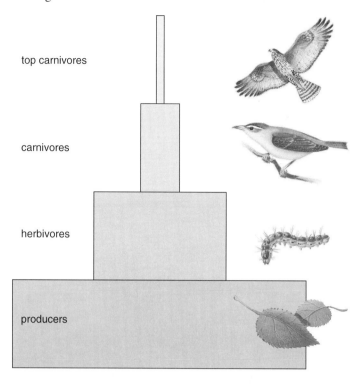

top carnivores

carnivores

herbivores

producers

8. a. In the ecological pyramid, write in two populations at the same trophic level as the caterpillar from the food web diagram. Write the names on this line. _____

   b. In the ecological pyramid, write in two populations at the same trophic level as the bird population. Write these names on this line. _____

9. a. Why is each higher trophic level smaller than the one preceding it? _____

   _____

   b. What is the so-called 10% rule of thumb? _____

   _____

## 34.3  GLOBAL BIOGEOCHEMICAL CYCLES (PP. 702–9)

The learning objectives for this section are:
- Construct and give a function for each part of a generalized biogeochemical cycle.
- Describe the water, carbon, nitrogen, and phosphorus biogeochemical cycles.
- Describe the human influence on each of these cycles and the resulting environmental problems that ensue.
- Explain the concept of biological magnification.

10. Examine the following diagram and then fill in the blanks.

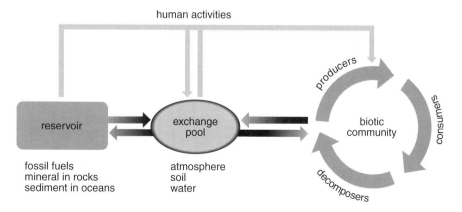

human activities

reservoir

exchange pool

producers

biotic community

consumers

decomposers

fossil fuels
mineral in rocks
sediment in oceans

atmosphere
soil
water

a. What is a reservoir? _____

b. What is an exchange pool? _____

c. What is a biotic community? _____

d. Explain the arrows labeled *human activities*. _____
   _____

11. Complete this diagram of the water cycle by filling in the boxes, using these terms:

    ice    $H_2O$ in the atmosphere    ocean    groundwaters

    Label the arrows, using these terms:

    precipitation (twice)    transpiration from plants and evaporation from soil    evaporation
    transport of water vapor by wind

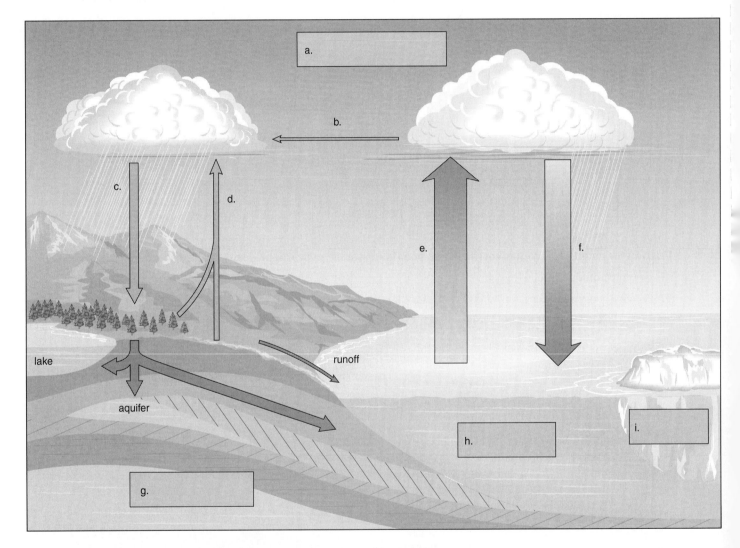

12. Place a check beside the statements that correctly describe the water cycle.
    a. _____ Water cycles between the land, the atmosphere, and the ocean, and vice versa.
    b. _____ We could run out of fresh water.
    c. _____ The ocean receives more precipitation than the land.
    d. _____ Water that is in the aquifers never reaches the oceans.

Fill in the blanks.

13. In the carbon cycle, carbon dioxide is removed from the atmosphere by the process of a. _____ but is returned to the atmosphere by the process of b. _____. Living things and dead matter in soil are carbon c. _____ and so are the d. _____ because of shell accumulation. In aquatic ecosystems, carbon dioxide from the air combines with water to produce e. _____ that algae can use for photosynthesis. In what way do humans alter the transfer rates in the carbon cycle? f. _____

14. Fill in the table to indicate the source of gases that cause the greenhouse effect.

| Gas | From |
| --- | --- |
| Carbon dioxide ($CO_2$) | a. _____ |
| Nitrous oxide ($N_2O$) | b. _____ |
| Methane ($CH_4$) | c. _____ |

d. Why are these gases called the greenhouse gases? _____

_____

15. Place a check beside all those statements that may be expected because of global warming.
   a. _____ a global temperature increase by as much as 4.5°C
   b. _____ melting of glaciers and a rise in sea level
   c. _____ massive fish kills and plant destruction
   d. _____ dryer conditions inland where droughts may occur
   e. _____ expansion of forests into arctic areas

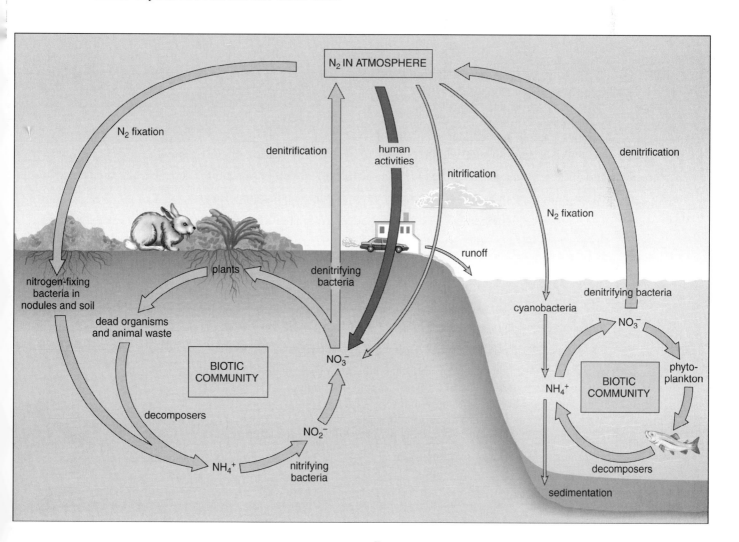

Questions 16 and 17 are based on the preceding diagram of the nitrogen cycle.

16. Match the descriptions to these types of bacteria:
    denitrifying bacteria
    nitrifying bacteria
    nitrogen-fixing bacteria
    a. _____ bacteria that convert nitrate to nitrogen gas
    b. _____ bacteria that convert ammonium to nitrate
    c. _____ bacteria in legume nodules that convert nitrogen gas to ammonium

17. Plants cannot utilize nitrogen gas. What are two ways in which plants receive a supply of nitrogen for incorporation into proteins and nucleic acids? _____

18. When humans produce fertilizers, the gas a._____ is removed from the atmosphere and changed to b._____, which enters the atmosphere. Acid deposition occurs when nitrogen oxides and c._____ in the atmosphere are converted to acids that return to earth.

19. Place a check beside those statements that may be expected because of acid deposition.
    a. _____ dying forests
    b. _____ lower agricultural yields
    c. _____ sterile lakes
    d. _____ corroded marble, metal, and stonework

20. Photochemical smog arises when a._____ and b._____ react with one another in the presence of sunlight. Smog contains the pollutants c._____ and d._____.

21. Place a check beside all those effects that may be expected from the occurrence of smog.
    a. _____ breathing difficulties
    b. _____ damage to plants
    c. _____ thermal inversions
    d. _____ cleaner air than usual

22. Place a check beside the statements that correctly describe the results when producers take up phosphate.
    a. _____ becomes a part of phospholipids
    b. _____ becomes a part of ATP
    c. _____ becomes a part of nucleotides
    d. _____ becomes a part of the atmosphere

23. Indicate whether these statements are true (T) or false (F). Rewrite all false statements to be true statements.

    a. _____ Excess phosphate in bodies of water may cause radiation poisoning. Rewrite: _____
    _____

    b. _____ Most ecosystems have plenty of phosphate. Rewrite: _____
    _____

    c. _____ The phosphorus cycle is a sedimentary cycle. Rewrite: _____
    _____

    d. _____ Phosphate enters ecosystems by being taken up by animals. Rewrite: _____
    _____

24. Place a check beside those items that correctly describe biological magnification.
    a. _____ pertains to organic poisons like pesticides and herbicides
    b. _____ affects other organisms but not humans
    c. _____ refers to an accumulation of foreign molecules in the body because they are not excreted
    d. _____ refers to an enlargement of certain parts of the body due to evolution

25. What is the ozone shield, and why is it important? _____
_____

26. Explain the significance of the following:

$$Cl + O_3 \rightarrow ClO + O_2$$

_____

27. What are some of the possible effects of increased ultraviolet radiation on humans and other organisms? _____

_____

Review key terms by using the following alphabetized list of terms to fill in the blanks below. Then complete the wordsearch.

```
R G L P T S Q Z I F A G A C N V
B E W D O O F R B A M N C G F D
G U P P Y L R O X A Q U I F E R
F T O D P G M U E F C T D H C L
L R E M U S N O C O R A D R O T
R O S A T D D E F U N N E L S I
P P A N V T N I C H E R P N Y M
N H T E N B T S A X L A O T S J
N I A H C D O O F Q S U S O T E
S C Q B E O R B A U F A I F E R
C A R N I V O R E E O C T I M U
I T F O R D O Q Y P O A I T A M
L I G L U N J U L P H N O L H Z
N O I T A C I F I R T I N E T F
I N N X C B L R Y T S F J R S R
```

*acid deposition*
*aquifer*
*carnivore*
*CFC*
*consumer*
*ecosystem*
*eutrophication*
*food chain*
*food web*
*nitrification*
*PAN*

a. _____ Organic compounds containing carbon, chlorine, and fluorine atoms.

b. _____ Rock layers that contain water and release it in appreciable quantities to wells or springs.

c. _____ The order in which one population feeds on another in an ecosystem.

d. _____ Organism that feeds on another organism in a food chain.

e. _____ Consumer in a food chain that eats other animals.

f. _____ Biological community together with the associated abiotic environment; characterized by a flow of energy and a cycling of inorganic nutrients.

g. _____ Type of chemical found in photochemical smog.

h. _____ Process by which nitrogen in ammonia and organic compounds is oxidized to nitrites and nitrates by soil bacteria.

i. _____ In ecosystems, complex pattern of interlocking and crisscrossing food chains.

j. _____ Enrichment of water by inorganic nutrients used by phytoplankton; over-enrichment often caused by human activities leading to excessive bacterial growth and oxygen depletion.

k. _____ The sulfate or nitrate salts of acids produced by commercial and industrial activities return to earth as rain or snow.

## OBJECTIVE QUESTIONS

Do not refer to the text when taking this test.

____ 1. A complex of interconnected food chains in an ecosystem is called a(an)
    a. ecosystem.
    b. ecological pyramid.
    c. trophic level.
    d. food web.

____ 2. Wolves and lions are at the same trophic level because they both
    a. are large mammals.
    b. eat only when they are hungry.
    c. eat primary consumers.
    d. live on land.
    e. Both *b* and *c* are correct.

____ 3. The biomass of herbivores is smaller than that of producers because
    a. herbivores are not as efficient in converting energy to biomass.
    b. some energy is lost during each energy transformation.
    c. the number of plants is always more than the number of herbivores.
    d. woody plants live longer than omnivores and herbivores.
    e. All of these are correct.

In questions 4–7, indicate whether the statements are true (T) or false (F).

____ 4. Energy flows through a food chain because it is constantly lost from organic food as heat.

____ 5. A food web contains many food chains.

____ 6. The weathering of rocks is one way that phosphate ions are made available to plants.

____ 7. Respiration returns carbon to the atmosphere.

____ 8. About _____ of the energy available at a particular trophic level is incorporated into the tissues at the next trophic level.
    a. 1%
    b. 10%
    c. 25%
    d. 50%
    e. 75%

Questions 9–11, refer to the following food chain: grass → rabbits → snakes → hawks

____ 9. Each population
    a. is always larger than the one before it.
    b. supports the next level.
    c. is an herbivore.
    d. is a carnivore.

____ 10. Rabbits are
    a. consumers.
    b. herbivores.
    c. more plentiful than snakes.
    d. All of these are correct.

____ 11. Hawks
    a. contain phosphate taken up by grass.
    b. give off $O_2$ that will be taken up by rabbits.
    c. die and decompose, and because of this they cannot contribute to a grazing food chain.
    d. All of these are correct.

____ 12. Which of the following contribute(s) to the carbon cycle?
    a. respiration
    b. photosynthesis
    c. fossil fuel combustion
    d. All of these are correct.

____ 13. The largest reserve of unincorporated carbon is in
    a. the soil.
    b. the atmosphere.
    c. the ocean.
    d. deep sediments.

____ 14. The greenhouse effect
    a. is caused by particles in the air.
    b. is caused in part by carbon dioxide.
    c. will cause temperatures to increase.
    d. will cause temperatures to decrease.
    e. Both *b* and *c* are correct.

____ 15. The form of nitrogen most plants make use of is
    a. atmospheric nitrogen.
    b. nitrogen gas.
    c. organic nitrogen.
    d. nitrates.

In questions 16–18, match the air pollutants to these conditions:
    a. ozone shield destruction
    b. global warming
    c. acid deposition
    d. photochemical smog

____ 16. CFCs

____ 17. $SO_2$

____ 18. $CO_2$

____ 19. UV radiation
    a. causes mutations.
    b. impairs crop growth.
    c. kills plankton.
    d. All of these are correct.

_____ 20. Nitrogen oxides and hydrocarbons react in the presence of sunlight to produce
   a. acid particles.
   b. ground level ozone.
   c. greenhouse gases.
   d. All of these are correct.

_____ 21. What may occur as a result of the greenhouse effect?
   a. coastal flooding
   b. loss of food
   c. excess plant growth
   d. Both *a* and *b* are correct.

_____ 22. What contributes to the greenhouse effect?
   a. nuclear power
   b. burning of fossil fuels
   c. geothermal energy
   d. Both *a* and *c* are correct.

_____ 23. Which is the cause of stratospheric ozone depletion?
   a. chlorine
   b. PANs
   c. nitrates
   d. Both *b* and *c* are correct.

_____ 24. Acid deposition is associated with
   a. dying lakes.
   b. dying forests.
   c. dissolving of copper from pipes.
   d. All of these are correct.

_____ 25. Biological magnification
   a. is more severe the longer the food chain.
   b. is more likely in aquatic food chains.
   c. involves hazardous wastes.
   d. All of these are correct.

## THOUGHT QUESTIONS

Answer in complete sentences.

26. Why is a food chain normally limited to four or five links?

27. How would the shortage of an element in the exchange pool affect an ecosystem? Explain.

**Test Results:** _____ number correct ÷ 27 = _____ × 100 = _____%

## ANSWER KEY

### STUDY QUESTIONS

**1. a.** decomposers **b.** carnivores **c.** autotrophs **d.** consumers **e.** herbivores **f.** omnivores   **2. a.** producers **b.** consumers **c.** decomposers **d.** inorganic nutrient pool **3. a.** producers **b.** consumers **c.** It dissipates. With every transformation, as when the energy in food is converted to ATP, there is always a loss of usable energy. Eventually, all solar energy taken in by plants becomes heat. **4. a.** carbon dioxide and water. **b.** consumer, decomposer **c.** carbon dioxide and water   **5. a.** three **b.** tree **c.** birds, chipmunks **d.** foxes, fishers **e.** tree → rabbits → snakes   **6.** old leaves and dead twigs → bacteria and fungi of decay → mice → hawks   **7.** Members of the grazing food web die and are decomposed by bacteria and fungi.   **8. a.** rabbits and deer **b.** foxes and snakes **9. a.** Less energy is available to be passed on. **b.** In general, only about 10% of the energy of one trophic level is available to the next trophic level.   **10. a.** a source that is usually available to the biotic community **b.** a source from which organisms do generally take chemicals, such as the atmosphere or soil. **c.** producers, consumers, and decomposers that interact through nutrient cycling and energy flow **d.** Humans remove elements from reservoirs and exchange pools and make them available to producers. For example, humans convert $N_2$ in the air to make fertilizer, and they mine phosphate to make fertilizer.   **11.** See Figure 34.8, page 703, in text. **12.** a, b, c   **13. a.** photosynthesis **b.** cellular respiration **c.** reservoirs **d.** oceans **e.** bicarbonate **f.** by burning fossil fuels that add carbon to the atmosphere   **14. a.** fossil fuel and wood burning **b.** fertilizer use and animal wastes **c.** biogas (guts of animals), in sediments, and in flooded rice paddies **d.** These gases are called greenhouse gases because, like the panes of a greenhouse, they allow solar radiation to pass through but hinder the escape of heat.   **15.** a, b, d, e   **16. a.** denitrifying **b.** nitrifying **c.** nitrogen fixing   **17.** nitrogen-fixing bacteria in nodules and nitrate in soil   **18. a.** $N_2$ **b.** $NO_3$ **c.** sulfur dioxide **19.** a, b, c, d **20. a.** No$_x$ **b.** HC **c.** PAN **d.** ozone **21.** a, b, c **22.** a, b, c **23. a.** F, . . . may cause algal bloom **b.** F, . . . have a limited supply of phosphate **c.** T **d.** F, . . . taken up by plants **24.** a, c **25.** Ozone is a layer within the stratosphere that protects the earth's surface from ultraviolet radiation. Organisms evolved in the presence of this ozone layer. **26.** The chlorine breaks down the ozone, and the UV radiation is not absorbed. **27.** It will increase the incidence of skin cancer and decrease the productivity of living systems. Loss of oceanic plankton will disrupt marine ecosystems.

## DEFINITIONS WORDSEARCH

```
                      A
B E W D O O F         C
  U             A Q U I F E R
  T                   D   C
  R E M U S N O C     D   O
  O           F       E   S
  P A N       C       P   Y
  H                   O   S
N I A H C D O O F     S   T
  C                   I   E
C A R N I V O R E     T   M
  T                   I
  I                   O
N O I T A C I F I R T I N
  N
```

**a.** CFC **b.** aquifer **c.** food chain **d.** consumer **e.** carnivore **f.** ecosystem **g.** PAN **h.** nitrification **i.** food web **j.** eutrophication **k.** acid deposition

## CHAPTER TEST

**1.** d **2.** c **3.** b **4.** F **5.** T **6.** T **7.** T **8.** b **9.** b **10.** d **11.** b **12.** d **13.** c **14.** e **15.** d **16.** a **17.** c **18.** b **19.** d **20.** b **21.** d **22.** b **23.** a **24.** d **25.** d **26.** By the laws of thermodynamics, energy conversion at each link of a food chain results in nonusable heat. Too little useful energy remains for more links. **27.** A shortage of an element such as nitrogen or phosphorus would reduce the biomass of the producer population. Therefore, the biomass of each succeeding population in the ecosystem would most likely be smaller than it otherwise would be.

# 35

# THE BIOSPHERE

Because the earth is a sphere, the sun's rays are vertical only at the equator, and temperature decreases from the equator to the poles. The tilt of the earth on its axis along with the earth's rotation about the sun creates the seasons. Because the oceans are warmer at the equator than at the poles, air rises at the equator and moves toward the poles; these air currents in turn cause ocean currents that affect **climate** about the world.

Warm air rising at the equator loses its moisture and then descends at about 30° north and south latitude and so forth to the poles. This movement of air in general accounts for different amounts of rainfall at different latitudes. Topography also plays a role in the distribution of rainfall.

Just south of the North Pole, the **tundra** has cold winters and short summers; the vegetation consists largely of short grasses and sedges and dwarf woody plants. Proceeding southward, the **taiga** is a coniferous forest, the **temperate deciduous forest** has seasons, and the **tropical rain forest** is a broad-leafed evergreen forest.

Among grasslands, which have less rainfall than forests, the **savanna** is a tropical grassland that supports the greatest number of different types of large herbivores. The prairie found in the United States has a limited variety of vegetation and animal life. In **deserts,** some plants, such as cacti, are succulents, and others are shrubs with thick leaves they often lose during dry periods.

Among aquatic biomes, freshwater communities include streams, rivers, lakes, and ponds. Lakes experience spring and fall overturns. Lakes and ponds have rooted plants in the littoral zone, plankton and fishes in the sunlit limnetic zone, and bottom-dwelling organisms in the profundal zone. **Estuaries** near the mouths of rivers are the nurseries of the sea. Marine communities include coastal communities and the oceans. An ocean has a **pelagic division** (open waters) and a **benthic division** (ocean floor). **Coral reefs** are productive communities found in shallow tropical waters.

Study the text section by section. Answer the study questions so that you can fulfill the learning objectives for each section.

## 35.1 CLIMATE AND THE BIOSPHERE (PP. 716–18)

The learning objectives for this section are:
- Explain how the distribution of solar energy affects climate.
- Describe other factors that result in climate differences in the biosphere.

1. Indicate whether these statements are true (T) or false (F).
   a. _____ Because the earth is a sphere, solar energy hitting earth is uniformly distributed.
   b. _____ The distribution of rainfall is partially due to topography.
   c. _____ Heat always passes from warm areas to colder areas.
   d. _____ The great deserts of the world lie at the equator.
   e. _____ Warm air moves from the equator to the poles.
   f. _____ Arctic winds across the Great Lakes produce lake-effect snows.
   g. _____ Rain shadows always form on the leeward side of the mountains.

The learning objective for this section is:
• Describe how climate affects the distribution of biomes.

2. Label this diagram that compares the effects of altitude and latitude on vegetation, using the following alphabetized list of terms. (Some terms are used more than once.)
   coniferous forest
   deciduous forest
   tropical forest
   tundra

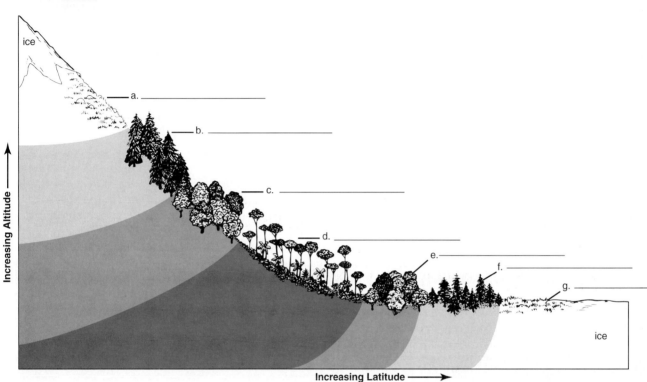

3. The diagram in question 2 emphasizes that vegetation is determined in part by ᵃ·_____.
   ᵇ·_____ also plays a major role, and therefore, tropical rain forests are found at the equator where both ᶜ·_____ and ᵈ·_____ are maximal.

4. For each biome listed, write a one- or two-word description for the temperature and rainfall.

| Biome | Temperature | Rainfall |
|---|---|---|
| Tundra | | |
| Desert | | |
| Grassland | | |
| Taiga | | |
| Temperate deciduous forest | | |
| Tropical rain forest | | |

5. Label the soil diagram, using the following alphabetized list of terms. Then indicate the layer in which *leaching* occurs.

    parent material
    topsoil
    subsoil

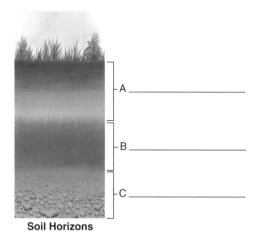

A _____

B _____

C _____

**Soil Horizons**

6. Because of limited leaching (due to limited rainfall), the A horizon is deep in ᵃ·_____, and this made the prairies of the United States good agricultural lands. Generally, in ᵇ·_____, both the A and B horizons supply inorganic nutrients for tree root growth. In tropical rain forests, however, because leaching is extensive, there is only a shallow ᶜ·_____ horizon; therefore, these forests ᵈ·_____ (can, cannot) support crops for many years.

## 35.3 TERRESTRIAL BIOMES (PP. 720–28)

The learning objective for this section is:
• Describe the major classes of terrestrial biomes.

7. From each pair, select the one that applies to the tundra, and write it on the answer line.
    a. _____ light—dark
    b. _____ cold—hot
    c. _____ short grasses—trees
    d. _____ musk ox—horses
    e. _____ epiphytes—permafrost

8. From each pair, select the one that applies to the taiga.
    a. _____ broad-leafed trees—narrow-leafed trees
    b. _____ cold—hot
    c. _____ cool lakes—pools and mires
    d. _____ zebras—moose

9. From each pair, select the one that applies to the temperate rain forest on the west coast of Canada and the United States.
    a. _____ short trees—tall trees
    b. _____ old trees—young trees
    c. _____ ferns and mosses—sedges and grasses

10. From each pair, select the one that applies to temperate deciduous forests.
    a. _____ conifers only—oak and maple trees
    b. _____ flowering shrubs—short grasses
    c. _____ caribou—white-tail deer
    d. _____ rabbits and skunks—lemmings and prairie chickens

11. From each pair, select the one that applies to tropical rain forests.
    a. _____ deciduous broad-leafed trees—evergreen broad-leafed trees
    b. _____ lianas and epiphytes—pine needles
    c. _____ few insects—many insects
    d. _____ colorful birds—drab birds
    e. _____ horses and zebras—monkeys and large cats

12. From each pair, select the one that applies to shrublands.
    a. _____ rainfall in summer—rainfall in winter
    b. _____ shrubs with thick roots—trees with shallow roots

13. From each pair, select the one that applies to United States prairies.
    a. _____ rabbits—prairie dogs
    b. _____ hawks—parakeets
    c. _____ trees—grasses

14. From each pair, select the one that applies to the African savanna.
    a. _____ elephants—moose
    b. _____ even rainfall—severe dry season
    c. _____ herds of herbivores—large primates

15. From each pair, select the one that applies to North American deserts.
    a. _____ cool days—cool nights
    b. _____ cacti—broad-leafed evergreen trees
    c. _____ lizards and snakes—elephants and zebras

## Hoop Dreams

Each biome is a hoop. The plants and animals are balls. Try to get the balls in the right hoops.

| Hoops (biomes) | Plants (balls) | Animals (balls) |
|---|---|---|
| desert | lichens | moose |
| taiga | spruce trees | beaver and muskrat |
| U.S. prairie | epiphytes | lemming |
| temperate deciduous forest | oak trees | monkey |
| tundra | grasses | lizard |
| African savanna | acacia trees | buffalo |
| tropical rain forest | cacti | wildebeest |

Possible number of baskets is 14. How many baskets did you make? _____

## 35.4 AQUATIC COMMUNITIES (PP. 729–37)

The learning objecitve for this section is:
• Describe the two major classes of aquatic biomes.

16. Aquatic communities can be divided into two major types: the a._____ communities that consist of lakes, ponds, rivers, and streams, and the b._____ communities along the coast and in the ocean itself.

17. Lakes occur as nutrient-poor or a._____ lakes and nutrient-rich or b._____ lakes. In the temperate zone, deep lakes are stratified. In the fall, as the top layer called c._____ cools, and in the spring as it warms, a (an) d._____ occurs. During this time, a mixing of e._____ and f._____ takes place.

18. List and describe the three life zones of a lake, noting the types of organisms found in each.

a. _____

b. _____

_____

c. _____

_____

19. Indicate whether these statements are true (T) or false (F).

a. _____ Salt marshes in the tropics and mangrove swamps in the temperate zone occur at the mouth of a river.

b. _____ Estuaries offer protection and nutrients to immature forms of marine life.

c. _____ Rocky coasts are protected, but sandy shores are bombarded by the seas as the tides roll in and out.

d. _____ There are different types of shelled and algal organisms at the upper, middle, and lower portions of the littoral zone of a rocky coast.

20. Place the appropriate letter next to each statement.

P—pelagic division    B—benthic division

a. _____ has the sublittoral and abyssal zones

b. _____ has the greater overall diversity of organisms

c. _____ includes neritic and oceanic provinces

d. _____ includes organisms living on the continental shelf and slope

e. _____ has organisms that depend on floating debris from above for food

f. _____ is penetrated by sunlight

21. Complete the following table by noting the amount of light present (bright/semidark/dark) and the types of organisms found (phytoplankton/strange-looking fish/filter feeders/carnivores/sea urchins) in each ocean zone.

| Ocean Zone | Amount of Light | Organisms |
|---|---|---|
| Epipelagic | | |
| Mesopelagic | | |
| Bathypelagic | | |
| Abyssal | | |

22. a. What supports life in the epipelagic zone? _____

b. What supports life in the abyssal zone? _____

## DEFINITIONS WORDMATCH

Review key terms by completing this matching exercise, selecting from the following alphabetized list of terms:

alpine tundra
arctic tundra
benthic division
biome
biosphere
climate
estuary
eutrophication
permafrost
rain shadow
savanna
taiga

a. _____ End of a river where fresh water and salt water mix as they meet.

b. _____ Major terrestrial community characterized by certain climatic conditions and dominated by particular types of plants.

c. _____ Ocean floor, which supports a unique set of organisms in contrast to the pelagic division.

d. _____ Terrestrial biome that is a grassland in Africa; characterized by few trees and a severe dry season.

e. _____ Weather condition of an area, including especially prevailing temperature and average daily/yearly rainfall.

f. _____ Permanently frozen ground usually occurring in the tundra, a biome of arctic regions.

## OBJECTIVE QUESTIONS

Do not refer to the text when taking this test.

For questions 1–10, indicate whether the statements are true (T) or false (F).

_____ 1. Climate determines the geographic location of a biome.

_____ 2. Grasslands usually receive a greater annual rainfall than deserts.

_____ 3. The taiga is the northernmost forested biome.

_____ 4. Temperate deciduous forests show the greatest species diversity of all forested biomes.

_____ 5. The leaves of tropical rain forest evergreen trees are needlelike.

_____ 6. The profundal zone of a lake is the zone closest to the shore.

_____ 7. An estuary acts as a nutrient trap, existing where a large river flows into an ocean.

_____ 8. The solid part of a coral reef consists of the skeletons of dead coral.

_____ 9. A food chain in the pelagic division could be: phytoplankton, zooplankton, small fish, herring.

_____ 10. The benthic division receives less light penetration than the pelagic division.

_____ 11. Which of the following phrases is NOT true of the tundra?
  a. low-lying vegetation
  b. northernmost biome
  c. few large mammals
  d. short growing season
  e. many different types of species

_____ 12. A temperate deciduous forest will
  a. be warm and moist.
  b. be hot and dry.
  c. be cold and have limited rain.
  d. have moderate temperatures and moderate rain.
  e. have moderate temperatures and little rain.

_____ 13. A tropical rain forest will typically
  a. be warm and moist.
  b. be hot and dry.
  c. be cold and have limited rain.
  d. have moderate temperatures and moderate rain.
  e. have moderate temperatures and little rain.

_____ 14. A desert will typically
  a. be warm and moist.
  b. be hot and dry.
  c. be cold and have limited rain.
  d. have moderate temperatures and moderate rain.
  e. have moderate temperatures and little rain.

_____ 15. The biome that best supports grazing animals is
  a. a tropical rain forest.
  b. a coniferous forest.
  c. a grassland.
  d. a desert.

_____ 16. Which biome has most of the animals living in trees?
  a. taiga
  b. temperate deciduous forest
  c. tropical rain forest
  d. savanna
  e. grassland

_____ 17. Which type of biome has succulent, leafless plants that have stems that store water and roots that can absorb great quantities of water in a brief period of time?
  a. tropical rain forest
  b. tundra
  c. temperate deciduous forest
  d. desert
  e. savanna

_____ 18. Large grazing animals are most numerous in which biome?
  a. tundra
  b. grassland
  c. coniferous forest
  d. deciduous forest
  e. tropical rain forest

_____ 19. Which zone of the ocean is the deepest?
  a. epipelagic
  b. mesopelagic
  c. bathypelagic
  d. abyssal
  e. estuarial

_____ 20. Which zone in the ocean receives the most sunlight?
  a. epipelagic
  b. mesopelagic
  c. bathypelagic
  d. abyssal
  e. estuarial

Answer in complete sentences.

21. Both the temperate rain forest and the chaparral occur in California. Explain various differences in climate and vegetation.

22. Explain why estuaries are the nurseries of the sea.

**Test Results:** _____ number correct ÷ 22 = _____ × 100 = _____ %

# ANSWER KEY

## STUDY QUESTIONS

**1. a.** F **b.** T **c.** T **d.** F **e.** T **f.** T **g.** T   **2.** See Figure 35.5, page 719, in text.   **3. a.** temperature **b.** Rainfall **c.** temperature **d.** rainfall
**4.**

| Temperature | Rainfall |
|---|---|
| cold | little |
| hot | little |
| moderate | limited |
| cool | moderate |
| moderate | rather high |
| hot | high |

**5. a.** topsoil, leaching **b.** subsoil **c.** parent material
**6. a.** temperate grasslands **b.** forests **c.** A **d.** cannot   **7. a.** dark **b.** cold **c.** short grasses **d.** musk ox **e.** permafrost   **8. a.** narrow-leafed trees **b.** cold **c.** cool lakes **d.** moose   **9. a.** tall trees **b.** old trees **c.** ferns and mosses   **10. a.** oak and maple trees **b.** flowering shrubs **c.** white-tail deer **d.** rabbits and skunks
**11. a.** evergreen broad-leafed trees **b.** lianas and epiphytes **c.** many insects **d.** colorful birds **e.** monkeys and large cats   **12. a.** rainfall in winter **b.** shrubs with thick roots   **13. a.** prairie dogs **b.** hawks **c.** grasses
**14. a.** elephants **b.** severe dry season **c.** herds of herbivores   **15. a.** cool nights **b.** cacti **c.** lizards and snakes
**Game: Hoop Dreams**—desert: cacti, lizard; taiga: spruce trees, moose; U.S. prairie: grasses, buffalo; temperate deciduous forest: oak trees, beaver and muskrat; tundra: lichens, lemming; African savanna: acacia trees, wildebeest; tropical rain forest: epiphytes, monkey   **16. a.** freshwater **b.** saltwater   **17. a.** oligotrophic **b.** eutrophic **c.** epilimnion **d.** overturn **e.** oxygen **f.** nu-

trients   **18. a.** *shore*—aquatic plants, microscopic organisms **b.** *limnetic zone*, sunlit main body—some surface organisms and plankton **c.** *profundal zone*, depths where sunlight does not reach—molluscs, crustaceans, worms   **19. a.** F **b.** T **c.** F **d.** T   **20. a.** B **b.** P **c.** P **d.** B **e.** B **f.** P
**21.**

| Amount of Light | Organisms |
|---|---|
| bright | phytoplankton |
| semidark | carnivores |
| dark | strange-looking carnivores |
| dark | filter feeders and sea urchins |

**22. a.** photosynthesis by algae **b.** debris floating down from above

## DEFINITIONS WORDMATCH

**a.** estuary   **b.** biome   **c.** benthic division   **d.** savanna **e.** climate   **f.** permafrost

## CHAPTER TEST

**1.** T   **2.** T   **3.** T   **4.** F   **5.** F   **6.** F   **7.** T   **8.** T   **9.** T
**10.** T   **11.** e   **12.** d   **13.** a   **14.** b   **15.** c   **16.** c   **17.** d
**18.** a   **19.** d   **20.** a   **21.** The temperate rain forest lies along the coast and has much rainfall; the old trees are covered by ferns and mosses and grow very tall. The chaparral occurs among hills and has limited rainfall; the shrubs that occur there are adapted to arid conditions and regrowth after fire.   **22.** An estuary, which is a partially enclosed body of water where fresh water and seawater meet, is a nutrient trap. Estuaries offer a protective environment where larval marine forms can mature before moving out to other coastal areas and the open sea.

# 36

# CONSERVATION OF BIODIVERSITY

**Conservation biology** is the scientific study of biodiversity and its management for sustainable human welfare. **Biodiversity** is the variety of life on earth; the exact number of species is unknown, but there are many more insects than other types of organisms. Biodiversity must be preserved at the genetic, community or ecosystem, and landscape levels of organization. Conservationists have discovered that biodiversity is not evenly distributed in the biosphere, and therefore saving particular areas may protect more species than saving other areas.

Biodiversity has both direct and indirect value. Direct value is the observable services of individual wild species. Wild species are the best source of new medicines. Wild species also meet other medical needs; for example, the bacterium that causes leprosy grows naturally in armadillos. Wild species also have agricultural value. Domesticated plants and animals are derived from wild species, which are a source of genes for the improvement of their phenotypes. Wild species can also be used as biological controls, and most flowering plants make use of animal pollinators. Additionally, wild species are still used for food (fish and shellfish, for instance). Hardwood trees from natural forests supply lumber for various purposes.

The indirect services provided by ecosystems are largely unseen but absolutely necessary for our well-being. These services include the workings of biogeochemical cycles, waste disposal, provision of fresh water, prevention of soil erosion, and regulation of climate. Additionally, people enjoy vacationing in natural settings.

**Habitat loss** is the most frequent cause of **extinction**, followed by the introduction of **alien species**, **pollution**, **overexploitation**, and **disease**. Habitat loss has occurred in all parts of the biosphere, but concern has now centered upon tropical rain forests and coral reefs where biodiversity is especially high. Alien species have been introduced into foreign ecosystems because of colonization, horticulture and agriculture, and accidental transport. Global warming is the form of pollution most responsible for extinction. Overexploitation can often be explained by the positive feedback cycle. For example, commercial fishing has become so efficient that fisheries of the world are collapsing.

Some researchers emphasize the need to preserve biodiversity hotspots because of their richness. When preserving ecosystems, it is wise to first determine which are the **keystone species**. Often it is necessary to save **metapopulations** because of habitat fragmentation. This requires saving **source populations** over **sink populations**.

Conservation biology is assisted by two types of computer analysis. A **gap analysis** tries for a fit between biodiversity concentrations and land still available to be preserved. A **population viability analysis** indicates the minimum size of a population needed to prevent extinction from happening. Since many ecosystems have been degraded, **habitat restoration** may be necessary before **sustainable development** is possible. Three principles of restoration are: (1) start before sources of wildlife and seeds are lost; (2) use simple biological techniques that mimic natural processes; and (3) aim for sustainable development so that the ecosystem fulfills the needs of humans.

## STUDY QUESTIONS

Study the text section by section. Answer the study questions so that you can fulfill the learning objectives for each section.

## 36.1 CONSERVATION BIOLOGY AND BIODIVERSITY (PP. 742–43)

The learning objectives for this section are:
- Describe the field of conservation biology and its values.
- Describe four types of biodiversity.
- Describe the distribution of diversity.

1. Place an X beside those fields involved in conservation biology:

   a. _____ genetics
   b. _____ physiology
   c. _____ behavior
   d. _____ veterinary science
   e. _____ range management

2. You can deduce from your answer to question 2 that conservation biology involves the development of a. _____ concepts and the b. _____ of these concepts to the everyday world.

3. Place an X beside the value principles that guide the field of conservation biology:

   a. _____ Extinctions due to human activities are undesirable.
   b. _____ Biodiversity is desirable to the health of the biosphere.
   c. _____ Biodiversity has no value in and of itself.
   d. _____ Extinctions due to human activities are of little concern.

4. Biodiversity involves the number of different organisms on earth. Name two other types of diversity:

   a. _____

   b. _____

5. Biodiversity is not evenly distributed, as exemplified by regions of the biosphere where a wide variety of species are found. These regions are called _____.

## 36.2 VALUE OF BIODIVERSITY (PP. 744–47)

The learning objective for this section is:
- Give examples of direct values and indirect values of biodiversity.

6. Give an example of the value of biodiversity to:

   a. medicine _____

   b. agriculture _____

   c. consumptive use _____

7. Place an X beside all those areas that are NOT at all dependent on biodiversity:

   a. _____ ecotourism
   b. _____ prevention of soil erosion
   c. _____ biogeochemical cycles
   d. _____ provision of fresh water
   e. _____ waste disposal
   f. _____ regulation of climate

8. Based on your answer to question 3, the conclusion is that biodiversity has _____ (much or little) indirect value.

## 36.3 CAUSES OF EXTINCTION (PP. 748–52)

The learning objectives for this section are:
- Identify the major factors responsible for loss of biodiversity.
- Explain how acid deposition, eutrophication, ozone depletion, organic chemicals, and global warming threaten the world's biodiversity.

9. Rank in order these threats to biodiversity from the one that is the greatest threat to the one that is the least threat.

    a. _____ pollution

    b. _____ habitat loss

    c. _____ disease

    d. _____ alien species

    e. _____ overexploitation

10. Indicate whether these statements are true (T) or false (F).

    a. _____ A species is often affected by more than one of the threats mentioned in question 9.

    b. _____ An alien specie is sometimes brought into a new area by horticulturists.

    c. _____ Tropical rain forest destruction should be associated with the introduction of disease into an area.

    d. _____ Although much of the interior of continents have suffered loss of habitat, the coastline has not suffered as much loss.

11. Match the concerns to the type of pollution:

    acid deposition    eutrophication    ozone depletion    organic chemicals    global warming

    a. _____ Forest and lakes are dying

    b. _____ Increased amount of UV radiation

    c. _____ Glaciers are melting and the sea level is rising

    d. _____ Pesticides and detergents can have hormonal effects

    e. _____ Massive fish kills

## 36.4 CONSERVATION TECHNIQUES (PP. 753–55)

The learning objectives for this section are:
- Describe methods of habitat preservation and principles that guide habitat restoration.
- Describe two types of computer analyses that help with conservation of biodiversity.

12. Sometimes it is necessary to decide the location and extent of an area to preserve. With that in mind what is a

    a. keystone species _____

    b. metapopulation _____

    c. sink population _____

    d. edge effect _____

13. A computer analysis called a gap analysis helps scientists find [a.] "_____", areas where biodiversity is high outside already preserved areas.
A population viability analysis helps researchers determine first how large a population has to be to maintain itself, and then how much [b.] _____ a species needs to remain at this size.

14. Sometimes it is necessary to restore an area so that it is suitable for the maintenance of species. What three principles guide restoration efforts?

    a. It's best to begin _____.

    b. It's best to use biological techniques that _____.

    c. The goal is _____ development so that an ecosystem can maintain itself.

Review key terms by completing this matching exercise, selecting from the following alphabetized list of terms:

*biodiversity hotspot*
*conservation biology*
*gap analysis*
*global warming*
*landscape*
*metapopulation*
*population*
*restoration ecology*
*sink population*
*source population*
*sustainable development*

a. _____ Use of computers to discover places where biodiversity is high outside of preserved areas.

b. _____ A number of interacting ecosystem fragments.

c. _____ Population subdivided into several small and isolated populations due to habitat fragmentation.

d. _____ Subdiscipline of conservation biology that seeks ways to return ecosystems to their former state.

e. _____ Species whose activities have a significant role in determining community structure.

f. _____ Region of the world that contains unusually large concentrations of species.

## CHAPTER TEST

### OBJECTIVE QUESTIONS

Do not refer to the text when taking this test.

_____ 1. What is defined as the scientific study of biodiversity and its management for human welfare?
  a. biology
  b. environmental biology
  c. conservation biology
  d. None of the above are correct.

_____ 2. A population viability analysis includes calculating the
  a. minimum population size needed to prevent extinction.
  b. amount of habitat needed to maintain this population size.
  c. how much pollution this species can withstand.
  d. Both *a* and *b* are correct.

_____ 3. What kind of change is aggravating the biodiversity crisis?
  a. genetic drift
  b. global environmental change
  c. excessive reproduction
  d. None of the above are correct.

_____ 4. One objective of _____ is to maintain the capacity of species to adapt to changing conditions through evolution.
  a. conservation biology
  b. global environmental change
  c. habitat fragmentation
  d. ecosystems

_____ 5. To maintain biodiversity, a gap analysis tells us
  a. which regions should be conserved.
  b. how large each preserved population is.
  c. if we are preserving the correct regions.
  d. Both *a* and *c* are correct.

_____ 6. The greatest majority of known species are:
  a. onychophorans.
  b. enchinoderms.
  c. insects.
  d. chordates.

_____ 7. Humans have
  a. not influenced extinction rates.
  b. reduced extinction rates.
  c. greatly increased extinction rates.
  d. reduced all activities that cause extinction.
  e. Both *b* and *d* are correct.

_____ 8. Biodiversity, a rich and valuable resource, includes
  a. genetic diversity.
  b. species richness.
  c. diverse communities and ecosystems.
  d. All of these are correct.

_____ 9. Which two ecosystems have the most species?
  a. tropical forests and coral reefs
  b. tropical forests and prairies
  c. coral reefs and the seashores
  d. rivers and lakes

_____10. Periodic mass extinctions have occurred due to
  a. lunar eclipses.
  b. geological events.
  c. astrophysical events.
  d. Both *b* and *c* are correct.

_____11. Humans threaten to cause a mass extinction when they participate in which habitat-destroying activities?
  a. tropical deforestation
  b. river damming and diversion
  c. lake pollution and eutrophication
  d. All of these are correct.

____12. Ecosystems are critically endangered in the United States largely due to
    a. pollution.
    b. habitat loss.
    c. alien species.
    d. overexploitation.

____13. Habitat fragmentation leads to
    a. demographic impacts.
    b. genetic consequences.
    c. extinctions.
    d. All of these are correct.

____14. Prehistoric humans caused many extinctions by
    a. hunting.
    b. habitat transformation.
    c. introduction of alien predators.
    d. All of these are correct.

____15. Climate shifts in terrestrial environments may
    a. eliminate habitats of some species.
    b. favor alien species.
    c. favor marine species.
    d. Both *a* and *b* are correct.

____16. Which environment has an uneven distribution of species?
    a. terrestrial environment
    b. freshwater environment
    c. marine environment
    d. All of these are correct.

____17. At the current species level, recent estimates suggest that the number of described species is about
    a. 1.4 – 1.6 million.
    b. 2.3 – 3.1 million.
    c. 4.1 – 5.2 million.
    d. 50 – 60 million.

____18. Which of these is an indirect value of biodiversity?
    a. medicinal value
    b. consumptive use value
    c. regulation of climate
    d. Both *a* and *b* are correct.

____19. Enforcement of biodiversity management must ultimately be defined by
    a. biological experts.
    b. the government.
    c. society.
    d. None of these are correct.

____20. The field of conservation biology became a separate field during
    a. the late 1890s.
    b. the 1990s.
    c. the 1950s.
    d. It has always been around.

## THOUGHT QUESTIONS

Answer in complete sentences.

21. Taking into consideration what you have learned about the value of biodiversity, what are the key reasons for saving the tropical rain forest?

22. What steps need to be taken if an ecosystem is to be restored?

**Test Results:** _____ number correct ÷ 22 = _____ × 100 = _____ %

## STUDY QUESTIONS

**1.** a,b,c,d,e   **2. a.** scientific **b.** application   **3.** a, b, c
**4. a.** genetic, community   **5.** hot spots   **6. a.** Rosy periwinkle is a source of a cancer medicine.   **b.** Crops are derived from wild species.   **c.** Fish are caught in the wild.
**7.** None should be checked because all are dependent on biodiversity.   **8.** much   **9.** b, d, a, e, c   **10. a.** T **b.** T **c.** F **d.** F   **11. a.** acid deposition **b.** ozone depletion **c.** global warming **d.** organic chemicals **e.** eutrophication   **12. a.** influence the viability of a community more than you would suspect from their numbers **b.** a population subdivided into several small and isolated populations **c.** a population that loses members because the environment is not favorable **d.** An edge reduces the amount of habitat typical of an ecosystem.   **13. a.** gap **b.** habitat **14. a.** right away **b.** mimic nature **c.** sustainable

## DEFINITIONS WORDMATCH

**a.** gap analysis   **b.** landscape   **c.** metapopulation
**d.** restoration ecology   **e.** keystone species   **f.** biodiversity hotspot

## CHAPTER TEST

**1.** c   **2.** d   **3.** b   **4.** a   **5.** d   **6.** c   **7.** c   **8.** d   **9.** a
**10.** d   **11.** d   **12.** b   **13.** d   **14.** d   **15.** d   **16.** d
**17.** a   **18.** c   **19.** c   **20.** b   **21.** Answers may vary but may include medicines yet to be discovered, ecotourism, formation of rain clouds, etc.   **22.** You must have sources of wildlife and seeds before they are lost. You should use simple biological techniques that mimic natural processes. You should aim for sustainable development so that the ecosystem still fulfills the needs of humans.

# Notes

# Notes

# Notes

# Notes

# Notes

# Notes

# Notes

# Notes

# Notes

# Notes

# Notes

# Notes

# Notes

# Notes

# Notes